AF894763

Computational Methods, Seismic Protection, Hybrid Testing and Resilience in Earthquake Engineering

GEOTECHNICAL, GEOLOGICAL AND EARTHQUAKE ENGINEERING

Volume 33

Series Editor

Atilla Ansal, *School of Engineering, Özyeğin University, Istanbul, Turkey*

Editorial Advisory Board

Julian Bommer, *Imperial College London, U.K.*
Jonathan D. Bray, *University of California, Berkeley, U.S.A.*
Kyriazis Pitilakis, *Aristotle University of Thessaloniki, Greece*
Susumu Yasuda, *Tokyo Denki University, Japan*

For further volumes:
http://www.springer.com/series/6011

Gian Paolo Cimellaro • Satish Nagarajaiah
Sashi K. Kunnath
Editors

Computational Methods, Seismic Protection, Hybrid Testing and Resilience in Earthquake Engineering

A Tribute to the Research Contributions of Prof. Andrei Reinhorn

Springer

Editors
Gian Paolo Cimellaro
Department of Structural, Geotechnical & Building Engineering (DISEG)
Politecnico di Torino
Turin, Italy

Satish Nagarajaiah
Civil & Env. Engineering and Mech. Engineering
Rice University
Houston, TX, USA

Sashi K. Kunnath
Civil and Environmental Engineering
University of California
Davis, CA, USA

ISSN 1573-6059 ISSN 1872-4671 (electronic)
ISBN 978-3-319-06393-5 ISBN 978-3-319-06394-2 (eBook)
DOI 10.1007/978-3-319-06394-2
Springer Cham Heidelberg New York Dordrecht London

Library of Congress Control Number: 2014949377

© Springer International Publishing Switzerland 2015
This work is subject to copyright. All rights are reserved by the Publisher, whether the whole or part of the material is concerned, specifically the rights of translation, reprinting, reuse of illustrations, recitation, broadcasting, reproduction on microfilms or in any other physical way, and transmission or information storage and retrieval, electronic adaptation, computer software, or by similar or dissimilar methodology now known or hereafter developed. Exempted from this legal reservation are brief excerpts in connection with reviews or scholarly analysis or material supplied specifically for the purpose of being entered and executed on a computer system, for exclusive use by the purchaser of the work. Duplication of this publication or parts thereof is permitted only under the provisions of the Copyright Law of the Publisher's location, in its current version, and permission for use must always be obtained from Springer. Permissions for use may be obtained through RightsLink at the Copyright Clearance Center. Violations are liable to prosecution under the respective Copyright Law.
The use of general descriptive names, registered names, trademarks, service marks, etc. in this publication does not imply, even in the absence of a specific statement, that such names are exempt from the relevant protective laws and regulations and therefore free for general use.
While the advice and information in this book are believed to be true and accurate at the date of publication, neither the authors nor the editors nor the publisher can accept any legal responsibility for any errors or omissions that may be made. The publisher makes no warranty, express or implied, with respect to the material contained herein.

Printed on acid-free paper

Springer is part of Springer Science+Business Media (www.springer.com)

Preface

One Contribution of Andrei Reinhorn that I Most Admire: Summary of After Lunch Remarks at the Reinhorn Symposium on September 30, 2013

After several years of discussion and consideration, the University at Buffalo made the decision and commitment in 1978 to develop an earthquake engineering research and education component in its new civil engineering building on the Amherst campus. Almost immediately following this decision, Professor Andrei Reinhorn was recruited as the first addition to the structural engineering faculty due to his expertise in structural dynamics and earthquake engineering. The new structural and geotechnical engineering laboratory housed a then state-of-the-art shaking table which was completed in 1982. This new simulator became fully operational in 1983, and Andrei was one of two faculty members responsible for the success of this major undertaking. Many years later, Andrei again co-led the team effort that secured another NSF award to expand the laboratory into a major NEES facility in 2004.

For 30 years, and particularly since 1986 when the NSF first awarded the University at Buffalo a center grant to establish the National Center for Earthquake Engineering Research (NCEER) followed by a subsequent NSF award to establish the Multidisciplinary Center for Earthquake Engineering Research (MCEER), Andrei has made many contributions in earthquake engineering education and research. He has become recognized by his peers for his important advances in structural control, experimental analysis, computational analysis, and resilience-based design. His achievements in any one of these areas alone would have been more than sufficient as one scholar's lifelong professional contribution. And in my humble opinion, Andrei's contribution to the earthquake engineering profession goes beyond even his scholarship and leadership in these four areas of earthquake engineering research and education.

For more than two centuries, engineering disciplines have been evolving and expanding, most rapidly in recent decades. Academic researchers in structural engineering or earthquake engineering were only able to concentrate on the in-depth pursuit of one or two specialty areas. But societal problems continued to grow in scope from rapid population growth, urbanization, globalization, and environmental/infrastructure degradation. These trends required that structural/earthquake engineering researchers not only be experts in one field, but also learn to be members of an integrated team of experts able to address large-scale societal problems. Developing this new generation of academic researchers is a great challenge for research-intensive universities; it will require many generations of new PhD graduates entering the academic environment to affect these changes.

Make no mistake, Reinhorn has helped move the bar in this grand effort. He has published some 175 refereed journal articles and book chapters; in those he has co-authored more than 100 of them with 11 different structures faculty members and their students at the University at Buffalo. This simultaneously demonstrates the breadth of his expertise and his ability to work with university colleagues in the same or closely-related disciplines. He has co-authored papers that were based on research carried out by students under the supervision of other faculty, and so has shown that he can contribute seriously to research in other groups. This is the essence of what we are trying to instill in the next generation of structural engineering researchers. Reinhorn has been a role model for this approach over the years. Most of his students and colleagues are here at this symposium. You have all experienced Andrei's leadership and structural engineering expertise. You have seen how he applies his many talents toward establishing earthquake resilient communities. This is the beginning of the fusion process that integrates the various sub-disciplines in civil engineering and that integrates the field of civil engineering structures with other emerging disciplines in electrical, material, mechanical and other knowledge domains to "Restore and Improve Urban Infrastructures." That ambitious vision is one of the Engineering Grand Challenges for the twenty-first century as identified by the National Academy of Engineering.

I have only mentioned a small portion of the many meaningful things that Andrei has done in his years of dedicated academic career. But let me end by reminding us all of the obvious: that the many successes and contributions he has made reflect the character of the man himself. He has exemplified the German Proverb that "individual effort is addition and team effort is multiplication".

The qualities that I admire most about Andrei are his vision, his wisdom, and his ability to build successful research teams in the highly individualized academic environment. He is an excellent role model for young academic researchers who are learning to engage in team efforts to re-shape structural engineering for a changing world!

Buffalo, NY, USA George C. Lee

Contents

Contributors

A.G. Ayala Instituto de Ingeniería, Universidad Nacional Autónoma de México, Delegación Coyoacán, México D.F., Mexico

Simonetta Baraccani Department DICAM, University of Bologna, Bologna, Italy

R. Barron Facultad de Ingeniería, Universidad Autónoma de Zacatecas, Zacatecas, Zac., Mexico

A. Belejo ICIST, DECivil, Instituto Superior Técnico, Universidade de Lisboa, Lisbon, Portugal

R. Bento ICIST, DECivil, Instituto Superior Técnico, Universidade de Lisboa, Lisbon, Portugal

Joseph M. Bracci Zachry Department of Civil Engineering, Texas A&M University, College Station, TX, USA

M. Bruneau Department of Civil, Structural and Environmental Engineering, University at Buffalo (SUNY), Buffalo, NY, USA

Gian Paolo Cimellaro, Ph.D., P.E. Department of Structural, Geotechnical & Building Engineering (DISEG), Politecnico di Torino, Turin, Italy

M.T. Cristofaro Department of Architecture (DiDA), University of Florence, Florence, Italy

M. De Stefano Department of Architecture (DiDA), University of Florence, Florence, Italy

Griffin Enyart City of Decatur, Decatur, IL, USA

Giada Gasparini Department DICAM, University of Bologna, Bologna, Italy

Chelsea Griffith Department of the Navy, Naval Sea Systems Command, Washington, DC, USA

Mircea Grigoriu Department of Civil and Environmental Engineering, Cornell University, Ithaca, NY, USA

Sashi K. Kunnath Civil and Environmental Engineering, University of California, Davis, CA, USA

O. Lavan Technion – Israel Institute of Technology, Technion City, Haifa, Israel

George C. Lee Department of Civil, Structural and Environmental Engineering, University at Buffalo (SUNY), Buffalo, NY, USA

Pauline Lopez Direction Innovation and R&D, Siniat, Avignon Cedex 9, France

Giuseppe Maddaloni Department of Engineering, University of Naples Parthenope, Naples, Italy

Gennaro Magliulo Department of Structures for Engineering and Architecture, University of Naples Federico II, Naples, Italy

Nicos Makris Department of Civil Engineering, Division of Structures, University of Patras, Patras, Greece

Gaetano Manfredi Department of Structures for Engineering and Architecture, University of Naples Federico II, Naples, Italy

D. Martinelli Department of Civil and Environmental Engineering, University of California at Berkeley, Berkeley, CA, USA

Adam Mueller Western Michigan University, Kalamazoo, MI, USA

Satish Nagarajaiah, Ph.D., FSEI Civil & Env. Engineering and Mech. Engineering, Rice University, Houston, TX, USA

Michele Palermo Department DICAM, University of Bologna, Bologna, Italy

Crescenzo Petrone Department of Structures for Engineering and Architecture, University of Naples Federico II, Naples, Italy

Marta Piqué Faculty ETSEIB, Universitat Politècnica de Catalunya, Barcelona, Spain

R. Pucinotti Dipartimento di Patrimonio, Architettura e Urbanistica, Università Mediterranea di Reggio Calabria, Reggio Calabria, Italy

C. Renschler Department of Geography, University at Buffalo (SUNY), Buffalo, NY, USA

Macarena Schachter Adaros Weidlinger Associates Inc., Washington, DC, USA

Xiaoyun Shao Western Michigan University, Kalamazoo, MI, USA

Stefano Silvestri Department DICAM, University of Bologna, Bologna, Italy

M. Tanganelli Department of Architecture (DiDA), University of Florence, Florence, Italy

Tomaso Trombetti Department DICAM, University of Bologna, Bologna, Italy

Michalis F. Vassiliou ETH Zürich, IBK, Zürich, Switzerland

S. Viti Department of Architecture (DiDA), University of Florence, Florence, Italy

Chapter 1
Introduction

Gian Paolo Cimellaro

The book wants to summarize the different research topics that Prof. Andrei Reinhorn tackled during his career with all his Ph.D. students and the one that he has inspired in other researchers from all over the world. Most of his research topics overlap each other, and a clear separation between them is quite difficult to be achieved, however we tried to organize the book in five main research parts

1. Computational methods
2. Experimental methods
3. Hybrid testing and control
4. Resilience-based Design
5. Seismic protection

Part 1 includes three chapters describing the genesis of 3D-Basis and IDARC and their impact on nonlinear dynamic analysis. It concludes with a novel method for solving random eigenvalues problem.

Part 2 describes experimental tests which were performed at Buffalo and in other earthquake laboratories in the world.

Part 3 introduces hybrid testing system and its application in developing hybrid simulation methods.

Part 4 introduces the concepts of resilience-based design and some applications in health care facilities and economy.

Part 5 describes different applications of seismic protection systems, such as viscously coupled shear walls, rocking systems, and the effect of nonlinear static procedures on the Seismic Assessment of asymmetric plan buildings.

G.P. Cimellaro, Ph.D., P.E. (✉)
Department of Structural, Geotechnical & Building Engineering (DISEG),
Politecnico di Torino, 10129 Turin, Italy
e-mail: gianpaolo.cimellaro@polito.it

G.P. Cimellaro et al. (eds.), *Computational Methods, Seismic Protection, Hybrid Testing and Resilience in Earthquake Engineering*, Geotechnical, Geological and Earthquake Engineering 33, DOI 10.1007/978-3-319-06394-2_1,
© Springer International Publishing Switzerland 2015

Fig. 1.1 Group picture during the Symposium on September 30th 2013

The book is a summary of the works presented by his former students during the Symposium which was organized to celebrate his work in the field of Earthquake Engineering on September 30th, 2013, at Ramada Hotel in Buffalo (Fig. 1.1).

This is book is to express the gratitude of his "*professional children*", the ambassadors of his research which has spread all over the world.

%%%

In honor of a
very special teacher and researcher

Prof. Andrei M. Reinhorn

The mediocre teacher tells,
The good teacher explains.
The superior teacher demonstrates.
The great teacher inspires.

-William Arthur Ward-

Thank you for inspiring so many around the world during 35 years of
Productive and dedicated teaching and research

From
Colleagues, students,
post-doctoral researchers and visiting
scholars from around the world

September 30, 2013

Part I
Computational Methods

Chapter 2
3D BASIS Origins, Novel Developments and Its Impact in Real Projects Around the World

Satish Nagarajaiah

Abstract Origins and development of **3D-BASIS** (**3-D**imensional **BAS**e **I**solated **S**tructures) was initially envisioned by the need for an efficient tool for nonlinear dynamic analysis of three-dimensional base isolated structures, particularly in solving the highly nonlinear bidirectional stick-slip hysteretic response of a collection of sliding isolation bearings and the resulting response of the superstructure, as this was not available at that time. The primary challenge was to solve the stick-slip behavior of friction bearings—modeled using a differential equation (Bouc-Wen Model) due to its efficiency in representing constant Coulomb friction or variable velocity depended friction by using a very small yield displacement during the stick phase resulting in very high tangential stiffness followed by a very small tangential stiffness during the sliding phase—and the resulting stiff differential equations. A challenge that is compounded when biaxial-friction is modeled, wherein even the traditional method of using Gear's method to solve stiff differential equations breaks down—a problem that was vexing the research team at University at Buffalo trying to solve the problem at that time. The answer was the development of the novel pseudo-force solution algorithm along with a semi-implicit Runge-Kutta method to solve the difficult problem. The efficient solution procedure is needed primarily for the nonlinear isolation system consisting of (1) sliding and/or elastomeric bearings, (2) fluid dampers, (3) other energy dissipation devices, while the superstructure is represented by three dimensional superstructure model appropriately condensed (where only master nodes at the center of mass of the floor are retained). This chapter describes the origins, development of 3D-BASIS and its impact.

S. Nagarajaiah, Ph.D, FSEI (✉)
Civil & Env. Engineering and Mech. Engineering, Rice University,
213 Ryon, MS 318, 6100 Main St., Houston, TX 77005, USA
e-mail: Satish.Nagarajaiah@rice.edu

G.P. Cimellaro et al. (eds.), *Computational Methods, Seismic Protection, Hybrid Testing and Resilience in Earthquake Engineering*, Geotechnical, Geological and Earthquake Engineering 33, DOI 10.1007/978-3-319-06394-2_2,
© Springer International Publishing Switzerland 2015

2.1 Introduction

Base isolation involves the introduction of isolation bearings and energy dissipating devices between the superstructure and its foundation. The laterally flexible isolation system shifts the fundamental period—considering an equivalent linear isolation system—of the structure beyond its fixed base period and the predominant periods of the ground motion. The period lengthening to typically 2–4 s is sufficient to reflect the earthquake energy. Energy dissipation in the isolation system is then useful in limiting the displacement response. The isolation bearings generally exhibit material nonlinearities and under certain conditions may also exhibit geometric nonlinearities. However, these nonlinearities are restricted to the isolation system. The superstructure is typically designed to exhibit elastic behavior.

This chapter presents a brief overview of the analytical modeling techniques used in the nonlinear dynamic analysis of base isolated structures. The localized nonlinearities at the base allow condensation of the linear superstructure to a small number of master degrees of freedom. All the nonlinear bearings and devices are explicitly modeled.

Mechanical properties of isolation bearings are described in detail. Material, friction, geometric and contact nonlinearities in the isolation system are discussed. Analytical models used for characterizing the behavior of isolation bearings and devices are presented. Formulation of the combined linear superstructure and nonlinear isolation system and solution procedure is presented. Computer programs that are most popularly used are described briefly.

2.2 Base Isolation Systems

Base isolation systems have gained wide acceptance (Buckle and Mayes 1990; Kelly 1997; Skinner et al. 1993; Soong and Constantinou 1994). The isolation bearings are typically connected between columns and foundation as shown in Fig. 2.1. The isolation system is designed to be very stiff in the vertical direction. The isolation system is designed to provide adequate initial stiffness under service loads, such as wind load, and to provide greater flexibility past yielding of the isolation bearings under strong ground motion or seismic loads.

There are two basic types of isolation bearings: elastomeric bearings and sliding bearings. Elastomeric bearings consist of laminated rubber layers and steel shim plates. Two types of elastomeric bearings that have been implemented in structures are the high damping rubber bearing and the lead rubber bearing. In both types the laminated rubber provides the lateral flexibility. The isolation system level displacements increase due to the lateral flexibility. Adding energy dissipation capacity reduces the isolation system displacements. The energy dissipation capacity is provided by the inherent damping capacity of the rubber in high damping bearings.

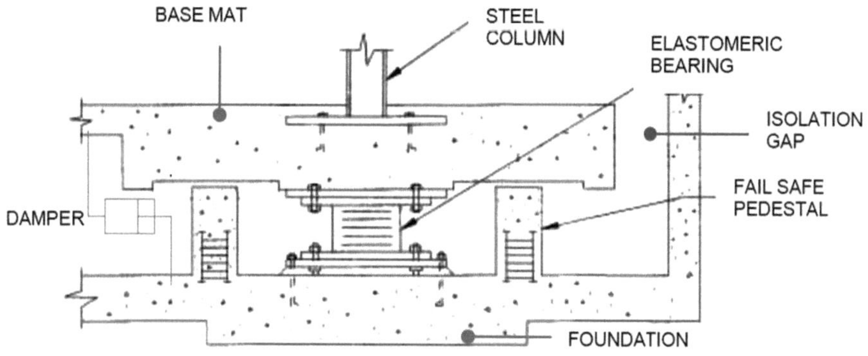

Fig. 2.1 Isolation system details including elastomeric bearing and damper

In lead-rubber bearings, which are typically manufactured with low damping rubber, the cylindrical lead plug within the rubber unit provides the energy dissipation capacity. Moreover, supplemental energy dissipating devices, primarily in the form of fluid viscous dampers, have been used in isolation systems to substantially enhance damping in applications in areas of very high seismicity.

Sliding bearings consist of Teflon or similar materials sliding on a stainless steel surface. Two types of sliding bearings that have been implemented in structures are the Friction Pendulum Sliding (FPS) bearings, spherically shaped sliding bearings, and the flat sliding bearings. Sliding bearings dissipate energy due to friction. Restoring force is provided by the spherical sliding surface in the FPS system or by added springs in the system with flat sliding bearings.

2.3 Material/Friction Nonlinearities of Base Isolation Bearings and Devices

2.3.1 Elastomeric Bearings

Elastomeric bearings are typically made of natural rubber and are classified into low damping and high damping bearings. The low damping bearings exhibit shear stiffness which is effectively linear to large shear strains (>100 %). The damping is in the range of 2–5 % of critical. Lead-rubber bearings are made up of low damping natural rubber with a lead core. The lead core is provided to increase the energy dissipation capacity to about 20–30 % of critical. The idealized force displacement behavior of a lead-rubber bearing can be characterized as bilinear hysteretic as shown in Fig. 2.2. The high initial stiffness offers rigidity under wind load and low level seismic load. The characteristic strength, $Q = A_p \sigma_{YL}$, where A_p is the lead plug

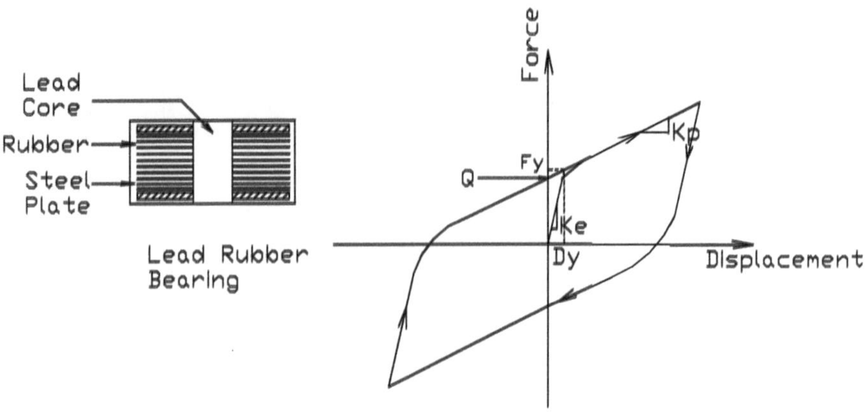

Fig. 2.2 Lead rubber bearing: bilinear force-displacement loop

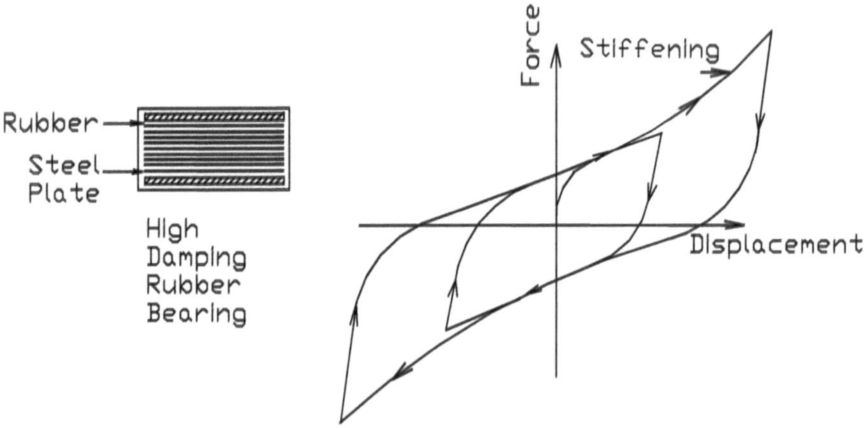

Fig. 2.3 High damping bearing: force displacement loop with stiffening

area and σ_{YL} is the effective shear yield stress of lead. The post yielding stiffness, K_p, is typically higher than the shear stiffness of the bearing without the lead core.

$$K_p = \frac{A_r G}{\sum t} f \tag{2.1}$$

Where, A_r is the bonded rubber area, $\sum t$ is the total rubber thickness, G is the shear modulus of rubber, and f is a factor larger than unity. Under proper conditions, f, may be equal to or less than 1.15. Moreover, the initial elastic stiffness, K_e, ranges between 6.5 and 10 times the post-yielding stiffness.

The stiffness and energy dissipation characteristics of high damping bearings are highly nonlinear and dependent on shear strain as shown in Fig. 2.3. The high damping bearings are made up of specially compounded rubber, which provides effective damping of 10–15 % of critical. The high damping bearings have high shear stiffness at low shear strains (<20 %) for rigidity under wind load and low level seismic load. The shear stiffness is typically lower in the range of 20–120 % shear strains. At large shear strains, the shear stiffness increases due to strain crystallization process in the rubber. The damping in high damping bearings is best characterized by a combination of hysteretic and viscous behavior. In the virgin stage and during the first cycle of movement, the bearings exhibit higher stiffness and damping than in the following cycles. The stiffness stabilizes by the third cycle, resulting in stable properties termed as scragged properties. Scragging of the bearings is the result of internal changes in the rubber. Recovery to the unscragged (virgin) properties occurs following sufficient time. The scragged state of the bearings can be modeled by a bilinear hysteretic model for shear strains of up to 200 %. The stiffening behavior (see Fig. 2.3) beyond this strain can also be modeled using more complex models (Constantinou et al. 2007; Tsopelas et al. 1994a; Kikuchi and Aiken 1997). The current technique used to model high damping bearings is to perform multiple analysis with bilinear hysteretic models; the parameters of the bilinear hysteretic models are determined at specific shear strain amplitudes. The bilinear model parameters can be established from test data of prototype bearings. These properties are the shear modulus, G, and the equivalent damping ratio, ξ (defined as the energy dissipated in a cycle of motion divided by 4π and by the maximum kinetic energy) under scragged conditions. G, is related to the post yielding stiffness $K_p^{\prime}$

$$K_p^{\prime} = \frac{GA_r}{\sum t} \tag{2.2}$$

The parameters of the model may be determined by use of the mechanical properties of G and ξ at a specific strain- for example parameters corresponding to the design displacement. The post yielding stiffness, $K_p^{\prime}$, is determined from (2.2), where as the characteristic strength, Q, may be related to the mechanical properties by assuming bilinear hysteretic behavior.

$$Q = \frac{\pi \ \xi \ K_p^{\prime} D^2}{(2 - \pi \ \xi)\, D - 2Dy} \tag{2.3}$$

Where, the yield displacement, Dy, is between 0.05 and 0.1 times the total rubber thickness and D is the design displacement. The yield force, Fy, is given by

$$Fy = Q + K_p^{\prime} Dy \tag{2.4}$$

and the post to pre-yielding stiffness ratio is given by

$$\alpha = \frac{K_p^{\prime} D y}{Fy} \tag{2.5}$$

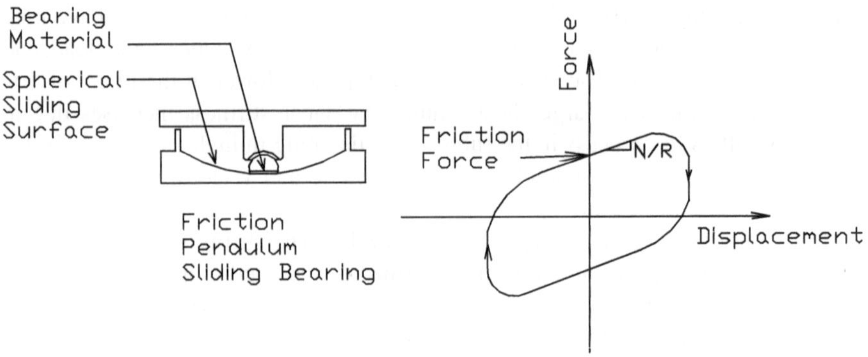

Fig. 2.4 Friction pendulum bearing: force-displacement loop (includes friction and recentering force)

Elastomeric bearings have finite vertical stiffness that affects the vertical response of the isolated structure. The vertical stiffness of an elastomeric bearing can be estimated as follows

$$k_v = \frac{E_c A_r}{\sum t} \tag{2.6}$$

Where, E_c is the compression modulus.

2.3.2 *Sliding Bearings*

Two types of sliding bearings are the flat sliding bearings with restoring force devices and the friction pendulum bearings (FPS) shown in Fig. 2.4. Flat sliding bearing is made up of Teflon sliding on a flat stainless steel surface. The recentering capability is provided by additional elastic springs. The FPS bearing, shown in Fig. 2.4, is made up of a composite material sliding on a spherical surface with radius of curvature R, which provides the recentering force. The behavior of FPS bearing can be represented by

$$F = \frac{N}{R}U + \mu_s N \,\mathrm{sgn}\left(\dot{U}\right) \tag{2.7}$$

Where, F is the force in the bearing, U and $\dot{U}$ are the displacement and velocity, respectively, μ_s is the coefficient of sliding friction (dependent on velocity and pressure) and N is the normal load on the bearing. It should be noted that for flat sliding bearings R is infinite. The coefficient of friction of sliding bearings depends on a number of parameters of which the composition of the sliding interface, bearing

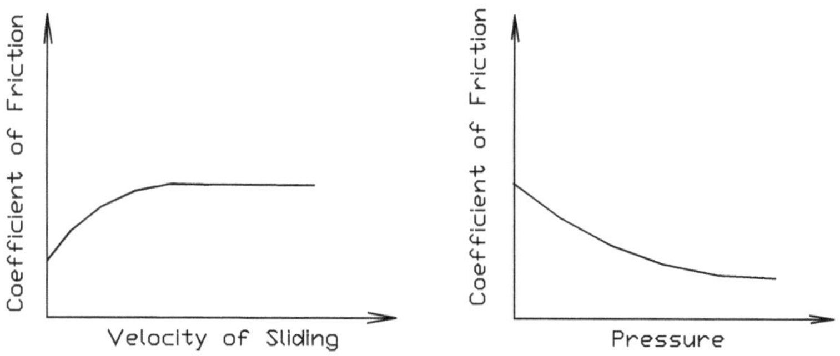

Fig. 2.5 Variation of coefficient of friction as a function of velocity of sliding and pressure

Fig. 2.6 Triple friction pendulum isolator

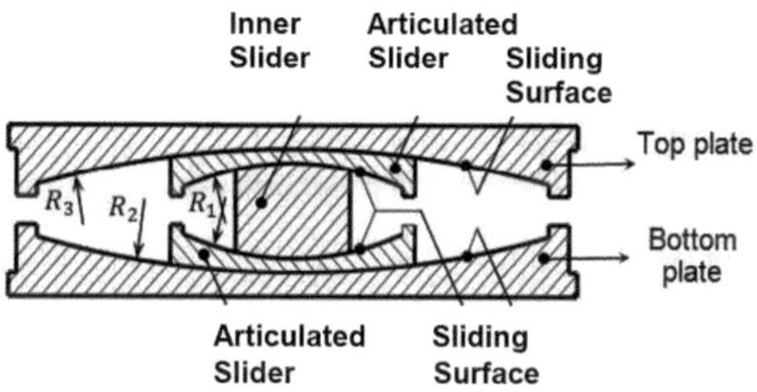

pressure and velocity of sliding (as shown in Fig. 2.5) are the most important. For interfaces consisting of polished stainless steel in contact with Teflon or composites the coefficient of friction may be described by Constantinou et al. (1990)

$$\mu_s = f_{\max} - (f_{\max} - f_{\min}) \exp\left(-a\left|\dot{U}\right|\right) \tag{2.8}$$

Where the parameters $f_{\min}$ and $f_{\max}$ describe, respectively, the coefficients of friction at essentially zero and large velocities of sliding and under constant pressure. Parameters $f_{\min}, f_{\max}$ and α depend on the bearing pressure, although only the dependency of $f_{\max}$ on pressure is of practical significance.

More recently Fenz and Constantinou (2008a, b), Morgan and Mahin (2011), Ray and Reinhorn (2012) and Dao et al. (2013) have studied the triple friction pendulum isolation bearing that has an inner slider and articulated sliders sliding inside concave sliding surfaces as shown in Fig. 2.6, and developed detailed analytical models with force-displacement behavior as shown in Fig. 2.7.

2.3.3 *Fluid Viscous Dampers*

Fluid dampers (Constantinou and Symans 1993) are used to enhance the damping in the isolation system and are connected between the base and foundation as shown in Fig. 2.1. Fluid viscous dampers produce force by forcing fluid (typically silicone

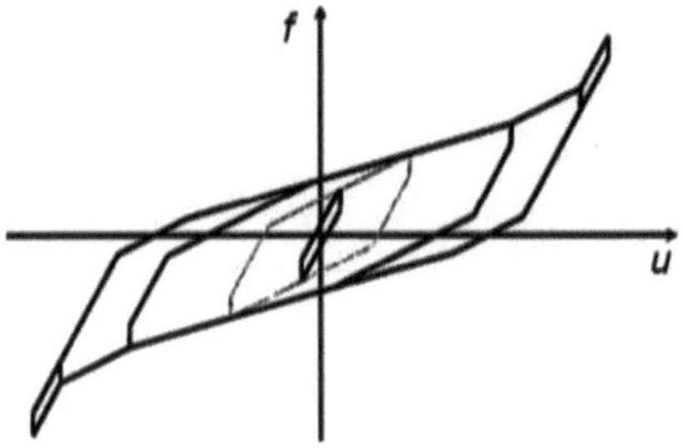

Fig. 2.7 Force (f)—displacement (u) behavior of triple friction pendulum isolator

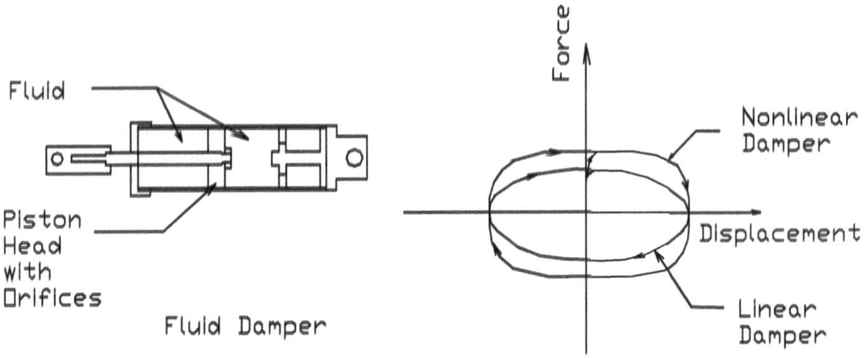

Fig. 2.8 Fluid damper: force displacement loop (velocity dependent damping force)

oil) through orifice passages as shown in Fig. 2.8. It is possible to shape the orifice passages (Constantinou and Symans 1993) in such a way as to produce an output force of the type

$$F = C \left| \dot{U} \right|^{\alpha} \operatorname{sgn}(\dot{U}) \tag{2.9}$$

Where C = damping coefficient, α is in the range of 0.5–1.0 and the representative force-displacement loops are shown in Fig. 2.8.

2.4 Modeling Material/Friction Nonlinearities of Isolation Bearings

Models with bilinear hysteretic characteristics can represent the behavior of elastomeric bearings. Several models have been used to represent the uniaxial and biaxial behavior of elastomeric isolation bearings. The uniaxial and biaxial behavior of elastomeric bearings have been modeled by Nagarajaiah (1990), Nagarajaiah et al. (1991a, b) using a viscoplasticity based modified Bouc-Wen model

(Bouc 1967; Wen 1976; Park et al. 1986). In the biaxial model forces F_X and F_Y are mobilized during the motion along the X and Y directions, respectively, of the elastomeric bearing:

$$F_X = \alpha \frac{Fy}{Dy} U_x + (1-\alpha)\, FyZ_x \qquad F_Y = \alpha \frac{Fy}{Dy} U_y + (1-\alpha)\, FyZ_y \quad (2.10)$$

in which α is the post-yielding to pre-yielding stiffness ratio, Fy is the yield force and Dy is the yield displacement. Z_x and Z_y are dimensionless variables governed by the following differential equations, which were proposed by (Park et al. 1986).

$$Dy\dot{Z}_x + \gamma \left|\dot{U}_x Z_x\right| Z_x + \beta \dot{U}_x {Z_x}^2 + \gamma \left|\dot{U}_y Z_y\right| Z_x + \beta \dot{U}_y Z_x Z_y - A\dot{U}_x = 0$$
$$Dy\dot{Z}_y + \gamma \left|\dot{U}_y Z_y\right| Z_y + \beta \dot{U}_y {Z_y}^2 + \gamma \left|\dot{U}_x Z_x\right| Z_y + \beta \dot{U}_x Z_y Z_x - A\dot{U}_y = 0 \quad (2.11)$$

Parameters A, β and γ are dimensionless—$A/(\beta+\gamma) = 1$ is chosen—and U_x,U_y and U_x,U_y represent, respectively, the displacements and velocities that occur at the isolation bearing. It can also be shown that the interaction curve of this biaxial model is circular.

In sliding bearings Teflon undergoes a small elastic shear deformation (less than 2 mm) before sliding commences. The small shear deformation of Teflon renders a finite but high elastic stiffness to the hysteretic loop, which can be captured by a hysteretic model. A hysteretic model for sliding bearings, which can account for the variation of coefficient of friction with velocity and bearing pressure observed in Teflon sliding bearings, has been presented by Tsopelas et al. (1994a), Constantinou et al. (1990), Nagarajaiah et al. (1991a), Mokha et al. (1993).

FPS bearings have been modeled using the hysteretic model in (2.11) with yield displacement, Dy, being very small (typically less than 2 mm).

$$F_X = \frac{N}{R} U_x + \mu_s N Z_x, \qquad F_Y = \frac{N}{R} U_y + \mu_s N Z_y \quad (2.12)$$

Where, μ_s is the coefficient of sliding friction and N is the normal load on the bearing. The normal load consists of gravity load, W, the effect of vertical ground acceleration, $\ddot{U}_v$, and the additional seismic load, P_{sl}, due to overturning moment.

$$N = W\left(1 + \frac{\ddot{U}_v}{g} + \frac{P_{sl}}{W}\right) \quad (2.13)$$

Where, g is the acceleration due to gravity. It should be noted that for flat sliding bearings R is infinite and (2.12) collapses to the model described in Constantinou

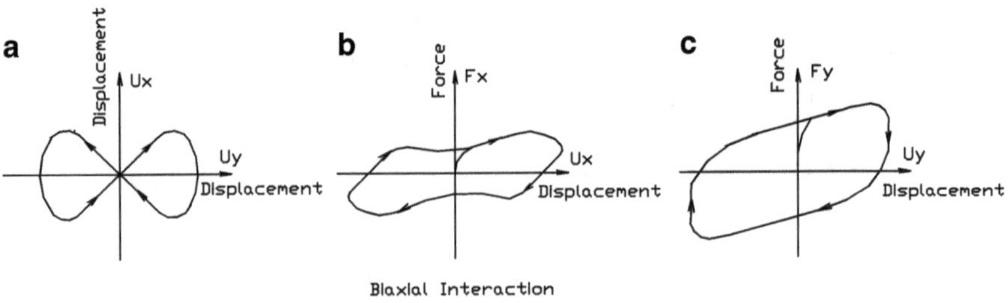

Fig. 2.9 Biaxial force-displacement behavior of flat sliding bearing: (**a**) Biaxial displacement profile, (**b**) Force-displacement response—X direction; (**c**) Force-displacement response—X direction

et al. (1990) and Nagarajaiah et al. (1991a) and experimentally verified by Mokha et al. (1993). A representative biaxial force-displacement behavior of flat slider is shown in Fig. 2.9. The eight shaped biaxial behavior is observed to have significant effect on the force-displacement behavior in the X direction.

Plasticity based models have been used to model isolation elements (SAP-ETABS 2014). Modified rate models have been used to represent the behavior of high damping bearings including stiffening (Tsopelas et al. 1994a; Kikuchi and Aiken 1997).

2.5 Geometric Nonlinearities of Base Isolation Bearings

2.5.1 Axial Load—Horizontal Displacement Effects

The elastomeric bearings when subjected to large axial forces and lateral horizontal displacements exhibit nonlinear and unstable behavior. This behavior is due to a combination of both geometric and material nonlinearities. The shear-force and horizontal displacement, F-u, curves are shown in Figs. 2.10 and 2.11.

The connection of elastomeric bearings can be either doweled or bolted to the column above and foundation below, which influences the behavior of the bearing. In doweled bearings rollover occurs and the displacement at which it commences can be calculated (Buckle and Kelly 1986). The rollover behavior can be modeled by including the P-Δ effects. In addition, since the bearing cannot sustain upward axial force because of doweled connection, uplift needs to be considered (as shown in Fig. 2.11).

Sliding bearings are stable even under large displacements. The change is axial load effects the coefficient of friction. Tsopelas et al. (1994a) and Nagarajaiah (1995) have modeled the influence of axial load on the coefficient of friction.

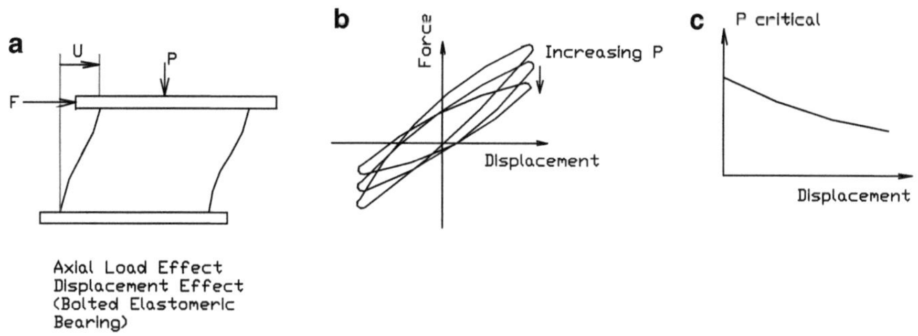

Fig. 2.10 Large axial load-horizontal displacement behavior of elastomeric bearings: (**a**) Bearing; (**b**) Horizontal force-displacement behavior; (**c**) Critical load as a function of horizontal displacement

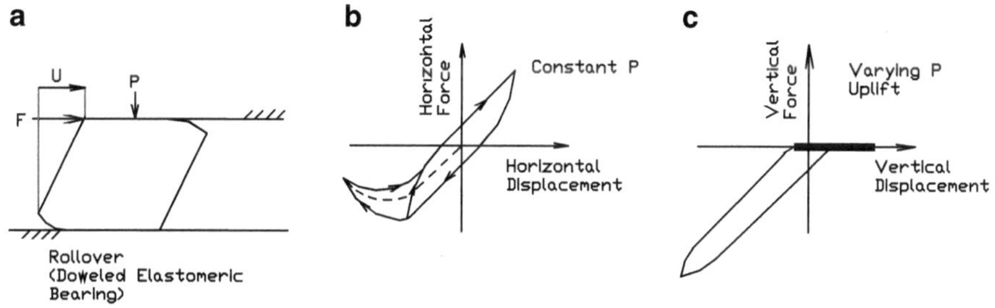

Fig. 2.11 Large axial load-horizontal displacement behavior of elastomeric bearings with rollover: (**a**) Bearing with rollover; (**b**) Horizontal force- displacement behavior with rollover; (**c**) Vertical force–vertical displacement behavior

2.6 Modeling Geometric Nonlinearities of Isolation Bearings

The axial load effect on elastomeric bearings has been modeled using linear models (Koh and Kelly 1986). A nonlinear analytical model has been developed by Nagarajaiah and Ferrell (1999), based on Koh-Kelly model, to include the effect of axial load and horizontal displacement. The large axial load horizontal behavior leads reduction in stiffness due to increasing axial load and reduction in critical axial load due to large horizontal displacement as shown in Fig. 2.10.

The critical load and horizontal displacement, P_{cr}-u, curves using the nonlinear analytical model developed by Nagarajaiah and Ferrell (1999) are shown in Fig. 2.10c which demonstrates that the bearing critical load drops with increasing horizontal displacement. The equilibrium paths demonstrate unstable post-critical behavior as observed in the experimental results. The critical load occurs at the limit point of each equilibrium path and horizontal tangential stiffness is zero at the limit point. The critical load drops with increasing horizontal displacement because the equilibrium paths are unstable. More details can be found in Constantinou et al. (2007).

2.7 Contact Nonlinearities of Base Isolation Systems

2.7.1 Uplift

Uplift occurs in doweled elastomeric bearings and in sliding bearings, due to loss of contact at the bearing as the column experiences axial tension forces (Fig. 2.11c). This loss of contact is reestablished as the cycle of motion reverses with compression forces in the column. Uplift is generally beneficial and reduces the base shear forces further (Nagarajaiah 1995). The effect of uplift can be modeled using a contact element with only compression stiffness (Nagarajaiah 1995; Nagarajaiah et al. 1992; Wilson 1993; Scheller and Constantinou 1998; SAP-ETABS 2014).

2.7.2 Pounding

The isolation gap (see Fig. 2.1) around the base isolated structure is provided to permit the maximum design displacement. Long period motions, as observed in 1994 Northridge earthquake and other recent earthquakes, could cause large base displacements. In such cases, unless appropriate measures are taken, pounding can occur accompanied by varying degree of damage to the superstructure. In the measured response of base isolated structures, subjected to Northridge earthquake, pounding has been observed when the isolation gap was not fully functional. The effect of pounding can be modeled by using gap elements (Nagarajaiah and Sun 2001).

2.8 Superstructure and Isolation System Modeling and Solution Procedures

The formulation and pseudoforce solution algorithm developed by Nagarajaiah (1990), Nagarajaiah et al. (1991a, b), which has been implemented in widely used computer program 3D-BASIS (Nagarajaiah et al. 1991a) for analyzing base isolated structures, is presented next. This is followed by brief description of the Ritz vector formulation and solution algorithm developed by Wilson (1993) for the widely used computer programs SAP-ETABS (2014).

2.8.1 Linear Superstructure and Nonlinear Isolation System

The superstructure, shown in Fig. 2.12, is treated as a linear elastic system. The superstructure and the base are modeled with three degrees of freedom per floor

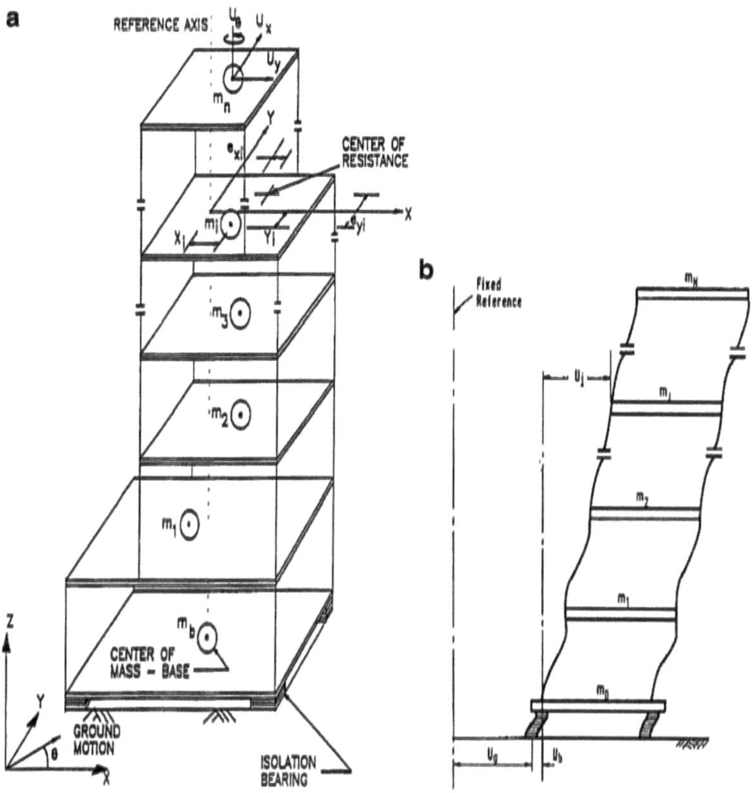

Fig. 2.12 (**a**) Asymmetric base isolated structure excited by bidirectional ground motion, (**b**) Displacement coordinates

at the center of mass as shown in Fig. 2.12a. The base and floors are assumed to be infinitely rigid in plane. The isolation system consisting of elastomeric or friction isolation bearings is modeled using discrete nonlinear biaxial elements. The equations of motion for the elastic superstructure are expressed in the following form.

$$M\ddot{u} + C\dot{u} + Ku = -M\ R\left(\ddot{u}_g + \ddot{u}_b\right) \tag{2.14}$$

Where, M, C, K are the superstructure mass, damping, and stiffness matrices in the fixed base condition, and R is the influence matrix. Furthermore, $\ddot{u}$, $\dot{u}$, and u represent the floor acceleration, velocity, and displacement vectors relative to the base, $\ddot{u}_b$ the vector of base acceleration relative to the ground, and $\ddot{u}_g$ the vector of ground acceleration (see Fig. 2.12b).

The equations of motion for the base are as follows

$$R^T M\left[\ddot{u} + R\left(\ddot{u}_b + \ddot{u}_g\right)\right] + M_b\left(\ddot{u}_b + \ddot{u}_g\right) + C_b\dot{u}_b + K_b u_b + f = 0 \tag{2.15}$$

Where, M_b = the diagonal mass matrix of the rigid base, C_b = the resultant damping matrix of viscous isolation elements, K_b = the resultant stiffness matrix of elastic isolation elements, and $\boldsymbol{f}$ = the global vector containing the forces mobilized in the nonlinear isolation elements with appropriate transformations. Employing modal reduction

$$u = \varphi\, u^* \tag{2.16}$$

Where, ϕ = the modal matrix, normalized with respect to the mass matrix, u^* = the modal displacement vector relative to the base. Combining Eqs. 2.14 to 2.16, the following matrix equation is obtained

$$\begin{bmatrix} \varphi^T M\varphi & \varphi^T MR \\ R^T M\varphi & R^T MR + M_b \end{bmatrix} \begin{Bmatrix} \ddot{u}^* \\ \ddot{u}_b \end{Bmatrix} + \begin{bmatrix} \varphi^T C\varphi & O \\ O & C_b \end{bmatrix} \begin{Bmatrix} \dot{u}^* \\ \dot{u}_b \end{Bmatrix} + \begin{bmatrix} \varphi^T K\varphi & O \\ O & K_b \end{bmatrix} \begin{Bmatrix} u^* \\ u_b \end{Bmatrix} + \begin{Bmatrix} O \\ f \end{Bmatrix} = -\begin{bmatrix} \varphi^T MR \\ R^T MR + M_b \end{bmatrix} \ddot{u}_g \tag{2.17}$$

Since the modal matrix ϕ is normalized with respect to mass, the following diagonal matrices are obtained $\varphi^T M\varphi = I$, $\varphi^T K\varphi = \omega^2$ and $\varphi^T C\varphi = 2\zeta\,\omega$, where ω = diagonal matrix of natural frequencies of the fixed base structure, ζ = diagonal matrix of damping ratios of the fixed base structure. The formulation in (2.17) developed by (Nagarajaiah 1990) facilitates efficient solution using pseudoforce algorithm.

2.8.2 Pseudoforce Formulation and Solution Algorithm

Equation 2.17 can be written as follows:

$$\tilde{M}\ddot{\tilde{u}}_t + \tilde{C}\dot{\tilde{u}}_t + \tilde{K}\tilde{u}_t + f_t = \tilde{P}_t \tag{2.18}$$

At time $t + \Delta t$

$$\tilde{M}\ddot{\tilde{u}}_{t+\Delta t} + \tilde{C}\dot{\tilde{u}}_{t+\Delta t} + \tilde{K}\tilde{u}_{t+\Delta t} + f_{t+\Delta t} = \tilde{P}_{t+\Delta t} \tag{2.19}$$

Written in incremental form

$$\tilde{M}\,\Delta\ddot{\tilde{u}}_{t+\Delta t} + \tilde{C}\,\Delta\dot{\tilde{u}}_{t+\Delta t} + \tilde{K}\,\Delta\tilde{u}_{t+\Delta t} + \Delta f_{t+\Delta t} = \tilde{P}_{t+\Delta t} - \tilde{M}\ddot{\tilde{u}}_t - \tilde{C}\dot{\tilde{u}}_t - \tilde{K}\tilde{u}_t - f_t \tag{2.20}$$

Where $\tilde{M}$, $\tilde{C}$, $\tilde{K}$, and $\tilde{P}$ represent the reduced mass, damping, stiffness, and load matrices in (2.20). Furthermore, the state of motion of modal superstructure and base is represented by vectors $\ddot{\tilde{u}}_t$, $\dot{\tilde{u}}_t$, $\tilde{u}_t$ in (2.17).

The incremental nonlinear global force vector $\Delta f_{t+\Delta t}$ in (2.20) is unknown. This global vector is brought on to the right hand side of (2.20) and treated as a pseudoforce vector. The two-stage solution algorithm developed by (Nagarajaiah 1990) involves the following steps: the solution of equations of motion using the unconditionally stable Newmark's constant average acceleration method, in the first stage and the solution of differential equations governing the behavior of the nonlinear isolation elements using the unconditionally stable semi-implicit Runge-Kutta method (Rosenbrock 1964), suitable for solutions of stiff differential equations, in the second stage. Furthermore, an iterative procedure consisting of corrective pseudoforces is employed within each time step until global equilibrium is achieved. The pseudoforce method with iteration is efficient due to the coefficient matrix of the equation of motion is formed and factorized only once at the beginning of the solution and repeatedly used. The method is particularly efficient due to the localized nonlinearities in the isolation elements that exist only at the base of the structure. The pseudoforce method converges to the correct solution even when severe nonlinearities such as planar sliding behavior along with biaxial effects are present. The method yields results of comparable accuracy of the predictor-corrector method (Nagarajaiah 1990).

The developed solution algorithm is as follows:

1. Initial Conditions.

   (a) Form stiffness matrix $\tilde{K}$, mass matrix $\tilde{M}$, and damping matrix $\tilde{C}$. Initialize $\tilde{u}_o, \dot{\tilde{u}}_o$ and $\ddot{\tilde{u}}_o$.

   (b) Select time step Δt, set parameters $\delta = 0.25$ and $\theta = 0.5$, and calculate the integration constants:

$$a_1 = \frac{1}{\delta(\Delta t)^2}, \quad a_2 = \frac{1}{\delta \Delta t}, \quad a_3 = \frac{1}{2\delta}$$
$$a_4 = \frac{\theta}{\delta \Delta t}, \quad a_5 = \frac{\theta}{\delta}, \quad a_6 = \Delta t \left(\frac{\theta}{2\delta} - 1\right)$$

   (c) Form the effective stiffness matrix

$$K^* = a_1 \tilde{M} + a_4 \tilde{C} + \tilde{K} \tag{2.21}$$

   (d) Triangularize K^* using Gaussian elimination (only if the time step is different from the previous step).

2. Iteration at each time step.

   (a) Assume the global pseudoforce vector

$$\Delta f^i_{t+\Delta t} = 0$$

   in iteration i = 1.

(b) Calculate the effective load vector at time $t + \Delta t$:

$$P^{*}_{t+\Delta t} = \Delta \tilde{P}_{t+\Delta t} - \Delta f^{l}_{t+\Delta t} + \tilde{M}\left(a_2 \dot{\tilde{u}}_t + a_3 \ddot{\tilde{u}}_t\right) + \tilde{c}\left(a_5 \dot{\tilde{u}}_t + a_6 \ddot{\tilde{u}}_t\right) \tag{2.22}$$

$$\Delta \tilde{P}_{t+\Delta t} = \tilde{P}_{t+\Delta t} - \left(\tilde{M}\ddot{\tilde{u}}_t + \tilde{C}\dot{\tilde{u}}_t + \tilde{K}\tilde{u}_t + f_t\right) \tag{2.23}$$

(c) Solve for displacements at time $t + \Delta t$:

$$K^{*}\Delta u^{l}_{t+\Delta t} = P^{*}_{t+\Delta t} \tag{2.24}$$

(d) Update the state of motion at time $t + \Delta t$:

$$\begin{aligned} \ddot{\tilde{u}}_{t+\Delta t} &= \ddot{\tilde{u}}_t + a_1 \Delta \tilde{u}^{l}_{t+\Delta t} - a_2 \dot{\tilde{u}}_t - a_3 \ddot{\tilde{u}}_t \\ \dot{\tilde{u}}_{t+\Delta t} &= \dot{\tilde{u}}_t + a_4 \Delta \ddot{\tilde{u}}^{l}_{t+\Delta t} - a_5 \dot{\tilde{u}}_t - a_6 \ddot{\tilde{u}}_t \\ \tilde{u}_{t+\Delta t} &= \tilde{u}_t + \Delta \tilde{u}^{l}_{t+\Delta t} \end{aligned} \tag{2.25}$$

(e) Compute the state motion at each bearing and solve for the nonlinear forces at each bearing using the semi-implicit Runge-Kutta method.
(f) Compute the nonlinear global force vector at the center of mass of the base $\Delta f^{l+1}_{t+\Delta t}$.
(g) Compute

$$error = \frac{\left|\Delta f^{l+1}_{t+\Delta t} - \Delta f^{l}_{t+\Delta t}\right|}{ref.\max f} \tag{2.26}$$

Where, $|\bullet|$ is the Euclidean norm.
(h) If error $\geq$ tolerance, further iteration is needed; iterate starting from step 2a and use $\Delta f^{l+1}_{t+\Delta t}$ as the pseudoforce vector and the state of motion at time t, $\tilde{u}_t$, $\dot{\tilde{u}}_t$ and $\ddot{\tilde{u}}_t$.

If error $\leq$ tolerance, no further iteration is needed; update the nonlinear global force vector $f_{t+\Delta t} = f_t + \Delta f^{l+1}_{t+\Delta t}$, reset time step if necessary, and go to step 2a if the time step is not reset or 1b if the time step is reset.

2.9 Semi-implicit Runge-Kutta Procedure in 3D-BASIS

The semi-implicit Runge-Kutta procedure in 3D-BASIS (Nagarajaiah 1990; Nagarajaiah et al. 1991a, b) was developed by the author as a part of his Ph.D. dissertation (Nagarajaiah 1990), based on the original work of Rosenbrock

(Rosenbrock 1964). The Bouc-Wen hysteretic model presented in Eq. (2.10) for uniaxial case (equation (11a-11b) for biaxial case) can be represented as follows:

$$\dot{z} = a_1 - a_2 z^{\eta} - a_3 z^{\eta} \tag{2.27}$$

The numerical procedure involves the following steps: Compute

$$z_{updated} = z_t + C_2 * k_r \tag{2.28}$$

Where,

$$k_r = \Delta t \left\{ \frac{1}{1 + \Delta t * C_1 * \left(-\frac{d\dot{z}_t}{dz} \right)} \right\} \dot{z}_t, \quad l_r = \Delta t \left\{ \frac{1}{1 + \Delta t * C_1 * \left(-\frac{d\dot{z}_t}{dz} \right)} \right\} \dot{z}_{updated}$$

Evaluate constants C^1 and C^2 using Rosenbrock's procedure (Rosenbrock 1964) and Taylor series [see Nagarajaiah (1990) for further details]. Calculate constants. Formulate

$$\dot{z}_t = z_P(1,1) = A_1 - A_2 (z_t)^{\eta} - A_3 (z_t)^{\eta} \tag{2.29}$$

With

$$\begin{aligned} z_{updated} &= z_t + C_2 \times RK \\ \dot{z}_{updated} &= A_1 - A_2 \left(z_{updated} \right)^{\eta} - A_3 \left(z_{updated} \right)^{\eta} \end{aligned} \tag{2.30}$$

Where,

$$RK = \frac{\dot{z}_t \Delta t}{\left\{ 1 + C_1 \Delta t \left(-\frac{d\dot{z}_t}{dz} \right) \right\}}, \quad -\frac{d\dot{z}_t}{dz} = \eta (z_t)^{\eta - 1} (A_2 + A_3)$$

Recalculate and then update

$$z_{t+\Delta t} = z_t + 0.75 \times RK + 0.25 \times RL \tag{2.31}$$

Where,

$$RL = \frac{\dot{z}_{updated} \Delta t}{\left\{ 1 + C_1 \Delta t \left(-\frac{d\dot{z}_t}{dz} \right) \right\}}, \quad -\frac{d\dot{z}_t}{dz} = \eta (z_t)^{\eta - 1} (A_2 + A_3)$$

This innovative solution procedure has been used by many other researchers at University at Buffalo, since its development by the author (Reinhorn 1994).

In summary the algorithm is summarized in the following table.

A. Initial Conditions:

1. Form stiffness matrix $\mathbf{K}^*$, mass matrix $\mathbf{M}^*$, and damping matrix $\mathbf{C}^*$. Initialize $\mathbf{u}_0^*$, $\dot{\mathbf{u}}_0^*$ and $\ddot{\mathbf{u}}_0^*$

2. Select time step Δt, set parameters $\delta = 0.25$ and $\theta = 0.25$, and calculate the integration constants:

$$\alpha_1 = \frac{1}{\delta(\Delta t)^2}; \alpha_2 = \frac{1}{\delta(\Delta t)}; \alpha_3 = \frac{1}{\delta};$$
$$\alpha_4 = \frac{\theta}{\delta(\Delta t)^2}; \alpha_5 = \frac{\theta}{\delta(\Delta t)}; \alpha_6 = \frac{\theta}{\delta}$$

3. Form the effective stiffness matrix

$$\mathbf{K}' = \alpha_1\mathbf{M}^* + \alpha_4\mathbf{C}^* + \mathbf{K}^*$$

4. Triangularize $\mathbf{K}'$ using Gaussian elimination(only if the time step is different from the previous step).

B. Iteration at each time step:

1 Assume the pseudo-force $\Delta\mathbf{f}^i_{t+\Delta t} = 0$ in iteration $i = 1$.

2. Calculate the effective load vector at time $t + \Delta t$:

$$\mathbf{P}'_{t+\Delta t} = \Delta\mathbf{P}^*_{t+\Delta t} - \Delta\mathbf{f}^*_{t+\Delta t} + \mathbf{M}^*(\alpha_2\dot{\mathbf{u}}^*_t + \alpha_3\ddot{\mathbf{u}}^*_t) + \bar{\mathbf{C}}(\alpha_5\dot{\mathbf{u}}^*_t + \alpha_6\ddot{\mathbf{u}}^*_t),$$
$$\Delta\mathbf{P}^*_{t+\Delta t} = \mathbf{P}^*_{t+\Delta t} - (\mathbf{M}^*\ddot{\mathbf{u}}^*_t + \mathbf{C}\dot{\mathbf{u}}^*_t + \mathbf{K}\mathbf{u}^*_t + \mathbf{f}_t)$$

3. Solve for displacements at time $t + \Delta t$:

$$\mathbf{K}'\Delta\mathbf{u}^i_{t+\Delta t} = \mathbf{P}'_{t+\Delta t}$$

4. Update the state of motion at time $t + \Delta t$:

$$\ddot{\mathbf{u}}^*_{t+\Delta t} = \ddot{\mathbf{u}}^*_t + \alpha_1\Delta\mathbf{u}^{*i}_{t+\Delta t} - \alpha_2\dot{\mathbf{u}}^*_t - \alpha_3\ddot{\mathbf{u}}^*_t;$$
$$\dot{\mathbf{u}}^*_{t+\Delta t} = \dot{\mathbf{u}}^*_t + \alpha_4\Delta\mathbf{u}^{*i}_{t+\Delta t} - \alpha_5\dot{\mathbf{u}}^*_t - \alpha_6\ddot{\mathbf{u}}^*_t;$$
$$\mathbf{u}^*_{t+\Delta t} = \mathbf{u}^*_t + \Delta\mathbf{u}^{*i}_{t+\Delta t}$$

5. Compute the state of motion at teach bearing and solve for the nonlinear force at each bearing using semi-implicit Runge-kutta method.

6. Compute the resultant nonlinear force vector at the center of mass of the base $\Delta\mathbf{f}^{i+1}_{t+\Delta t}$.

7 Compute

$$Error = \frac{|\Delta\mathbf{f}^{i+1}_{t+\Delta t} - \Delta\mathbf{f}^{i}_{t+\Delta t}|}{\text{Ref}\cdot Max\cdot Moment}$$

Where $|\cdot|$ is the euclidean norm.

8. If Error$\geq$ tolerance, further iteration is needed, iterate starting form step B-1 and use $\Delta\mathbf{f}^{i+1}_{t+\Delta t}$ as the pseudo-force and the state of motion at time t, $\mathbf{u}^*_t$, $\dot{\mathbf{u}}^*_t$ and $\ddot{\mathbf{u}}^*_t$.

9 If Error$\leq$ tolerance, no further iteration is needed, update the nonlinear force vector: $\mathbf{f}_{t+\Delta t} = \mathbf{f}_t + \Delta\mathbf{f}^{i+1}_{t+\Delta t}$ reset time step if necessary, go to step B-1 if the time step is not reset or A-2 if the time step is reset.

2.10 3D-BASIS Suite of Computer Programs

The presented formulation and solution algorithm has been implemented in—the most widely used software for analyzing base isolated structures—the class of computer programs 3D-BASIS (Nagarajaiah et al. 1991a, b), 3D-BASIS-M (Tsopelas et al. 1991), 3D-BASIS-ME (Tsopelas et al. 1994b), and 3D-BASIS-

TABS (Nagarajaiah et al. 1993; Reinhorn et al. 1994). 3D-BASIS-TABS is a combination of 3D-BASIS and ETABS—a widely used building analysis software (Wilson et al. 1975). 3D-BASIS-TABS offers the advantage of modeling the superstructure using linear elastic beam, column, and shear wall elements, while 3D-BASIS, 3D-BASIS-M, and 3D-BASIS-ME can only model the superstructure using the condensed version with 3 degrees of freedom per floor. 3D-BASIS-M and 3D-BASIS-ME offer the advantage of analyzing multiple buildings on a common isolation basemat, with the isolation system below, while only a single building on isolation system can be analyzed using 3D-BASIS and 3D-BASIS-TABS. Most recent in the series is 3D-BASIS-ME-MB, which includes the capability of model uplift (Tsopelas et al. 2005). 3D-BASIS class of programs are distributed through the Multidisciplinary Center for Earthquake Engineering, Buffalo, and National Information Service for Earthquake Engineering, University of California, Berkeley.

2.11 ETABS and SAP

The formulation in the widely used ETABS is similar to that of 3D-BASIS, but for the use of Ritz modal vectors instead of eigenvectors. Time history analysis is performed by mode superposition method. The modal equations are integrated by a method, which is exact for a linear variation of the load during the time step. The forces in the nonlinear elements are calculated at the end of each time step. The forces are treated as pseudoforces and brought on to the right hand side of the equations of motion. Iteration is then performed within the time step until convergence is achieved.

The most recent version of the widely used SAP series and ETABS series (Wilson et al. 1975) are computer programs SAP and ETABS (SAP-ETABS 2014). ETABS is a linear building analysis computer program and SAP is a finite element computer program. The programs have linear elastic beam, shell, plane, and solid elements. Their latest versions ETABS and SAP have discrete nonlinear elements. The nonlinear elements in the program include uniaxial and biaxial plasticity element, viscous damper element with nonlinear exponent on velocity term, gap (compression only) and hook (tension only) element, biaxial plasticity element, biaxial element for friction and/or pendulum behavior (Scheller and Constantinou 1998). These elements allow analysis of complete three dimensional superstructure models with localized nonlinear elements such as elastomeric and friction isolators and damping devices. ETABS and SAP use a similar solution procedure to that used in 3D-BASIS. SAP and ETABS (SAP-ETABS 2014) are proprietary computer programs of Computers and Structures Inc., Berkeley, CA.

2.12 Key Innovations

The acronym 3D-BASIS stands for **3D-BAS**e **I**solated **S**tructures, coined by the author in 1989. Several key innovations in formulation and computational techniques, needed in the development of 3D-BASIS, have been summarized in this chapter. Briefly,

1. New Nonlinear/Inelastic Analytical Models for Elastomeric Bearings, Sliding Bearings, Three Dimensional Sliding and Elastomeric Base Isolated Structures
2. Formulation of the Pseudo-force Solution Algorithm, Semi-Implicit Algorithm, and proof of Convergence using Closed Form Solutions (Nagarajaiah 1990)
3. Due to highly nonlinear biaxial interaction for sliding isolation bearings, the differential equations become very stiff and even Gear's predictor-corrector method for stiff differential equation fails
4. Formulated a new stable semi implicit integration method based on Rosenbrock's (1964) method
5. Integration Coefficients 0.78886751 and −1.1547005 were derived originally by Nagarajaiah (1990) to maintain a fourth order truncation error
6. Verification using extensive shake table test results.

2.13 3D-BASIS Used for Analysis of Important Base Isolation Projects Around the World

3D-BASIS for Nonlinear Dynamic Analysis of Base Isolated Structures has been cited in several important code related documents [FEMA 273/274 (FEMA 273/274 1997), ATC 33, NEHRP, NIST]. "The most widely used computer program for analyzing base isolated structures today is the 3D-BASIS suite of programs . . ." is a direct quote from the book on "Earthquake Resistant Design with Rubber" by Professor James M. Kelly (1997) —see page 234.

3D-BASIS has been used for analysis and design of numerous projects around the world; the most important of which are listed below.

- U. S. Court of Appeals, San Francisco, CA, 1990–1991
- LNG Tanks, Greece, 1994
- San Francisco International Airport, CA, 1996
- ATATURK International Airport in Istanbul, Turkey, 2000
- Statue of Hermes, Museum at Olympia, Greece, 2004
- Mills Peninsula Hospital, Burlingame, CA, 2005-Currently Complete
- Washington Hospital, Fremont, CA, 2005-Currently Complete
- Stanford University Hospital, CA, 2008—Currently complete
- Lunskoye and Piltun Offshore Oil Platforms, Sakhalin, Russia, 2008—Currently Complete
- San Francisco General Hospital, CA, 2011—under construction
- Arkundagi Offshore Oil Platform, Sakhalin, Russia, 2013—Under Construction

2.14 Concluding Remarks

3D-BASIS release was significant milestone in nonlinear dynamic analysis of three-dimensional base isolated structures, particularly in solving the highly nonlinear bidirectional stick-slip hysteretic response of a collection of sliding isolation bearings and the resulting response of the superstructure. In this chapter techniques used in the nonlinear dynamic analysis of base isolated structures in 3D-BASIS have been presented. These techniques developed by the author and other researchers have been implemented in the latest NEHRP Guidelines/ Commentary for the Seismic Rehabilitation of Buildings (Ballot Version) FEMA 273/274 (1997). The presented nonlinear dynamic techniques and computer programs have been widely used in the analysis and design of many new and retrofit base isolation projects around the United States and also around the world. In summary,

- 3D-BASIS has had a unique impact on nonlinear dynamic analysis of base isolated structures around the world
- SAP uses the similar formulation as in 3D-BASIS; for more detail refer to Wilson et al. (1993) and SAP-ETABS (2014).
- The latest version of 3D-BASIS—3D-BASIS-MB-ME—Includes Triple Pendulum Model, uplift and new response prediction and display features.

Results from OpenSees (2014) and SAP-ETABS (2014), which are widely used currently in earthquake engineering simulation, are verified using 3D-BASIS suite of programs by many researchers and practitioners (Sarkisian and Lee 2011). In addition now that Structural Health Monitoring is being adopted steadily more real measured data will become available to validate and verify any future developments (Nagarajaiah and Sun 2000, 2001).

Acknowledgments 3D-BASIS suite of computer programs would not have been possible without the vision of Professor Andrei Reinhorn. 3D-BASIS suite of computer programs has been realized by the sustained efforts of the author, Professor Andrei Reinhorn, and his collaborators at University at Buffalo.

References

Bouc R (1967) Forced vibration of mechanical systems with hysteresis. In: Proceedings 4th conference on nonlinear oscillations, Prague. See also Mathematical model for hysteresis. Report to Centre de Recherches Phy-siques, Marseille, France, pp 16–25

Buckle IG, Kelly J (1986) Properties of slender elastomeric isolation bearings during shake table studies of a large scale model bridge deck. Jt Seal Bearing Syst ACI 1:247–269

Buckle IG, Mayes RL (1990) Seismic isolation: history, application, and performance—a world overview. Earthq Spectra 6(2):161–202

Constantinou MC, Symans MD (1993) Experimental study of seismic response of buildings with supplemental fluid dampers. J Struct Des Tall Build 2:93–132

Constantinou MC, Mokha A, Reinhorn AM (1990) Teflon bearings in base isolation II: modeling. J Struct Eng ASCE 16(4):455–474

Constantinou MC, Whittaker AS, Kalpakidis Y, Fenz DM, Warn GP (2007) Performance of seismic isolation hardware under service and seismic loading, Technical report MCEER-07-0012

Dao ND, Ryan KL, Sato E, Sasaki T (2013) Predicting the displacement of triple pendulum™ bearings in a full-scale shaking experiment using a three-dimensional element. Earthq Eng Struct Dyn. doi:10.1002/eqe.2293

FEMA 273/274 (1997) NEHRP guidelines/commentary for the seismic rehabilitation of buildings, Federal Emergency Management Agency

Fenz DM, Constantinou MC (2008a) Spherical sliding isolation bearings with adaptive behavior: theory. Earthq Eng Struct Dyn 37(2):163–183

Fenz DM, Constantinou MC (2008b) Modeling triple friction pendulum bearings for response-history analysis. Earthq Spectra 24(4):1011–1028

Kelly JM (1997) Earthquake-resistant design with rubber, 2nd edn. Springer, New York

Kikuchi M, Aiken ID (1997) An analytical hysteresis model for elastomeric seismic isolation bearings. Earthq Eng Struct Dyn 26(2):214–231

Koh CG, Kelly JM (1986) Effects of axial load on elastomeric bearings, Report no. UCB/EERC-86/12. Earthquake Eng. Res. Ctr., University of California, Berkeley

Mokha A, Constantinou MC, Reinhorn AM (1993) Verification of friction model of teflon bearings under triaxial load. J Struct Eng ASCE 119(1):240–261

Morgan TA, Mahin SA (2011) The use of innovative base isolation systems to achieve complex seismic performance objectives, PEER-Center-Report 2011/06. UC Berkeley

Nagarajaiah S (1990) Nonlinear dynamic analysis of three dimensional base isolated structures. Ph.D. dissertation, University at Buffalo

Nagarajaiah S (1995) Seismic response of multistory sliding isolated structures with uplift. In: Proceedings of structures congress '95. ASCE, Boston, pp 1044–1047

Nagarajaiah S, Ferrell K (1999) Stability of elastomeric seismic isolation bearings. J Struct Eng ASCE 125(9):946–954

Nagarajaiah S, Sun X (2000) Response of base isolated USC hospital building in Northridge earthquake. J Struct Eng ASCE 126(10):1177–1186

Nagarajaiah S, Sun X (2001) Base isolated FCC building: impact response in Northridge earthquake. J Struct Eng ASCE 127(9):1063–1074

Nagarajaiah S, Reinhorn AM, Constantinou MC (1991a) 3D-BASIS nonlinear dynamic analysis of three dimensional base isolated structures: Part II, Report no. 91-0005. National Ctr. for Earthquake Eng. Res., University of Buffalo, Buffalo

Nagarajaiah S, Reinhorn AM, Constantinou MC (1991b) Nonlinear dynamic analysis of three dimensional base isolated structures. J Struct Eng ASCE 117(7):2035–2054

Nagarajaiah S, Reinhorn AM, Constantinou MC (1992) Experimental study of sliding isolated structures with uplift restraint. J Struct Eng ASCE 118(6):1666–1682

Nagarajaiah S, Li C, Reinhorn AM, Constantinou MC (1993) 3D-BASIS-TABS computer program for nonlinear dynamic analysis of three dimensional base isolated structures, Report NCEER-93-0011. National Ctr. for Earthquake Eng. Res., Buffalo

OpenSees (2014) Open system for earthquake engineering simulation, An open-source program available at http://opensees.berkeley.edu

Park YJ, Wen YK, Ang AHS (1986) Random vibration of hysteretic systems under bidirectional ground motions. Earthq Eng Struct Dyn 11(6):749–770

Ray T, Reinhorn A (2012) Enhanced smooth hysteretic model with degrading properties. J Struct Eng. doi:10.1061/(ASCE)ST.1943-541X.0000798

Reinhorn A (1994) Personal communication with professor Reinhorn

Reinhorn AM, Nagarajaiah S, Constantinou MC, Tsopelas P, Li R (1994) 3D-BASIS-TABS: V2.0 computer program for nonlinear dynamic analysis of 3D base isolated structures, Technical report no. 94-0018. National Ctr. for Earthquake Eng. Res., Univ. of Buffalo, Buffalo

Rosenbrock HH (1964) Some general implicit processes for numerical solution of differential equations. Comput J 18(1):50–64

SAP-ETABS (2014) Structural analysis and Design Software. Computers and Structures Inc., Berkeley

Sarkisian M, Lee P (mark.sarkisian@som.com, peter.lee@som.com) (2011) Personal communication (S.M.-Engineering Director and L.P., Associate Director), Skidmore, Owings & Merrill, San Francisco, Mar 7, 2011

Scheller J, Constantinou MC (1998) Response history analysis of structures with seismic isolation and energy dissipation systems: verification examples for program SAP2000, Technical report no. 99-0002. Multidisciplinary Ctr. for Earthquake Eng. Res., Univ. of Buffalo, Buffalo

Skinner RI, Johnson H, McVerry H (1993) Introduction to seismic isolation. Wiley, New York

Soong TT, Constantinou MC (1994) Passive and active structural vibration control in civil engineering. Springer, New York

Tsopelas P, Nagarajaiah S, Constantinou M C, Reinhorn AM (1991) 3D-BASIS-M: nonlinear dynamic analysis of multiple building base isolated structures, Report no. 91-0014. National Ctr. for Earthquake Eng. Res., Univ. of Buffalo, Buffalo

Tsopelas P, Nagarajaiah S, Constantinou MC, Reinhorn AM (1994a) Nonlinear dynamic analysis of multiple building base isolated structures. J Comp Struct 50(1):47–57

Tsopelas P, Constantinou MC, Reinhorn AM (1994b) 3D-BASIS-ME: nonlinear dynamic analysis of seismically isolated single and multiple structures, Report no. 94-0010. National Ctr. for Earthquake Eng. Res., Univ. of Buffalo, Buffalo

Tsopelas P, Roussis PC, Constantinou MC, Buchanan R, Reinhorn M (2005) 3D-BASIS-ME-MB: computer program for nonlinear dynamic analysis of seismically isolated structures, Technical report no. MCEER-05-0009. Multidisciplinary Ctr. for Earthquake Eng. Res., Univ. of Buffalo, Buffalo

Wen YK (1976) Method of random vibration of hysteretic systems. J Struct Eng ASCE 117(7):2035–2054

Wilson EL (1993) An efficient computational method for the base isolation and energy dissipation analysis of structural systems. Proc Semin Seism Isolation Passive Energy Dissipation ATC 17-1:365–376

Wilson EL, Hollings JP, Dovey HH (1975) ETABS—three dimensional analysis of building systems. Report no. UCB/EERC-75/13. Earthquake Eng. Res. Center, Univ. of California, Berkeley

Chapter 3
The Genesis of IDARC and Advances in Macromodeling for Nonlinear Analysis of RC Structures

Sashi K. Kunnath

Abstract The development of IDARC (Park et al., IDARC: Inelastic Damage Analysis of Reinforced Concrete frame shear-wall structures. Technical report NCEER 87-0008, State University of New York, Buffalo, 1987) (Inelastic Damage Analysis of Reinforced Concrete structures) was initially motivated by the need for an efficient computer program to aid the design of shaking table experiments but then evolved into a tool for seismic assessment of reinforced concrete structures. The primary modeling technique employed in the IDARC computational platform is the representation of the overall behavior of components through macromodels. The effectiveness of the macro elements is enhanced through the introduction of distributed flexibility models that account for the effects of spread plasticity. Nonlinear material behavior is specified by means of a generic hysteretic force-deformation model that incorporates stiffness degradation, strength deterioration and pinching or bond-slip effects. Solution modules for nonlinear static, monotonic, quasi-static cyclic and transient seismic loads were implemented. The final response quantities are expressed in terms of damage indices that provide engineers with a qualitative interpretation of the structural response.

3.1 Introduction

The primary objective of any analysis, linear or nonlinear, is to estimate force and deformation demands at critical regions and to assess the likely performance of the structure due to the imposed loading. When the loading is deterministic and the system is expected to respond linearly, the modeling tasks are significantly simplified. However, for transient non-deterministic loads and structural systems that

S.K. Kunnath (✉)
Civil and Environmental Engineering, University of California, Davis, CA 95616, USA
e-mail: skkunnath@ucdavis.edu

G.P. Cimellaro et al. (eds.), *Computational Methods, Seismic Protection, Hybrid Testing and Resilience in Earthquake Engineering*, Geotechnical, Geological and Earthquake Engineering 33, DOI 10.1007/978-3-319-06394-2_3,
© Springer International Publishing Switzerland 2015

exhibit degrading inelastic behavior, the variables and uncertainties associated with selecting model variables or choosing an analysis method are more challenging. The consequences of choosing a simple model to carry out a complex nonlinear analysis or using a complex model to conduct a simple linear static analysis can be far-reaching. It is always important to ascertain the level of complexity required to achieve a desired solution.

The range of modeling options available to a designer is diverse. If the purpose of an analysis is to determine the state of stress or strain at a particular location in a structural component or connection, then it is necessary to resort to finite element models that incorporate detailed constituent material behavior. However, if the quantities of interest are more global in nature such as member rotation or inter-story drift in a building, then the use of approximate member models may be sufficient.

Since the inelastic behavior of RC structures, particularly under seismic loads, is mostly concentrated at known locations in an element, a macromodel approach can often be used with remarkable reliability. Macromodels are computationally efficient and offer a great deal of flexibility in modeling. Since it is possible to account for a variety of behavior patterns in an equivalent sense, they can be used to effectively and efficiently model the global response of reinforced concrete structures.

Material nonlinearities typically account for most of the inelastic behavior in reinforced concrete and is usually incorporated in one of the following ways: (i) through the use of constitutive stress-strain relations at a microscopic level (micromodels) as is done in the finite element or fiber model analyses; or (ii) a macromodel approach wherein the inelastic material behavior is described using force-deformation rules (also called hysteresis rules) which reflect member behavior as a whole.

3.2 Development of Force-Deformation Models

A typical load-displacement relationship of a reinforced concrete member is characterized by three stages: an elastic uncracked stage, a stage characterized by crack propagation, and a yielded or plastic stage. Of these, nonlinearity is evident only in the second and third stages and are a result of two major material effects: cracking of concrete and plasticity of the reinforcement. Other sources of non linearity also exist, such as bond-slip, aggregate interlock in cracked concrete and dowel action of the reinforcing steel, all of which are essentially independent of time, and long-term behavior such as creep and shrinkage which are time-dependent phenomena.

The primary difference between a macromodel approach and a finite element or fiber model representation is that, for macromodels, no constitutive equations are used. Instead, the inelastic behavior is described using force-deformation rules which reflect member behavior as a whole. The basis for development of force-deformation models is experimental testing.

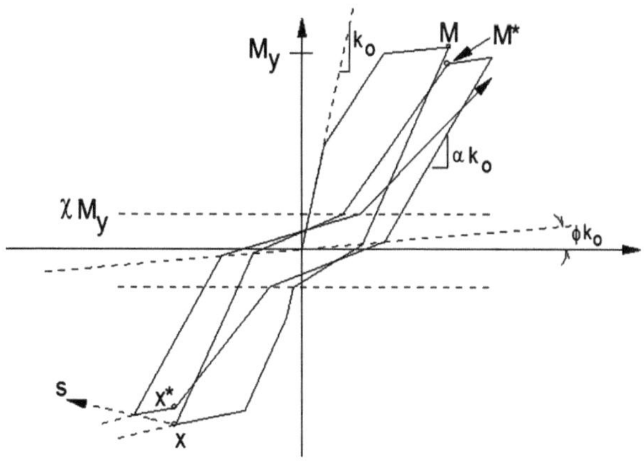

Fig. 3.1 Multi-parameter degrading hysteresis model

3.2.1 Uniaxial Models

Numerous models to reproduce a range of force-deformation behavior can be found in the literature. Prior to the development of stiffness degrading models, most nonlinear programs used simple bilinear force-deformation representations. Part of the reason for this was the fact that non-degrading bilinear models were adequate for analysis of steel structures and complexities associated with numerical implementation were minimal. The first notable improvement of the bilinear model was provided by Clough and Johnston (1966). The next significant contribution in RC analysis came from Takeda et al. (1970) that also laid the foundation of so-called "degrading" hysteresis models. Since then numerous enhancements have occurred in hysteresis model development with considerations of strength deterioration, pinching, bond-slip becoming rather commonplace in structural modeling.

A typical model with the versatility to incorporate most of the desired effects in reinforced concrete is shown in Fig. 3.1. This basic model, which is implemented in the IDARC series of programs, uses several control parameters to establish the rules under which inelastic loading reversals take place. A variety of hysteretic properties can be achieved through the combination of a non-symmetric trilinear curve and these control parameters which characterize stiffness degradation, strength deterioration and pinching, respectively. For example, α, which can be expressed as a function of the deformation, controls the amount of stiffness loss; ϕ and χ control the initiation and degree of pinching; and the slope **s** and the change in expected peak strength (M to M*) controls the softening due to system deterioration. A sample simulation of observed behavior using this model is shown in Fig. 3.2. The hysteresis curves in this case was obtained from tests of a precast concrete connection with a hybrid combination of mild steel and post-tensioning steel (Cheok et al. 1998). However, the specification of hysteretic rules in an actual analysis is rather empirical and should be based on available experimental data. In the absence of such data, a parametric study to evaluate the sensitivity of these parameters is necessary.

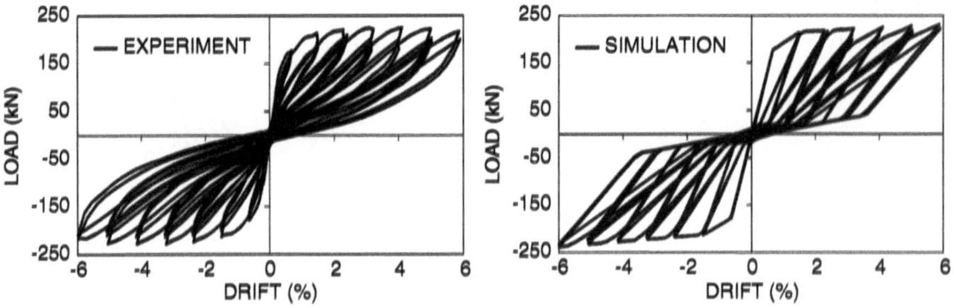

Fig. 3.2 Simulation of observed hysteretic behavior

3.2.2 Bidirectional Models

Experimental evidence suggests that stiffness and strength degradation in reinforced concrete is accelerated in the presence of biaxial moments and varying axial loads. Yield surface models and fiber models are generally the only means of considering multiaxial material behavior in the inelastic range. However, in the context of macromodels, a convenient scheme for expressing bi-directional moments with possible interaction effects is needed for implementation in an analytical framework. One such scheme, which is an extension of the Bouc-Wen formulation (Bouc 1967; Wen 1976), was proposed by Kunnath and Reinhorn (1990). An expression between moment and curvature is derived, as follows:

$$\begin{Bmatrix} M_x \\ M_y \end{Bmatrix} = \alpha \begin{bmatrix} (EI)_x & 0 \\ 0 & (EI)_y \end{bmatrix} \begin{Bmatrix} \phi_x \\ \phi_y \end{Bmatrix} + (1-\alpha) \begin{bmatrix} M_x^y & 0 \\ 0 & M_y^x \end{bmatrix} \begin{Bmatrix} Z_x \\ Z_y \end{Bmatrix} \tag{3.1}$$

M_x, M_y, ϕ_x and ϕ_y are the moments and curvatures in the x and y directions respectively; M_x^y and M_y^x are the corresponding yield moments in the respective directions; α is the post-yield stiffness ratio; and Z_x, *and* Z_y are the non-dimensional hysteretic force components given by:

$$\begin{Bmatrix} \dot{Z}_x \\ \dot{Z}_y \end{Bmatrix} = (A\,[I] - B\,[\Omega])\ \ [\Phi] \begin{Bmatrix} \dot{\phi}_x \\ \dot{\phi}_y \end{Bmatrix} \tag{3.2}$$

$$[\Phi] = \begin{bmatrix} 1/\phi_x^y & 0 \\ 0 & 1/\phi_y^y \end{bmatrix}; \quad [\Omega] = \begin{bmatrix} Z_x^2\left[\operatorname{sgn}(\dot{\phi}_x Z_x)+1\right] & Z_x Z_y\left[\operatorname{sgn}(\dot{\phi}_y Z_y)+1\right] \\ Z_x Z_y\left[\operatorname{sgn}(\dot{\phi}_x Z_x)+1\right] & Z_y^2\left[\operatorname{sgn}(\dot{\phi}_{yx} Z_y)+1\right] \end{bmatrix} \tag{3.3}$$

$$\operatorname{sgn}(\dot{\phi}\, Z) = 1 \quad \text{if} \quad \dot{\phi}\, Z > 0 \quad \text{or} \quad = -1 \quad \text{if} \quad \dot{\phi}\, Z < 0 \tag{3.4}$$

Details of the implementation of the above model which also incorporates stiffness and strength degradation are described in Kunnath and Reinhorn (1990). However, it must be mentioned that it is essential to select the base parameters of the model to eliminate rate-effects (Constantinou and Adnane 1987). The Bouc-Wen model and its derivatives also exhibit erroneous drift accumulation during constant-amplitude cycling in the same quadrant. The numerical implementation of this model, therefore, requires certain modifications to eliminate or minimize this error prior to use in a nonlinear analysis program.

3.3 Element Models

The basic objective in developing a member model for nonlinear analysis is to adequately characterize the varying stiffness properties of the element during a cyclic loading event.

3.3.1 Beam-Column Elements

To adequately represent spread plasticity due to distributed cracking in a beam-column element, IDARC utilizes a distributed flexibility approach wherein the flexibility matrix, including shear distortions, relating moments and rotations at the ends of the element is typically expressed as:

$$\begin{Bmatrix} \theta_A \\ \theta_B \end{Bmatrix} = \begin{bmatrix} f_{AA} & f_{AB} \\ f_{BA} & f_{BB} \end{bmatrix} \begin{Bmatrix} M_A \\ M_B \end{Bmatrix} \tag{3.5}$$

In the above equation, θ_A and θ_B are the rotations at the ends, M_A and M_B are the moments at the ends of the element. The flexibility coefficients are obtained from:

$$f_{ij} = \int_0^L \left[\frac{m_i(x)m_j(x)}{EI(x)}\right] dx + \int_0^L \left[\frac{v_i(x)v_j(x)}{GA_Z}\right] dx \tag{3.6}$$

where $m_i(x)$ and $m_j(x)$ are the moment distributions due to a virtual unit moment at end "*i*" or "*j*", respectively; $v_i(x)$ and $v_j(x)$ are the corresponding shear distributions. In such a formulation, the flexibility and resulting stiffness matrix is a function of the assumed curvature distribution across the member. Figure 3.3 shows the moment distribution in a typical beam element subjected to combined gravity and lateral loads. A pair of linear distributions (one which considers only yielding and the other which considers both cracking and yielding) is shown in Fig. 3.3. More complex nonlinear distributions can also be considered.

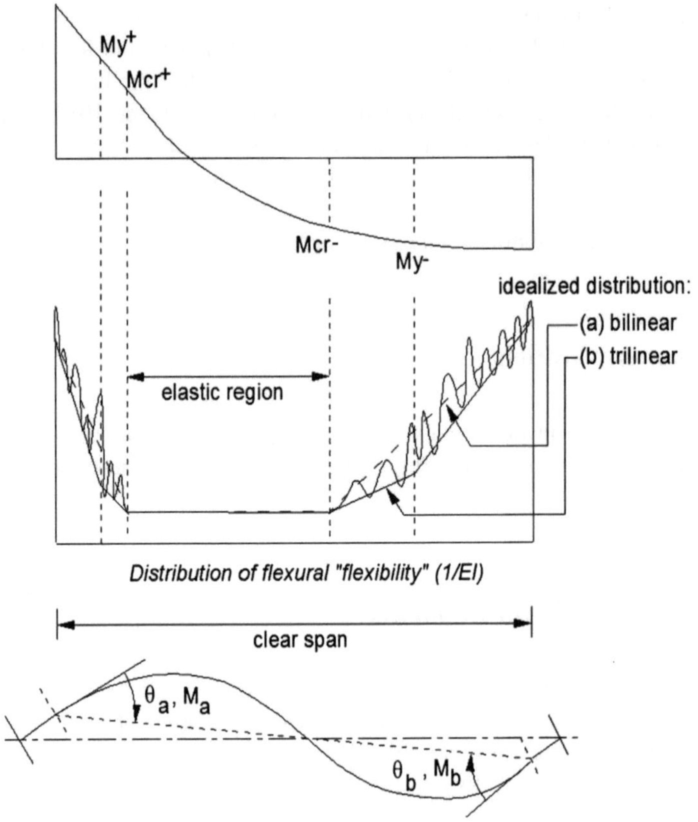

Fig. 3.3 Moment and curvature distribution in a flexural member bent in double curvature

3.3.2 *Shear Wall Elements*

Despite their enormous stiffness and strength, walls can respond in a stable ductile fashion under flexural conditions. In the past, walls were modeled as a single vertical element located at the center of the wall. If the wall was part of a building frame, this central vertical element was connected to adjoining beams by rigid links. An enhanced macro element model of a shear wall is the multiple-vertical-line-element (MVLE) model. A representation of this model is displayed in the general building model shown later in Fig. 3.5. It consists of a central vertical line to model flexural and shear behavior in the wall unit as a single entity and additional axial elements along the wall length to model axial stresses resulting from flexural deformations. Such a model was first used by Kabayesawa et al. (1983) in an attempt to capture the response of a seven story shear wall building in a full-scale pseudo-dynamic test conducted in Japan. Enhancements and variations of the model have also been proposed (Vulcano et al. 1989; Fajfar and Fischinger 1990). Stacking a set of MVLE to represent a single wall panel appears to produce the best results.

One drawback of the MLVE model lies in ascertaining the correct failure mode of the wall. The properties of the shear spring in the MVLE model are dependent on the flexural strength and axial force. Since the properties of the spring have to be assigned prior to the analysis, the potential interaction of these forces is not properly accounted. Separate formulations to compute the shear vs. strain characteristics of the wall panel can be derived from strut-and-tie modeling.

3.3.3 Beam-Column Joints

Since seismic design requires the total column capacity to be in excess of 20 % of the total girder capacity at the joint, the intersection of large column cross-sections with beams results in a sizable joint region that can deform in shear and contribute to the overall joint rotations. Moreover, the shear capacity of the joint may be exceeded in certain inadequate or nonductile connections leading to additional inelastic distortion and rapid deterioration of the connection. When the joint shear forces are high, and the beam column panel zones are not adequately reinforced, panel-zone distortions can have measurable impact on the story drift.

Existing procedures in frame analysis assume either (a) the panel to be rigid in which case the angle between the panel zone and the adjacent members (beams and columns) are assumed to remain at right angles even after the panel zone has undergone severe shear deformation, or (b) a linear, elastic relationship to exist between the shearing forces and panel-zone distortions. In the former case, a single moment and associated joint rotation is used at the center of the panel. The latter approach recognizes the significance of joint deformations but does not account for large inelastic rotations that may occur if the shear capacity of the joint is exceeded. A simple macro-model that can be used effectively to represent joint deformations is shown in Fig. 3.4. A panel zone mechanism based on pure shear deformations form the basis of the model. In such a model, it is assumed that column moments can differ from beam moments. The panel zone connecting the beams and columns behave as rigid bars and can deform independent of one another. With reference to the figure, M_b, θ_b, M_c, θ_c are the moments and rotations of beams and columns, respectively. The shear distortion of the panel, γ_p is the relative change in the rotations of the beam and column element, as follows:

$$\gamma_p = (\theta_b - \theta_c) \tag{3.7}$$

A relationship of the following form can be derived:

$$\begin{Bmatrix} M_c \\ M_b \end{Bmatrix} = V_p G \begin{bmatrix} 1 & -1 \\ -1 & 1 \end{bmatrix} \begin{Bmatrix} \theta_c \\ \theta_b \end{Bmatrix} \tag{3.8}$$

Here, V_p is the volume of the panel and G is the shear modulus of the material. The shear versus strain behavior is specified by means of a trilinear envelope with

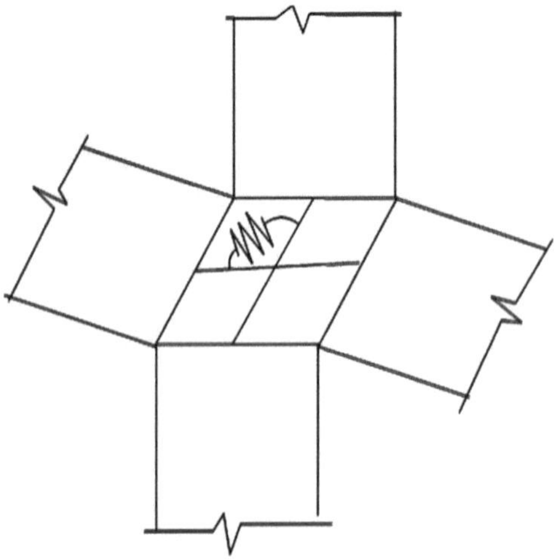

Fig. 3.4 Modeling joint deformations in beam-column joints

degrading cyclic characteristics. This macromodel was developed and implemented in IDARC by Alath and Kunnath (1995) and has been successfully used by Bracci et al. (1998) to correlate the observed behavior of steel-concrete hybrid connections undergoing significant inelastic joint distortions.

3.4 Modeling RC Frame-Wall Buildings

DRAIN-2D (Kannan and Powell 1973) and SAKE (Otani 1974) were among the earliest non-commercial nonlinear two-dimensional frame analysis programs. DRAIN-2D was developed as a general-purpose frame analysis program with an array of element types (beam-column, panel, and truss) and used a two-component model to incorporate inelastic action. SAKE, on the other hand, was directed more towards reinforced concrete and was based entirely on a single component beam-column element with a Takeda-type hysteresis model.

A frame-wall system is modeled in IDARC as an assemblage of two-dimensional frames interconnected by transverse beams whose stiffness contributions in the two-dimensional plane of analysis is considered. Though torsional modes of vibration and bidirectional interaction are not accounted for, properly constructed two dimensional models can yield acceptable results for the large majority of regular structures. Three degrees-of-freedom (d-o-f) are used at each node however all lateral floor degrees-of-freedom are slaved to a floor master d-o-f based on the assumption that floor slabs act as rigid diaphragms.

Enhanced idealizations can be achieved by incorporating certain aspects of three-dimensional interaction within the two-dimensional framework. For example, the effect of transverse elements and in-plane floor flexibility can be included as shown in Fig. 3.5. The addition of other two-dimensional elements such as infill panels,

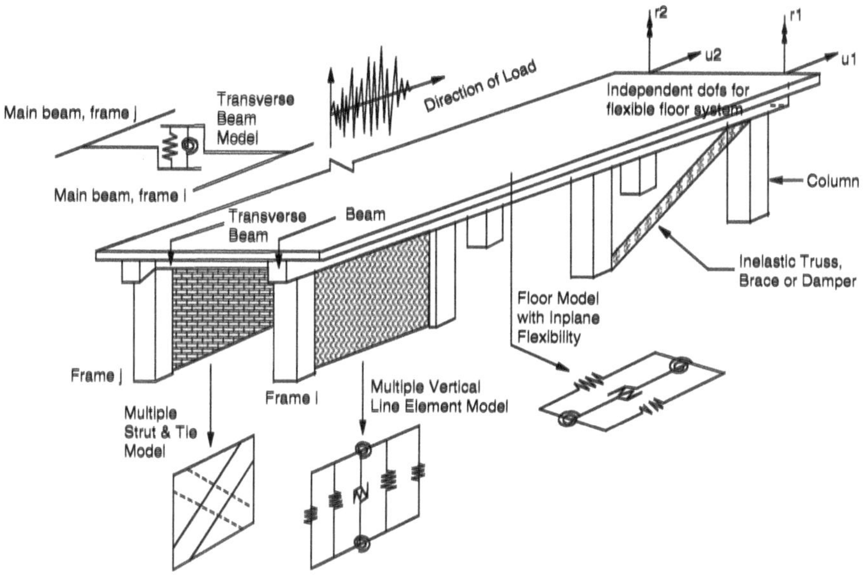

Fig. 3.5 Representation of a RC building as an assemblage of macro-elements

interstory braces or energy-dissipating elements, can provide the user with a variety of modeling options. These enhancements were incorporated into the later versions of IDARC.

3.5 Damage Modeling

The current release of IDARC incorporates three models for damage quantification:

Modified Park-Ang Damage Model: The Park-Ang damage model (Park et al. 1984) was incorporated in the original release of the program (Park et al. 1987). Furthermore, the Park and Ang damage model is also an integral part of the three parameter hysteretic model since the rate of strength degradation is directly related to the parameter β described below. The Park-Ang damage index for a structural element is defined as:

$$DI_{P\&A} = \frac{\delta_m}{\delta_u} + \frac{\beta}{\delta_u P_y} \int dE_h \tag{3.9}$$

where δ_m is the maximum experienced deformation; δ_u is the ultimate deformation of the element; P_y is the yield strength of the element; $\int dE_h$ is the hysteretic energy absorbed by the element during the response history; and β is a model constant parameter. Three damage indices are computed using this damage model: (1) Element damage index; (2) Story damage index for vertical and horizontal components; and (3) Overall building damage. The combination of damage indices is accomplished through weighting factors which are presently expressed as a function of the dissipated hysteretic energy. Since the above model returns a finite damage value for elastic response, the model was modified to subtract the recovered elastic deformations during enhancements added to Version 3 of the program (Kunnath et al. 1992).

Fatigue Based Damage Model: The fatigue based damage model was introduced by Reinhorn et al. (2009). The damage model was developed based on basic structural response considerations, and a low-cycle fatigue rule. The damage index is:

$$DI = \frac{\delta_a - \delta_y}{\delta_u - \delta_y} \frac{1}{\left(1 - E_h / \left\{4\left(\delta_u - \delta_y\right) F_y\right\}\right)} \tag{3.10}$$

where δ_a is the maximum experienced deformation, rotation, or curvature; δ_y is the yield deformation capacity; δ_u is the ultimate deformation capacity; F_y is the yield force capacity; and E_h is the cumulative dissipated hysteretic energy. Note that simplifying the above damage model for the case when the ratio $(\delta_a - \delta_y)/(\delta_u - \delta_y)$ is close to unity, the above expression simplifies to:

$$DI = \frac{\delta_a - \delta_y}{\delta_u - \delta_y} + \frac{E_h}{4\left(\delta_u - \delta_y\right) F_y} \tag{3.11}$$

This is similar to the Park-Ang formulation for $\beta = 0.25$.

Global Damage Model: Another measure of structural damage is the variation in the fundamental period of vibration of the structure. This history is related to the overall stiffness loss in the structure due to inelastic behavior. DiPasquale and Cakmak (1988) defined the softening in the structure as:

$$DI = 1 - \frac{(T_0)_{initial}}{(T_0)_{equivalent}} \tag{3.12}$$

A possible interpretation of the computed damage estimates is summarized in Table 3.1.

3.6 Sample Simulation Using IDARC

Next, the response of a seven-story reinforced concrete building that was instrumented during the 1994 Northridge earthquake is analyzed. The objective herein is to ascertain if the two-dimensional modeling framework and analysis techniques, as presented in earlier sections, are adequate to predict the inelastic response of concrete structures under strong ground motions.

Table 3.1 Damage classification

Damage state[a]	Usability	Damage description[a]	Park-Ang damage index	Inter-story drift demand (%)	
				Flexure	Shear
None		No visible damage, cosmetic or structural	0.0	<0.25	<0.05
Insignificant	Usable	Only cosmetic repair required; no structural repair	0.0–0.2	0.25–0.50	0.05–0.20
Moderate	Temporarily unusable	Repairable structural damage	0.2–0.4	0.5–2.0	0.25–0.50
Heavy	Unusable	Extensive damage requiring demolition or replacement	>0.5	>2.0	>1.0

[a]ATC-33 classification

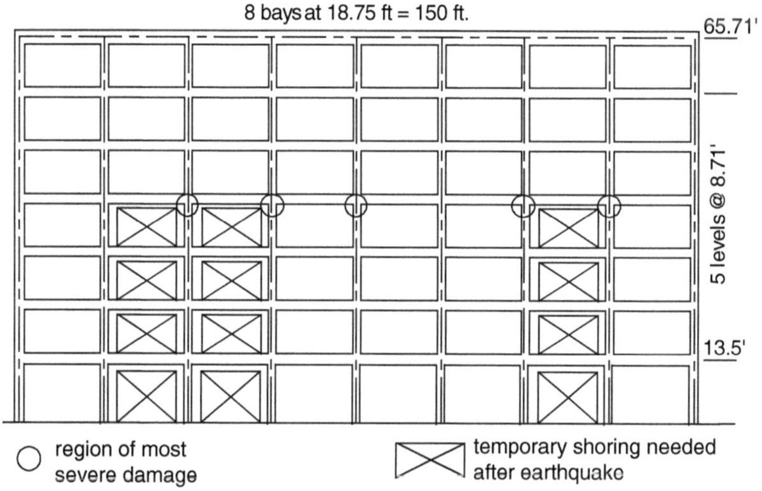

Fig. 3.6 South perimeter frame of the Holiday Inn building

Detailed building information, damage data and estimates of section properties is reported in Islam (1996). The structural damage was primarily confined to the longitudinal perimeter frames with the most severe damage between fourth and fifth floors of the south perimeter frame (Fig. 3.6). The south perimeter frame, which sustained the most severe damage, was selected for evaluation. Material and section property data were adjusted to include the effects of concrete aging and probable damage from the 1971 earthquake. These are reported in Islam (1996).

An incremental time-history analysis was performed on this frame using the recorded accelerations at the base level of the building as the input motion. Figure 3.7 shows the comparison of recorded and calculated roof displacement time-history. The recorded peak displacement at the roof level in this direction

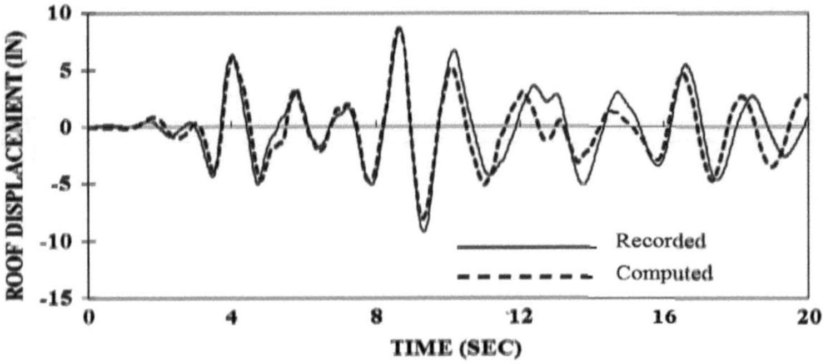

Fig. 3.7 Comparison of observed response with numerical prediction

was 9.20″ @ 9.34 s compared to approximately 9.0 in. predicted by the computer program at almost the same instant. The building period, based on the recorded roof displacement response, appears to be about 1.5 s during the early part of the earthquake, 2.1 s between 10 and 20 s and about 2.4 s thereafter. The program-generated time history compares well with the observed response.

3.7 Some Modeling and Numerical Implementation Issues

All nonlinear computer programs are based on "incremental" or "iterative" linear methods. Nonlinearity is a consequence of stiffness changes occurring during (as in iterative schemes) or at the end (as in incremental schemes) of each linear step. Nonlinear constitutive models (either stress-strain or force-deformation) are specified as continuous or multi-linear paths. For continuously varying stiffness properties, it is necessary to carry out an iterative analysis to ensure equilibrium at the end of each step.

The event-to-event strategy is suitable for multi-linear models since it is possible to predict the required force (or deformation) to cause the next change in state. However, for degrading models with numerous stiffness changes in a single pass (loading or unloading) such as the incorporation of pinching, bond-slip, strength and/or stiffness deterioration, the number of events that need to be checked may become excessive, particularly for large structural systems.

The stiffness (or flexibility) matrix for a member needs to be updated for one or more of the following reasons: (a) a transition in stiffness as prescribed by the hysteresis model or the stress-strain curve; (b) a shift in the contraflexure point; and (c) a change in the plastic hinge length. All such changes lead to unbalanced forces between two solution steps. Item (a) can be dealt with using an event-to-event strategy. The changes resulting from either varying contraflexure location or hinge length are not associated with any predefined event change. Hence, an iterative approach to ensure stability of the final solution is necessary.

Given the complexity of carrying out an iterative or event-to-event analysis, it is sometimes prudent to consider approximate ways to treat convergence issues. A compensation procedure, for example, can be adopted to minimize equilibrium errors. The unbalanced force in any given time step is applied as a corrective force in the subsequent time step. Alternatively, the velocities or accelerations can be adjusted to satisfy dynamic equilibrium at the end of each step. Either approach is not physically accurate but serves to limit numerical errors from increasing without bounds.

3.8 Concluding Remarks

The release of IDARC was a significant milestone in nonlinear analysis of RC structures. Among other features it was the first non-commercial special-purpose computer program to introduce spread plasticity formulations for beam-column elements, the first to include a shear-wall element, the first to consider the effects of transverse (orthogonal) elements within a two-dimensional framework, the first to incorporate a versatile hysteretic model that was capable of simulating stiffness degradation, strength deterioration and pinching behavior, the first to provide a pushover curve as a standard nonlinear static analysis option, and the first to incorporate a damage model to post-process the results of the seismic analysis.

Since the first release of IDARC in 1987, the program has seen numerous upgrades and enhancements and is currently distributed as Version 7.0. Nonlinear analysis of reinforced concrete structures in general has also seen significant advances in the past two decades with fiber-based modeling becoming more commonplace. Ultimately, the efforts of modeling must find their way into computational tools. Currently, there are several available computer programs which, to varying degrees, are capable of performing nonlinear analysis of RC structures. Of these, OpenSees (2013) is the most widely used program within the seismic research community.

It is conceivable that high-fidelity finite-element based simulations will become more reliable and used more routinely as computational facilities become even more powerful and affordable. However, instrumentation of existing buildings and bridges and full-scale structural testing must continue so that computational tools used in sophisticated evaluation are continually validated and enhanced.

Acknowledgments The development and release of the initial version of IDARC and its subsequent enhancements would not have been possible without the support, guidance and vision of my mentor and friend Andrei Reinhorn.

References

Alath S, Kunnath SK (1995) Modeling inelastic shear deformations in RC beam-column joints. In: Proceedings, 10th ASCE engineering mechanics conference, Boulder

Bouc R (1967) Forced vibration of mechanical systems with hysteresis. In: Proceedings 4th conference on nonlinear oscillations, Prague. See also "Mathematical model for hysteresis." Report to Centre de Recherches Physiques, Marseille, pp 16–25

Bracci JM, Atahan AO, Kunnath SK (1998) Analytical modeling of beam-column joints in composite SRC frame systems. ACI Special Publication, American Concrete Institute, Farmington

Cheok G, Stone WC, Kunnath SK (1998) Seismic response of precast concrete frames with hybrid connections. ACI Struct J 95(5):527–539

Clough RW, Johnston SB (1966) Effects of stiffness degradation on earthquake ductility requirements. In: Proceedings Japan earthquake engineering symposium, Tokyo, pp 227–232

Constantinou MC, Adnane MA (1987) Dynamics of soil-base-isolated-structure systems: evaluation of two models for yielding systems. Report to NSF. Department of Civil Engineering, Drexel University

DiPasquale E, Cakmak AS (1988) Identification of the serviceability limit state and detection of seismic structural damage. Technical report NCEER-88-0022. National Center for Earthquake Engineering Research, Princeton University

Fajfar P, Fischinger M (1990) Mathematical modeling of reinforced concrete structural walls for nonlinear seismic analysis. In: Proceedings, European conference on structural dynamics. A.A. Balkema Publishers, Bochum, pp 471–478

Islam S (1996) Analysis of Northridge earthquake response of a damaged nonductile concrete frame building. Struct Des Tall Build J (Wiley) 5(3):151–182

Kabayesawa T, Shiohara H, Otani S, Aoyama H (1983) Analysis of the full-scale seven story reinforced concrete test structure. J Fac Eng 37(2):431–478

Kannan AE, Powell GH (1973) DRAIN-2D: a general purpose computer program for dynamic analysis of inelastic analysis of inelastic plane frames. EERC report 73-6 and 73-22. University of California, Berkeley

Kunnath SK, Reinhorn AM (1990) Model for inelastic biaxial bending interaction of RC beam columns. ACI Struct J 86(3):284–291

Kunnath SK, Reinhorn AM, Lobo RF (1992) IDARC version 3.0: a program for the inelastic damage analysis of reinforced concrete structures. Report no. NCEER-92-0022. National Center for Earthquake Engineering Research, State University of New York at Buffalo, Buffalo

OpenSees (2013) Open system for earthquake engineering simulation, An open-source program. Available at http://opensees.berkeley.edu

Otani S (1974) SAKE: a computer program for inelastic response of R/C frames to earthquakes, SRS report no. 413. Civil Engineering Studies, University of Illinois at Urbana-Champaign, Urbana

Park YJ, Ang AH-S, Wen YK (1984) Seismic damage analysis and damage-limiting design of R/C buildings, Technical report no. SRS 516. Civil Engineering Studies, University of Illinois, Urbana

Park YJ, Reinhorn AM, Kunnath SK (1987) IDARC: inelastic damage analysis of reinforced concrete frame shear-wall structures. Technical report NCEER 87-0008. State University of New York, Buffalo

Reinhorn AM, Roh H, Sivaselvan M, Kunnath SK, Valles R, Madan A, Li C, Lobo RF, Park YJ (2009) IDARC2D version 7.0: a program for the inelastic damage analysis of structures. Technical report MCEER-09-0006. State University of New York at Buffalo

Takeda T, Sozen MA, Nielsen NN (1970) Reinforced concrete response to simulated earthquakes. J Struct Div ASCE 96(ST12):2557–2573

Vulcano A, Bertero VV, Colotti V (1989) Analytical modeling of R/C structural walls. In: Proceedings, 9WCEE, Tokyo, vol 6, pp 41–46

Wen YK (1976) Method for random vibration of hysteretic systems. J Eng Mech Div ASCE 102(EM2):249–263

Chapter 4
A Novel Method for Solving Random Eigenvalue Problems

Mircea Grigoriu

Abstract Matrices with random entries are encountered in finite element/difference formulations of a broad range of mechanics problems. Monte Carlo simulation, the only general method for solving this class of problems, is usual impractical when dealing with realistic problems.

A new method is presented for solving stochastic problems with random matrices that is based on the representation of the entries of random matrices by stochastic reduced order models (SROMs) and surrogate models. SROMs are random elements with finite numbers of samples that are selected from the samples of target random elements in an optimal manner. Surrogate models are approximations for quantities of interest with known expressions. Numerical examples are used to illustrate the implementation and the performance of the SROM method. The examples include inverses and eigenvalues/eigenvectors of random matrices.

4.1 Introduction

Physical systems have uncertain properties so that their states satisfy stochastic equations, that is, equations with random coefficients, input, and/or end conditions. Analytical solutions of stochastic equations are available in special cases that may be limited practical interest. Numerical methods need to be employed to solve this class of equations. Their implementation requires to discretize both the physical and the probability spaces. Finite element/finite difference and other methods can be used to discretize the physical space. For example, the discrete version of a linear partial differential equation with random coefficients is a stochastic algebraic

---

M. Grigoriu (✉)
Department of Civil and Environmental Engineering, Cornell University,
Room 363 Hollister Hall, Ithaca, NY 14853, USA
e-mail: mdg12@cornell.edu

G.P. Cimellaro et al. (eds.), *Computational Methods, Seismic Protection, Hybrid Testing and Resilience in Earthquake Engineering*, Geotechnical, Geological and Earthquake Engineering 33, DOI 10.1007/978-3-319-06394-2_4,
© Springer International Publishing Switzerland 2015

equation, that is, an algebraic equation with random matrix. The discretization of the probability space is usually accomplished by approximating the random functions in the definition of a stochastic problem by parametric random functions, that is, deterministic functions of space and/or time that depend on a finite number of random variables (Grigoriu 2012a, Section 6.3.3). The stochastic dimension of the resulting equation is finite and equal to the dimension of the random vector Z collecting the random variables in the expression parametric model.

Monte Carlo simulation, the only general method for solving stochastic problems, has attractive features. It is conceptually simple and non-intrusive, in the sense that it constructs probabilistic descriptions for solutions of stochastic equations by solving deterministic versions of these equations corresponding to samples of their random entries. A significant limitation of the method is the computational demand that can be prohibitive in applications. This limitation has promoted the development of approximate methods for solving general stochastic problems, for example, the stochastic Galerkin and collocation methods if the uncertainty in Z is arbitrary and the Taylor, perturbation, Neumann series methods if the uncertainty in Z is small (Grigoriu 2012a, Chapter 8). The focus of many of the approximate methods dealing with random matrices is the random eigenvalue problem. For example, some of the solutions of this problem are based on perturbation (Adhikari 2004), polynomial chaos expansions (Ghosh et al. 2005), crossing theory (Grigoriu 1992), and dimension decomposition (Rahman 2007).

We proposed a new method for solving stochastic problems with random matrices. The method can be viewed as a smart Monte Carlo simulation. Like Monte Carlo it uses samples of Z, i.e., samples of random matrices, to construct approximate solutions. In contrast to Monte Carlo simulations that selects a large number of samples of Z at random, we (1) represent Z by stochastic reduced order models (SROMs), that is, random vectors $\tilde{Z}$ that have finite numbers of samples selected in an optimal manner, and (2) construct piecewise linear approximations for the mapping from Z to the solution of stochastic problems, referred to as surrogate models. The SROM method is conceptually simple, accurate, efficient, non-intrusive, and simple to implement. Numerical examples are presented to illustrate the proposed method. The examples include inverses and eigenvalues/eigenvectors of random matrices.

4.2 SROM Method

Denote by $\{A_{ij}(Z),\ i,j = 1,\dots,n\}$ the entries of an (n,n)-matrix $A(Z)$ assumed to be measurable functions of an $\mathbb{R}^d$-valued random variable $Z = (Z_1,\dots,Z_d)$ defined on a probability space $(\Omega,\mathscr{F},P)$. Accordingly, these entries are random variables on $(\Omega,\mathscr{F},P)$. Our objectives are to calculate statistics of quantities of interest $Q(Z)$ for $A(Z)$, e.g., the inverse $B(Z) = A(Z)^{-1}$ and the eigenvalues/eigenvectors of $A(Z)$.

We construct surrogate models $Z \mapsto \tilde{Q}(Z)$ for the mapping $Z \mapsto Q(Z)$ in two steps. First, Z is represented by a stochastic reduced order model (SROM) $\tilde{Z}$, that is, an $\mathbb{R}^d$-valued random variable that has a finite number of samples $\{\tilde{z}_k\}$, $k = 1, \ldots, m$, selected from the samples of Z. The model size m needs to be selected and constitutes a tradeoff between accuracy and computational demand. Second, an approximation $\tilde{Q}(Z)$ is constructed for $Q(Z)$ by using the output of mapping $Z \mapsto \tilde{Q}(Z)$ and local properties of this mapping at the samples $\{\tilde{z}_k\}$ of $\tilde{Z}$.

Resulting surrogate models can be used to generate samples of quantities of interest efficiently since the mapping $Z \mapsto \tilde{Q}(Z)$ is available analytically. The samples are used to find approximately moments, distributions, and other properties for quantities of interest.

4.2.1 SROMs for Random Elements

Let $\Gamma = Z(\Omega)$ denote the range of Z and let $\{\Gamma_k\}$ be a measurable partition of Γ. The mapping $\tilde{Z}(\omega) = \sum_{k=1}^{m} \tilde{z}_k \, 1\big(\omega \in Z^{-1}(\Gamma_k)\big)$ from $(\Omega, \mathscr{F})$ to $(\mathbb{R}^d, \mathscr{B}(\mathbb{R}^d))$ is measurable so that $\tilde{Z}$ is an $\mathbb{R}^d$-valued random variable defined on the same probability space as Z. Following considerations in Grigoriu (2012b), $\{\Gamma_k\}$ are the cells of the Voronoi tessellation in Γ with centers the samples $\{\tilde{z}_k\}$ of a SROM $\tilde{Z}$ of Z, i.e., $\Gamma_k = \{z \in \Gamma : \|z - \tilde{z}_k\| < \|z - \tilde{z}_l\|,\ l \neq k\}$.

An arbitrary pair $\{\tilde{z}_k, \Gamma_k\}$, $k = 1, \ldots, m$, with the above properties defines completely a simple random vector $\tilde{Z}$ on $(\Omega, \mathscr{F}, P)$. For example, the marginal moments of order $q \geq 1$, marginal distributions, and correlations of $\tilde{Z}$ are $\tilde{\mu}_i(q) = \sum_{k=1}^{m} p_k \, \tilde{z}_{k,i}^q$, $\tilde{F}_i(u) = \sum_{k=1}^{m} p_k \, 1(\tilde{z}_{k,l} \leq u)$, and $\tilde{r}_{ij} = \sum_{k=1}^{m} p_k \, \tilde{z}_{k,l} \, \tilde{z}_{k,j}$, $i, j = 1, \ldots, d$, where $\tilde{z}_{k,i}$ denotes the coordinate i of $\tilde{z}_k$ and $p_k = P(Z \in \Gamma_k)$. We are interested in members of this family of random vectors that are optimal in some sense. The optimal $\tilde{Z}$, referred to as a SROM for Z, can be obtained by minimizing, for example, the objective function

$$
\begin{aligned}
e(\{\tilde{z}_k, p_k\}) = {} & \max_{1 \leq l \leq d} \sum_{q=1}^{\bar{q}} \big(\mu_l(q) - \tilde{\mu}_l(q)\big)^2 + \max_{1 \leq l \leq d} \max_{u} \big(F_l(u) - \tilde{F}_l(u)\big)^2 \\
& + \max_{1 \leq l, j \leq d} \big(r_{lj} - \tilde{r}_{lj}\big)^2,
\end{aligned}
\tag{4.1}
$$

where $\mu_l(q) = E[Z_l^q]$, $F_l(u) = P(Z_l \leq u)$, and $r_{lj} = E[Z_l \, Z_j]$ denote target marginal moments, marginal distributions, and correlations of Z. The objective function in Eq. 4.1 quantifies the discrepancy between properties of $\tilde{Z}$ and Z. Other objective functions can be used. Several optimization algorithms have been developed for constructing SROMs (Grigoriu 2012b; Warner et al. 2013).

4.2.2 Surrogate Models

Let $\tilde{Z}$ be a SROM of Z with samples $\{\tilde{z}_k\}$ and Voronoi cells $\{\Gamma_k\}$, $k = 1, \ldots, m$. Consider a quantity of interest $Q(Z)$, i.e., a measurable function of Z. We approximate $Q(Z)$ by

$$\tilde{Q}(Z) = \sum_{k=1}^{m} 1(Z \in \Gamma_k) \left[Q(\tilde{z}_k) + \sum_{r=1}^{d} \frac{\partial Q(Z)}{\partial Z_r} |_{(Z=\tilde{z}_k)} \left(Z_r - \tilde{z}_{k,r} \right) \right], \tag{4.2}$$

where $\{Z_r\}$ and $\{\tilde{z}_{k,r}\}$ are the components of Z and $\tilde{z}_k$. This approximation consists of hyperplanes tangent to $Q(Z)$ at $\{Q(\tilde{z}_k)\}$ that extend over the cells $\{\Gamma_k\}$ of the Voronoi tessellation. It is referred to as a surrogate model for $Q(Z)$.

The surrogate model $\tilde{Q}(Z)$ depends on the values $\{Q(\tilde{z}_k)\}$ of the quantity of interest $Q(Z)$ and of its partial derivatives $\{\partial Q(Z)/\partial Z_r\}$ with respect to the components of Z at the samples $\{\tilde{z}_k\}$ of $\tilde{Z}$. The implementation of $\tilde{Q}(Z)$ involves $m + d\,m = (d+1)\,m$ deterministic solutions, m for finding $\tilde{Q}(Z)$ at the samples $\{\tilde{z}_k\}$ of $\tilde{Z}$ and $d\,m$ for finding the partial derivatives $\partial Q(Z)/\partial Z_r$ at each $\tilde{z}_k$. Once $\{Q(\tilde{z}_k)\}$ and $\{\partial Q(Z)/\partial Z_r\}$ have been calculated, samples of $\tilde{Q}(Z)$ can be obtained by elementary calculations from samples of Z.

We give without proofs two properties of the surrogate model $\tilde{Q}(Z)$ in Eq. 4.2. First, $\tilde{Q}(Z)$ constitutes a piecewise first order Taylor series approximation for $Q(Z)$ with expansion points the samples $\{\tilde{z}_k\}$ of $\tilde{Z}$. Second, the discrepancy between $Q(Z)$ and $\tilde{Q}(Z)$ is equal to

$$Q(Z) - \tilde{Q}(Z) = \frac{1}{2} \sum_{k=1}^{m} 1(Z \in \Gamma_k) \, \|Z - \tilde{z}_k\|^2 \left(\sum_{r=1}^{d} \alpha^{(r)}(\tilde{z}_k, Z) \frac{\partial}{\partial Z_r} \right)^{(2)} Q(Z_k^*), \tag{4.3}$$

where $\{\alpha^{(r)}(\tilde{z}_k, Z) = (Z - \tilde{z}_k)/\|Z - \tilde{z}_k\|\}$ are the components of the unit vector along the line $L(\tilde{z}_k, Z)$ defined by $\tilde{z}_k$ and Z, the operator $\left(\sum_{r=1}^{d} \alpha^{(r)}(\tilde{z}_k, Z)\, \partial/\partial Z_r\right)^{(2)}$ defines the second directional derivative along $L(\tilde{z}_k, Z)$, and $Z^* \in L(\tilde{z}_k, Z)$ is interior to the segment with ends $\tilde{z}_k$ and Z.

4.3 Inverse of Random Matrices

Let $A(Z)$ be an (n,n)-random matrix and denote by $B(Z) = A(Z)^{-1}$ its inverse. It is assumed that (1) the random matrix $B(Z) = A(Z)^{-1}$ exists almost surely, so that the matrices $\{B(\tilde{z}_k) = A(\tilde{z}_k)^{-1}\}$ are defined, and (2) the partial derivatives $\{\partial B(Z)/\partial Z_r\}$, $r = 1, \ldots, d$, exist for $Z = \tilde{z}_k$. The quantity of interest in this case is the inverse $B(Z)$ of $A(Z)$.

The representation in Eq. 4.2 gives

$$\tilde{A}(Z) = \sum_{k=1}^{m} 1(Z \in \Gamma_k) \left[A(\tilde{z}_k) + \sum_{r=1}^{d} \frac{\partial A(Z)}{\partial Z_r} |_{(Z=\tilde{z}_k)} \left(Z_r - \tilde{z}_{k,r} \right) \right] \quad \text{and}$$

$$\tilde{B}(Z) = \sum_{k=1}^{m} 1(Z \in \Gamma_k) \left[B(\tilde{z}_k) + \sum_{i=1}^{d} \frac{\partial B(Z)}{\partial Z_r} |_{(Z=\tilde{z}_k)} \left(Z_r - \tilde{z}_{k,r} \right) \right], \tag{4.4}$$

for random matrices $A(Z)$ and $B(Z)$, respectively, where $A(\tilde{z}_k)$ and $B(\tilde{z}_k)$ are the matrices $A(Z)$ and $B(Z)$ for $Z = \tilde{z}_k$. The partial derivatives $\partial A(Z)/\partial Z_r$ and $\partial B(Z)/\partial Z_r$ in the previous equation can be obtained from

$$\frac{\partial B(Z)}{\partial Z_r} = -B(Z) \frac{\partial A(Z)}{\partial Z_r} B(Z) \quad \text{and}$$

$$\frac{\partial A(Z)}{\partial Z_r} = -A(Z) \frac{\partial B(Z)}{\partial Z_r} A(Z) \tag{4.5}$$

for $Z = \tilde{z}_k, k = 1, \ldots, d$.

Example. Consider the random matrix

$$A(Z) = \begin{bmatrix} Z_1 + Z_2 & -Z_2 & 0 \\ -Z_2 & Z_2 + Z_3 & -Z_3 \\ 0 & -Z_3 & Z_3 \end{bmatrix}, \tag{4.6}$$

where $Z_i = F^{-1} \circ \Phi(G_i)$, F is a Beta distribution with range $[a, b]$ and shape parameters $\{p, q\}$, $G_i \sim N(0, 1)$, $E[G_i\, G_j] = \rho^{|i-j|}$, $i, j = 1, 2, 3$, and $\rho \in (-1, 1)$.

The surrogate models in Eq. 4.4 depend on the partial derivatives of $A(Z)$ with respect to the coordinates $\{Z_r\}$ of Z, the inverses of $A(Z)$, and the partial derivatives of $B(Z)$ with respect to the coordinates $\{Z_r\}$ of Z at the samples $\{Z = \tilde{z}_k\}$ of $\tilde{Z}$. These quantities can be obtained explicitly in this case and are

$$\frac{\partial A(Z)}{\partial Z_1} = \begin{bmatrix} 1 & 0 & 0 \\ 0 & 0 & 0 \\ 0 & 0 & 0 \end{bmatrix}, \quad \frac{\partial A(Z)}{\partial Z_2} = \begin{bmatrix} 1 & -1 & 0 \\ -1 & 1 & 0 \\ 0 & 0 & 0 \end{bmatrix}, \quad \text{and} \quad \frac{\partial A(Z)}{\partial Z_3} = \begin{bmatrix} 0 & 0 & 0 \\ 0 & 1 & -1 \\ 0 & -1 & 1 \end{bmatrix}, \tag{4.7}$$

and

$$B(Z) = \begin{bmatrix} 1/Z_1 & 1/Z_1 & 1/Z_1 \\ 1/Z_1 & (Z_1 + Z_2)/(Z_1\, Z_2) & (Z_1 + Z_2)/(Z_1\, Z_2) \\ 1/Z_1 & (Z_1 + Z_2)/(Z_1\, Z_2) & (Z_1\, Z_2 + Z_2\, Z_3 + Z_3\, Z_1)/(Z_1\, Z_2\, Z_3) \end{bmatrix}.$$

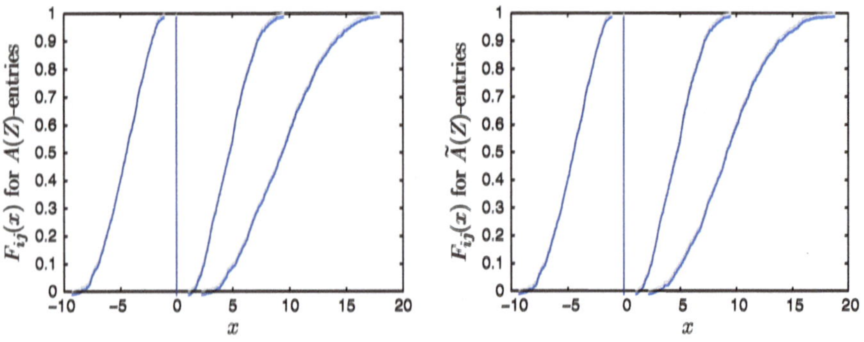

Fig. 4.1 Estimates of the distributions of the entries of $A(Z)$ (*left panel*) and of $\tilde{A}(Z)$ with $m = 5$ (*right panel*) for $a = 1, b = 10, p = 2, q = 3$, and $\rho = 0.7$

The partial derivatives $\partial B(Z)/\partial Z_r$ of $B(Z)$ result from the above equation by elementary calculations.

Numerical results are for $a = 1$, $b = 10$, $\rho = 0.7$, shape parameters $(p, q) = (2, 3)$. The left and right panels in Fig. 4.1 show estimates of the distributions of the entries of $A(Z)$ and $\tilde{A}(Z)$ in Eq. 4.4 with $m = 5$, that are calculated from samples of $A(Z)$ and $\tilde{A}(Z)$ corresponding to 1,000 independent samples of Z. Although the representation of $\tilde{A}(Z)$ is based on a SROM $\tilde{Z}$ with $m = 5$, the estimated distributions of the entries of $A(Z)$ and $\tilde{A}(Z)$ are indistinguishable at the figure scale. There are only five lines since some of the entries of $A(Z)$ coincide. From left to right, the lines in the figure are estimates of the distributions of $A_{12}(Z) = A_{23}(Z)$, $A_{12}(Z)$, $A_{33}(Z)$, and $A_{11}(Z) = A_{22}(Z)$.

The solid lines in the top left, top right, bottom left, and bottom right panels in Fig. 4.2 are estimates of the distributions of the entries of $\tilde{B}(Z)$ for SROMs with $m = 5$, $m = 10$, $m = 20$, and $m = 30$ samples. The dash lines are estimates of the distributions of the entries of $B(Z)$. All estimates are obtained from 1,000 independent samples of Z. From left to right, the lines in these plots are the distributions of $B_{11}(Z) = B_{12}(Z) = B_{13}(Z)$, $B_{22}(Z) = B_{23}(Z)$, and $B_{33}(Z)$. In contrast to results in the previous figure, larger SROMs are needed to approximate satisfactorily the distributions of the entries of $B(Z)$. This is expected since the entries of $A(Z)$ are linear functions of Z while those of $B(Z)$ depend in a rather complex manner on Z. The distributions of $\tilde{B}_{ij}(Z)$ improve significantly as the size of SROMs for Z is increased from $m = 5$ to $m = 10$. This improvement continues as m is further increased but a slower rate.

Numerical results in Figs. 4.1 and 4.2 suggest that the distributions of the entries of $\tilde{A}(Z)$ and $\tilde{B}(Z)$ approach those of $A(Z)$ and $B(Z)$ as m increases. The rate of convergence depends on features of the quantity of interest and the functional form of the distribution of the random element Z.

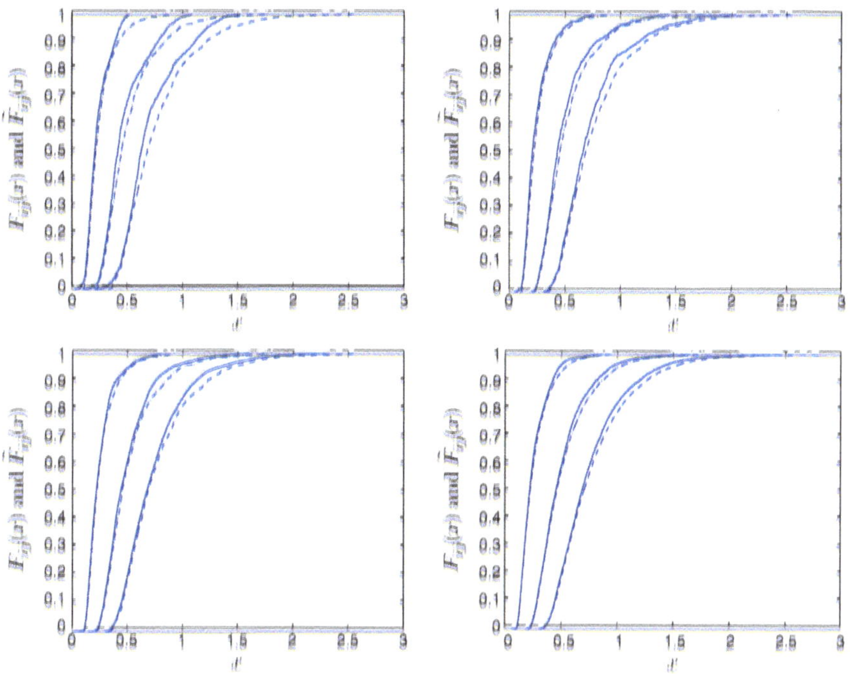

Fig. 4.2 Estimates of the distributions of the entries of $\tilde{B}(Z)$ for $m = 5$ (*top left panel*), $m = 10$ (*top right panel*), $m = 20$ (*bottom left panel*), and $m = 30$ (*bottom left panel*) are shown in *solid lines*. *Dotted lines* are corresponding estimates for $B(Z)$

4.4 Random Eigenvalue Problem

The quantities of interest $Q(Z)$ in this case are the eigenvalues and eigenvectors of an (n,n)-random matrix $A(Z)$ whose entries are, as previously, measurable functions of an $\mathbb{R}^d$-valued random variable Z defined on a probability space $(\Omega, \mathscr{F}, P)$. It is assumed that the determinant of $A(Z)$ is not zero with probability one. As previously, $\tilde{Z}$ denotes a SROM for Z with defining parameters $\{\tilde{z}_k, \Gamma_k\}$, $k = 1, \ldots, m$. For simplicity, we assume that the eigenvalues $\{\Lambda_i(Z)\}$ of $A(Z)$ are distinct. The case of multiple eigenvalues is treated elsewhere.

4.4.1 Eigenvalues

The surrogate model in Eq. 4.2 has the form

$$\tilde{\Lambda}_i(Z) = \sum_{k=1}^{m} \left[\tilde{\lambda}_{i,k} + \sum_{r=1}^{d} \tilde{\lambda}_{i,k}^{(r)} (Z_r - \tilde{z}_{k,r}) \right] 1(Z \in \Gamma_k), \tag{4.8}$$

where $\tilde{\lambda}_{i,k} = \Lambda_i(\tilde{z}_k)$,

$$\lambda_{i,k}^{(r)} = \frac{\partial \Lambda_i(Z)}{\partial Z_r} \, |_{Z=\tilde{z}_k} = -\frac{c_{1,k}^{(r)} \lambda_{i,k}^{n-1} + \cdots + c_{n-1,k}^{(r)} \lambda_{i,k} + c_{n,k}^{(r)}}{n \lambda_{i,k}^{n-1} + (n-1) c_{1,k} \lambda_{i,k}^{n-2} + \cdots + c_{n-1,k}}, \quad k = 1, \ldots, m, \tag{4.9}$$

$\{c_{i,k}\}$ are the coefficients $\{C_i\}$ of the characteristic equation $\det(A - \Lambda I) = \Lambda^n + C_1 \Lambda^{n-1} + \cdots + C_{n-1} \Lambda + C_n = 0$ for $Z = \tilde{z}_k$, and $c_{i,k}^{(r)} = \partial C_i(Z)/\partial Z_r$ at $Z = \tilde{z}_k$ (Grigoriu 2002, Section 8.3.2.3). The coefficient $\{C_i\}$ can be calculated recursively from the relationship

$$C_k = -\big(C_{k-1} D_1 + C_{k-2} D_2 + \cdots + C_1 D_{k-1} + D_k\big)/k, \quad k = 1, \ldots, n, \tag{4.10}$$

with $C_0 = 1$ and $D_k(Z) = \mathrm{tr}\big(A(Z)^k\big)$. The surrogate model in Eq. 4.8 and independent samples of Z can be used to estimate properties of the eigenvalues of $A(Z)$.

The partial derivatives $\{\lambda_{i,k}^{(r)}\}$ in Eq. 4.9 result by differentiating the characteristic equation $\det(A - \Lambda I) = 0$ with respect to the coordinates $\{Z_r\}$ of Z. This operation gives

$$n \Lambda^{n-1} \frac{\partial \Lambda}{\partial Z_r} + \frac{\partial C_1}{\partial Z_r} \Lambda^{n-1} + (n-1) C_1 \Lambda^{n-2} \frac{\partial \Lambda}{\partial Z_r} + \cdots + \frac{\partial C_{n-1}}{\partial Z_r} \Lambda + C_{n-1} \frac{\partial \Lambda}{\partial Z_r} + \frac{\partial C_n}{\partial Z_r} = 0$$

or

$$\left(n \Lambda^{n-1} + (n-1) C_1 \Lambda^{n-2} + \cdots + C_{n-1}\right) \frac{\partial \Lambda}{\partial Z_r} = -\left(\frac{\partial C_1}{\partial Z_r} \Lambda^{n-1} + \cdots + \frac{\partial C_{n-1}}{\partial Z_r} \Lambda + \frac{\partial C_n}{\partial Z_r}\right),$$

which yields Eq. 4.9 for $Z = \tilde{z}_k$.

Example. Let $A(Z)$ be the random matrix in Eq. 4.6, where Z is the $\mathbb{R}^3$-valued Beta random variable defined following this equation. The coefficients of the characteristic equation $\det(A - \Lambda I) = 0$ of A are functions of Z and have the expressions $C_1(Z) = -Z_1 - 2 Z_2 - 2 Z_3$, $C_2(Z) = Z_1 Z_2 + 2 Z_1 Z_3 + 3 Z_2 Z_3$, and $C_3(Z) = -Z_1 Z_2 Z_3$, so that $\partial C_1(Z)/\partial Z_r$ is $\{-1, -2, -2\}$, $\partial C_2(Z)/\partial Z_r$ is $\{Z_2 + 2 Z_3, Z_1 + 3 Z_3, 2 Z_1 + 3 Z_2\}$, and $\partial C_3(Z)/\partial Z_r$ is $\{-Z_2 Z_3, -Z_1 Z_3, -Z_1 Z_2\}$ for $r = 1, 2, 3$ (Grigoriu 2012b). These expressions and their derivatives with respect to the components $\{Z_r\}$ of Z for $Z = \tilde{z}_k$ give the coefficients $\{c_{i,k}\}$ and $\{c_{i,k}^{(r)}\}$ that are used in Eq. 4.9 to calculate the partial derivatives $\{\lambda_{i,k}^{(r)}\}$.

The solid lines in the left and right panels of Fig. 4.3 show estimates of the distributions of $\tilde{\Lambda}_i(Z)$ obtained from Eq. 4.8 for SROMs $\tilde{Z}$ with $m = 5$ and $m = 10$ and 1,000 independent samples of Z. The dash lines are estimates of the distributions of $\Lambda_i(Z)$ inferred from samples of these random variables calculated by solving 1,000 deterministic eigenvalue problems for $A(Z)$ with Z replaced by independent samples of it. Estimates of the distributions of $\{\Lambda_i(Z)\}$ are satisfactory even when based on a SROM of Z with $m = 5$. Monte Carlo and SROM-based

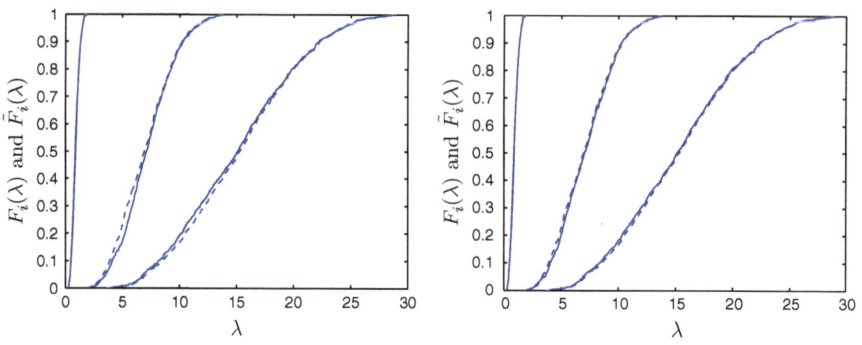

Fig. 4.3 Estimates of the distribution $F_i(\lambda)$ of $\Lambda_i(Z)$ by Monte Carlo in *dash lines* and SROM in *solid lines* for $m = 5$ (*left panel*) and $m = 10$ (*right panel*) by the first method

estimates for $\Lambda_1(Z)$ coincide at the figure scale. The construction of SROM-based estimates for $\{\Lambda_i(Z)\}$ involves solutions of m deterministic eigenvalue problems and the elementary calculations specified by Eqs. 4.8 and 4.9.

4.4.2 Eigenvectors

Let $\{U_i(Z)\}$ and $\{V_i(Z)\}$ denote the right and left eigenvectors of $A(Z)$, so that they satisfy the equations $A(Z)\,U_i(Z) = \Lambda_i(Z)\,U_i(Z)$ and $A(Z)'\,V_i(Z) = \Lambda_i(Z)\,V_i(Z)$, $i = 1, \ldots, n$. The eigenvectors are measurable functions of Z so that they are $\mathbb{R}^n$-valued random variables defined on the same probability space as Z. The surrogate model in Eq. 4.2 for $U_i(Z)$ has the form

$$\tilde{U}_i(Z) = \sum_{k=1}^{m} 1(Z \in \Gamma_k) \left[\tilde{u}_{i,k} + \sum_{r=1}^{d} \tilde{u}_{i,k}^{(r)} (Z_r - \tilde{z}_{k,r}) \right], \tag{4.11}$$

where $\{\tilde{u}_{i,k} = U_i(\tilde{z}_k)\}$ are the right eigenvectors of the deterministic matrix $\tilde{a}_k = A(\tilde{z}_k)$ and the sensitivity factors $\{\tilde{u}_{i,k}^{(r)}\}$ are the partial derivatives $\{\partial U_i(Z)/\partial Z_r\}$ calculated at $Z = \tilde{z}_k$. Similar representations hold for the left eigenvectors $\{V_i(Z)\}$ of $A(Z)$.

It is not possible to construct equations for the gradients of the eigenvectors of $A(Z)$ by differentiating their defining equations with respect to the coordinates of Z. For example, the derivative of $A\,U_i - \Lambda_i\,U_i = 0$ with respect to an arbitrary coordinate Z_r of Z is

$$\frac{\partial A(Z)}{\partial Z_r} U_i(Z) + A(Z) \frac{\partial U_i(Z)}{\partial Z_r} - \frac{\partial \Lambda_i(Z)}{\partial Z_r} U_i(Z) - \Lambda_i \frac{\partial U_i(Z)}{\partial Z_r} = 0, \tag{4.12}$$

which, for $Z = \tilde{z}_k$, gives

$$\left(\tilde{a}_k - \tilde{\lambda}_{l,k}\, I\right) \tilde{u}_{l,k}^{(r)} = \left(\tilde{\lambda}_{l,k}^{(r)} - \tilde{a}_k^{(r)}\right) \tilde{u}_{l,k} \tag{4.13}$$

where $\tilde{a}_k = A(\tilde{z}_k)$, $\tilde{\lambda}_{l,k} = \Lambda_l(\tilde{z}_k)$, $\tilde{u}_{l,k} = U_l(\tilde{z}_k)$, $\tilde{u}_{l,k}^{(r)} = \partial U_l(\tilde{z}_k)/\partial Z_r$, $\tilde{a}_k^{(r)} = \partial A(\tilde{z}_k)/\partial Z_r$, and I denotes the identity matrix. Since $\tilde{a}_k - \tilde{\lambda}_{l,k}\, I$ is not invertible, $\tilde{u}_{l,k}^{(r)}$ cannot be calculated directly from Eq. 4.13 directly. These vectors can be obtained from $\tilde{u}_{l,k}^{(r)} = \sum_{j=1}^{n} b_{lj}^{(r)}\, \tilde{u}_{j,k}$, where

$$b_{lj}^{(r)} = \begin{cases} 0, & l = j \\ \tilde{v}_{j,k}' \left(\tilde{\lambda}_{l,k}^{(r)} I - \tilde{a}_k^{(r)}\right) \tilde{u}_{l,k} / \left(\tilde{\lambda}_{j,k} - \tilde{\lambda}_{l,k}\right), & l \neq j, \end{cases} \tag{4.14}$$

where $\tilde{v}_{j,k} = V_j(\tilde{z}_k)$. If matrices $\{\tilde{a}_k\}$ are symmetric, their left and right eigenvectors coincide and Eq. 4.14 holds with $\tilde{v}_{j,k} = \tilde{u}_{j,k}$.

The implementation of the surrogate model in Eq. 4.11 involves (1) solutions of m deterministic eigenvalue problems delivering the eigenvalues $\{\tilde{\lambda}_{l,k}\}$, right eigenvectors $\{\tilde{u}_{l,k}\}$, and left eigenvectors $\{\tilde{v}_{l,k}\}$, $l = 1, \ldots, n$, for matrices $\{\tilde{a}_k\}$, $k = 1, \ldots, m$, (2) determination of the partial derivatives $\{\tilde{\lambda}_{l,k}^{(r)}\}$ in Eq. 4.9 for the eigenvalues $\{\tilde{\lambda}_{l,k}\}$, and (3) calculations in Eq. 4.14 involving the eigenvalues and eigenvectors of matrices $\{\tilde{a}_k\}$.

Example. Consider the random matrix $A(Z)$ in Eq. 4.6 depending on an $\mathbb{R}^3$-valued Beta random variable with parameters specified in the first example. The SROM-based eigenvectors $\tilde{U}_l(Z)$ of $A(Z)$ are given by Eq. 4.11 with $d = 3$. Since $A(Z)$ is symmetric, its right and left eigenvectors coincide so that the vectors $\tilde{u}_{l,k}^{(r)}$ in the expression of $\tilde{U}_l(Z)$ can be calculated from Eq. 4.14 with $\tilde{u}_{j,k}$ in place of $\tilde{v}_{j,k}$. As previously stated, the construction of $\tilde{U}_l(Z)$ involves solutions of m deterministic eigenvalue problems for matrices $\tilde{a}_k$ to obtain the eigenvalues $\{\tilde{\lambda}_{l,k}\}$ and the eigenvectors $\{\tilde{u}_{l,k}\}$, calculation of the sensitivity factors $\{\tilde{\lambda}_{l,k}^{(r)}\}$ in Eq. 4.9 for the eigenvalues $\{\tilde{\lambda}_{l,k}\}$, and determination of the projections $\{b_{lj}^{(r)}\}$ of $\{\tilde{u}_{l,k}^{(r)}\}$ on the eigenvectors $\{\tilde{u}_{l,k}\}$ of $\tilde{a}_k$.

Following numerical results are for the random matrix in Eq. 4.6 depending on the $\mathbb{R}^3$-valued Beta random variable Z with $a = 1$, $b = 10$, $p = 2$, $q = 3$, and $\rho = 0.7$. The left and right panels in Fig. 4.4 show 1,000 independent samples of the eigenvectors $U_l(Z)$ and $\tilde{U}_l(Z)$ with $m = 20$. The plots provide a qualitative description of the exact and approximate eigenvectors of $A(Z)$. The construction of the approximate eigenvectors $\tilde{U}_l(Z)$ requires to solve only $m \ll n$ deterministic eigenvalue problems in additional computational effort involved in the construction of a SROM $\tilde{Z}$ for Z. Once the expression of $\tilde{U}_l(Z)$ has been found, the generation of samples of this random variable takes very little time.

Errors of moments of $\tilde{U}_l(Z)$ relative to Monte Carlo estimates of corresponding moments of $U_l(Z)$, $l = 1, 2, 3$, have been calculated for SROMs $\tilde{Z}$ with $m = 10$ and $m = 20$. The absolute values of these errors for first, second, third, and fourth

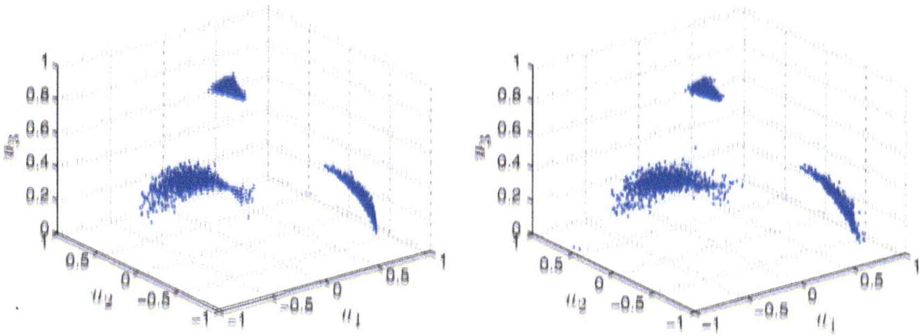

Fig. 4.4 Samples of the eigenvectors of $U_i(Z)$ (*left panel*) and $\tilde{U}_i(Z)$ with $m = 20$ (*right panel*)

moments in percentages are less than 1.49, 2.95, 4.7, and 8.4 for $m = 10$ and 0.99, 1.93, 4.6, and 6.91 for $m = 20$.

4.5 Conclusions

A new method has been developed for calculating statistics of inverses and eigenvalues/eigenvectors of random matrices. The method is conceptually simple, general, efficient, accurate, and non-intrusive. Its implementation involves two steps. First, stochastic reduced order models (SROMs) are developed for the random elements in the definition of random matrices. Second, surrogate models are constructed for quantities of interest and used to find statistics of these quantities by Monte Carlo simulation. SROMs are random elements with finite numbers of samples that are selected from the samples of target random elements in an optimal manner. Surrogate models provide approximations for quantities of interest.

Numerical examples have been used to illustrate the implementation and the performance of the SROM method. The examples include inverses of random matrices and solutions of random eigenvalue problems. The proposed method is satisfactory in all numerical examples.

Acknowledgements The work reported in this paper has been supported by the National Science Foundation under grand CMMI-0969150. This support is gratefully acknowledged.

References

Adhikari S (2004) Complex modes in stochastic systems. Adv Vib Eng 3(1):1–11

Ghosh D, Ghanem RG, Red-Horse J (2005) Analysis of eigenvalues and model interaction of stochastic systems. AIAA J 43(10):2196–2201

Grigoriu M (1992) A solution of the random eigenvalue problem by crossing theory. J Sound Vib 158(1):58–80

Grigoriu M (2002) Stochastic calculus. Applications in science and engineering. Birkhäuser, Boston
Grigoriu M (2012a) Stochastic systems. Uncertainty quantification and propagation. Springer series in reliability engineering. Springer, London/Heidelberg/New York/Dordrecht. ISBN 978-1-4471-2326-9, ISBN 978-1-4471-2327-9 (eBook)
Grigoriu M (2012b) A method for solving stochastic equations by reduced order models and local approximations. J Comput Phys 231(19):6495–6513
Rahman S (2007) Stochastic dynamic systems with complex-valued eigensolutions. Int J Numer Methods Eng 71:963–986
Warner J, Grigoriu M, Aquino W (2013, in press) Stochastic reduced order models for random vectors. Applications to random eigenvalue problems. Probab Eng Mech 31:1–11

Part II
Experimental Methods

Chapter 5
Three Dimensional Formulation of Large Displacement Problems: The Zipper Frame Example

Macarena Schachter Adaros

Abstract A zipper frame is a conventional braced frame with chevron type braces at all floors, connected through a zipper column linking their midpoints. Under lateral force, a 3 stories frame is expected to behave as follows: after the 1st floor braces buckle, the zipper column transmits the unbalance vertical force to the 2nd story, forcing them to buckle. The new unbalanced vertical force is then transmitted to the third story, which is designed to remain elastic. This is known as the zipper mechanism. Three shaking table tests were performed to evaluate the seismic response of a 3 story suspended zipper frame model. In every shaking table test performed, the braces buckled out of plane but the zipper mechanism did not always develop. Following the test results, a new analytical tool was developed to capture the 3 dimensional phenomena of buckling. It is based on the well known corotational formulation. Every element is described independently. The rigid body modes are removed by using a corotated coordinate system. Within the deformed shape, the element can undergo large rotations and large strains, which are evaluated respect to the corotated origin. The structural problem is solved in the state space, where displacements as well as velocities are variables. The final set of equations is a combination of differential and algebraic equations that is solved using the IDA software. This paper presents the main results of the shaking table tests as well as some general description of the proposed formulation and its capability to verify the shaking table findings.

M. Schachter Adaros (✉)
Weidlinger Associates Inc., 1825 K St. NW, Washington, DC 20006, USA
e-mail: macarena.schachter@gmail.com

G.P. Cimellaro et al. (eds.), *Computational Methods, Seismic Protection, Hybrid Testing and Resilience in Earthquake Engineering*, Geotechnical, Geological and Earthquake Engineering 33, DOI 10.1007/978-3-319-06394-2_5,
© Springer International Publishing Switzerland 2015

5.1 Introduction

In a traditional K braced (Chevron) frame design, the braces resist the lateral earthquake load transmitted by the beam as axial forces. Under a strong ground motion one of the braces is expected to buckle under compression while the adjacent one is expected to yield in tension. Due to differences in compression and tension capacities, there will be a non-null vertical resultant transmitted to the beam as a shear force, which makes its design very costly. In 1988 Khatib et al. suggested adding a column between the beams connected at same joints where the braces connect. This column will transmit the vertical unbalanced load to the upper floor, redistributing it to the braces, thus the compression brace will be subjected to an even greater compression, triggering buckling. A new unbalanced vertical force will be created at the upper floor beam. If this floor is linked to the upper one, the process would repeat. The propagation of buckling and yielding to the upper floors seems similar to a "zipper" from which this system derived the name of "zipper frame". The solution is appealing; however if all the braces at all floors are allowed to buckle, then collapse of the frame is likely to occur. To overcome this problem, Leon and Yang (2003) devised a new mechanism defined as "Suspended zipper frame". In this mechanism, the upper story of a frame is purposely designed to remain elastic thus preventing the collapse of the frame.

The suspended zipper frame was selected as the structural system to be tested in a collaborative NEES project, in order to validate the system behavior and develop method of analysis and evaluation. A series of tests was performed on the new 7 m × 7 m shake table at the NEES equipment site at University at Buffalo. The model utilized is an incomplete similitude one third scale model derived from a prototype designed by Leon and Yang (2003) at Georgia Institute of Technology. The test results and conclusions are presented in this paper.

The behavior of the braces while buckling is highly nonlinear and three-dimensional. A new macromodel formulation that captures this behavior is developed. The formulation uses information from the nodes and from some limited number of points along the member, with no need for a finite element or fiber description of the section. It is based on the concept of corotated frames to separate rigid body motions from deformations allowing for large rotations and displacements at both stages. It is formulated in the state space where second order differential equations are transformed into a set of first order differential and algebraic equations that are solved using the IDA software. Nonlinearity in the material is introduced by using the constitutive law in rate form developed by Simeonov (1999).

5.2 Test Setup

The test setup and the test specimen are shown in Fig. 5.1 while details of the section components are presented in Schachter 2008. The setup consists of two independent three stories gravity frames sandwiching the braced frame specimen. Each gravity

Fig. 5.1 Test setup – *top left*: mass to specimen connection; *bottom left*: hinge at the gravity frame; *right*: specimen

frame is composed of double hinged columns and story masses weighing 8.5 kips (~4.0 mtons) each. The braced frame provides the only lateral resistance in the system when connected together.

The model has dimensions scaled to 1:3 of the prototype. However, due to limitations in the shaking table capacity and the independence of the lateral and vertical resistance systems, a different acceleration scale was determined form the similitude conditions as 2:1. This means that a value for the acceleration in the model has to be scaled by 2 to get the equivalent acceleration in the prototype.

5.3 Test Protocol

The ground motion selected was LA22yy (from the SAC set of ground motions for Los Angeles area). This record is 1.15 times the recorded Kobe earthquake. Its original peak ground acceleration is 0.92 g. The acceleration was gradually increased from 15 % of the original acceleration up to 100 % or until failure.

For the first test, the time scaled record was scaled down to its 15, 30, 45, 60, 80 and 100 % of the PGA. For test #2 the sequence was 30, 80, 100 and 120 %. For the test #3 the sequence was: 30, 80, 100, 120, 140 %.

5.4 Results of Shake Table Test #1

The structure responded elastically, as expected, during the 15 and 30 % of test ground motion. During the 45 % scaled test, the west brace in the 1st floor buckled elastically, while the east brace remained elastic. Braces in the 2nd floor remained elastic. No measurement of importance was registered in the zipper column. During the 60 % scaled test, both braces at the first floor buckled. The Zipper column did not record any axial strains of the magnitude expected. During this ground motion, the second floor east brace buckled at the same time as the first floor east brace did. The third story zipper column was not engaged. The behavior during this test is consistent with a chevron braced frame but not with a zipper frame. During the 80 and 100 % scaled tests, some of the strain gauges at first story level failed to record information and both second story braces buckled.

5.5 Results of Shaking Table Test #2

A lateral restraint was added in the middle of the beams as shown in Fig. 5.2. The restraint was designed to keep the beam straight but a Teflon interface ensures the free movement in the direction of motion.

In a first attempt to perform 30 % scaled ground motion test, problems with the control of the shaking table created an input greater than the desirable, triggering buckling at the west brace. The behavior was similar to the one found during the 45 % scaled motion in Test #1. The east brace remained elastic. During the 80 % scaled test, both braces in the first floor buckled and yielded. The Krypton coordinate tracking system recorded clearly the buckling at the first story braces. This was one of the first measurements of buckled braces in a dynamic test. As expected, the

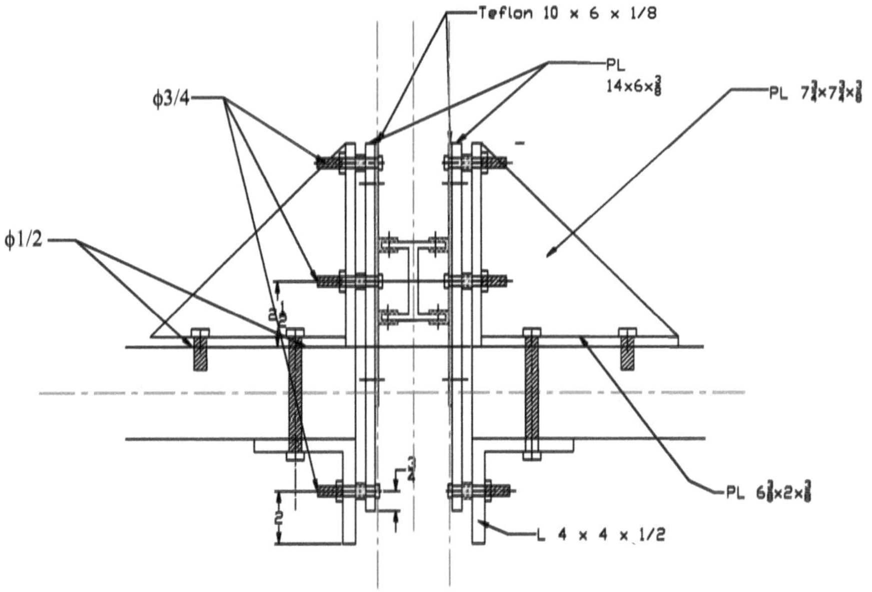

Fig. 5.2 Lateral restraint at the beam

zipper column was engaged and remained elastic, after the braces buckled. The braces at the second floor remained elastic. From data analysis, it was observed that the beam was being subjected to torsion although the values were very small. Since the sides of the flanges are in contact with the Teflon surface of the restraint, it is possible that a large amount of the torsional moment was being transmitted to the plates rather than to the beam. In conclusion, during Test #2, the zipper mechanism developed and the frame behaved as expected.

5.6 Results of Shaking Table Test #3

This test was a repetition of test #1. The lateral restraints were not used. The frame was heavily instrumented to record torsion in the beam and redundant instrumentation on the zipper columns and braces was provided. During the 30 % scaled motion, the frame remained elastic. During the 80 % scaled motion, both braces in the first floor buckled, the second story zipper column was engaged and no torsion was recorded. During the 100 % scaled motion, the second story braces buckled and the third story zipper column was engaged. During the 120 % scaled motion, the braces did not have much capacity in compression, therefore the tension brace carried most of the load and the force in the zipper column was maximum, however it was not big enough to yield. During the 140 % scaled motion, the braces were fatigued and one of them broke in tension after buckling.

The behavior of the frame was excellent. There was no sign of torsion or yielding at the beam. At the end of the test, the structure was leaning to one side and most of the damage was concentrated on the first story.

The Krypton tracking system recorded the 3D displacement of the braces. Figure 5.3 shows the in plane vs. out of plane movement of every probe placed along the East brace. Probe 1 was at the lower end, probes 4 and 5 at the middle and probe 7 at the upper end.

5.7 Comparison of Test Results

The difference in behavior in the second test as compared with the first is substantial. The transverse restraint of the beam held until the 100 % scaled motion was applied. This allowed the engagement of the 2nd floor zipper column as early as in the 30 % scaled test. The third floor zipper column helped to resist the vertical unbalanced force coming from the first floor and is also engaged (with a small amount of force) at low levels of ground shaking. Therefore an important part of the vertical force is transmitted to the third floor braces as intended in the suspended "zipper" frame suggested by Leon and Yang (2003). The beams at 1st and 2nd floors deflect downward when the braces at the respective levels buckle. These results demonstrate that restraining the beam laterally prevents undesirable torsion in the beam.

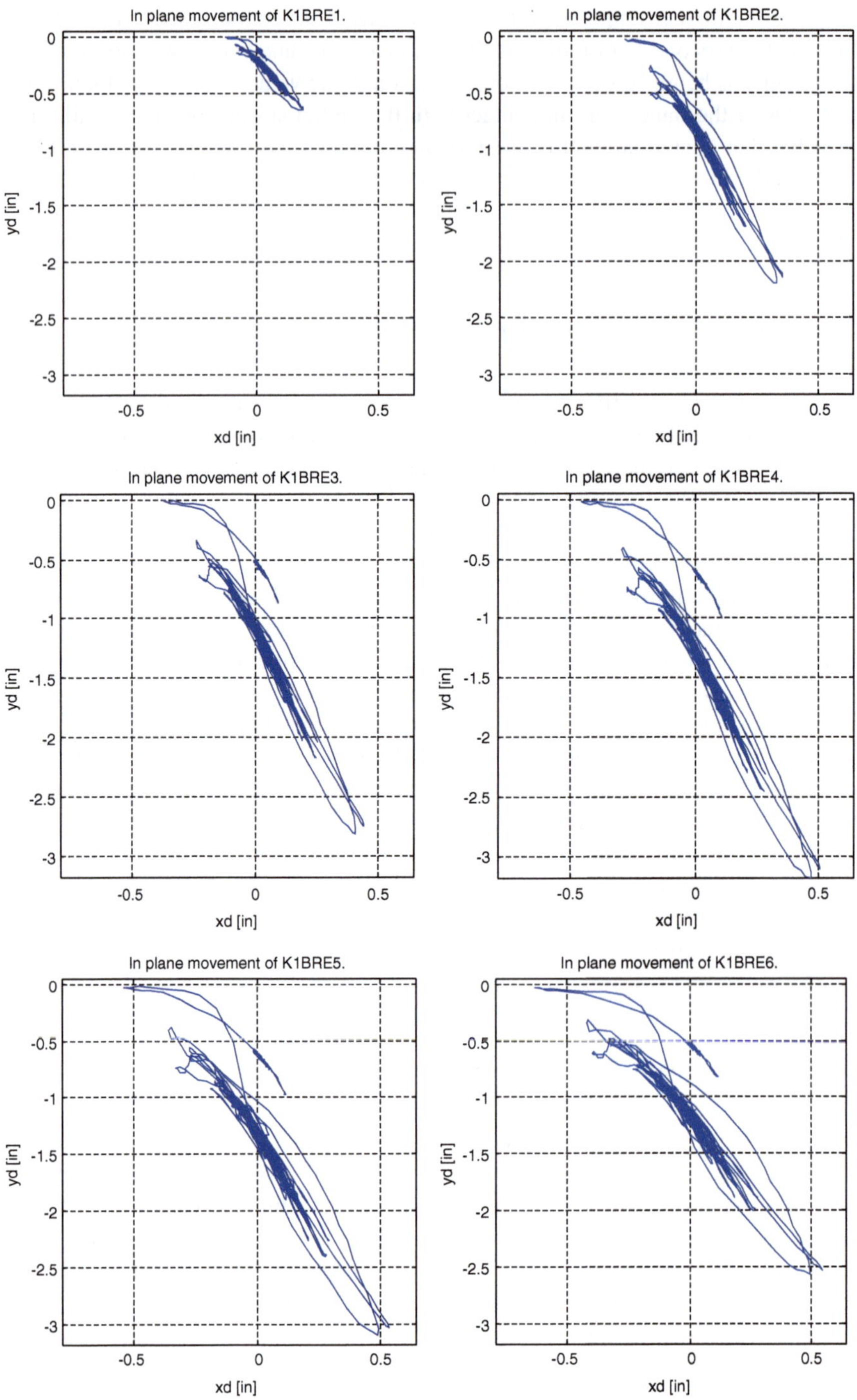

Fig. 5.3 In plane vs. out of plane movement as recorded by Krypton tracking system

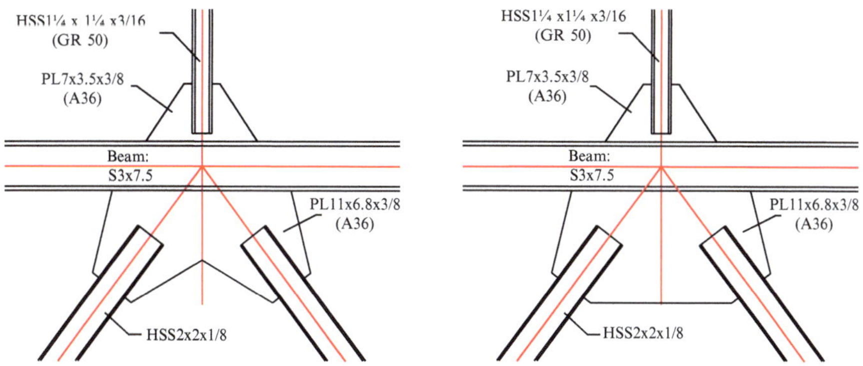

Fig. 5.4 Detail of designed gusset (*left*) and tested gusset (*right*)

The difference in behavior between the third test and the first test is explained by the brace to beam gusset plate geometry. The plate was designed to have a notch in the middle as shown in Fig. 5.4 (AISC 2002). The frames tested in Test #1 and Test #2 did not have this detail due to a manufacturing error while the frame tested in Test #3 was built correctly. When the braces buckle out of plane, the stiffness of the gusset determines how much moment gets transferred to the beam as torsion (Bruneau et al. 1998; Goel 1992). A very flexible gusset absorbs the energy through deformation while a very rigid gusset transfers the full moment to the beam above.

5.8 Analytical Model

A new formulation capable of geometric and material nonlinearities was developed to try to reproduce the results of the zipper frames tests. A general description of the proposed formulation is presented next with some examples.

The new formulation is based on the well known Corotational formulation (Crisfield 1990; Simo and Vu-Quoc 1986). This methodology separates the total displacements of an element into rigid body motion and deformation. A rigid body motion does not generate stresses, just re-orients them, therefore all the forces come from the deformational part of the motion. However, the rigid body motions can be very large, thus second order effects can be very important.

The proposed formulation is solved in the state space (Sivaselvan 2003) where a set of parameters and their rates are chosen as the global and local variables. The differential equations of equilibrium, compatibility and material nonlinearity are solved simultaneously which allows the equilibrium equations to be written a as a set of first order differential equation, thus simplifying the numerical solution.

The state space requires:

- The definition of global variables and local variables.
- Evolution equations for global variables and local variables.
- Equations that link global and local variables.

5.9 Coordinate Systems

In a structural member, the rigid body motions are represented by the movement of the chord of the element with respect to a given coordinate system. In a macromodel, an element is defined by 2 end nodes and the chord is then a 3D straight line linking them. In an initial condition, the element is undeformed and the chord coincides with the element itself. In a deformed configuration, the chord of the element can be described as a translation and a rotation of the initial chord. The same result is obtained through the vectorial composition of the positions of the nodes in respect to a global coordinate system. To calculate the position and rotations of any point along the member, it is sufficient to know its position and rotation with respect to the chord since the location of the chord in respect to a global system has already been determined. The corotational method defines a new additional coordinate system located along the chord of the element in its deformed configuration. This new coordinate system is called corotational because it follows the movement of the chord through time and space. From the point of view of any point along the deformed shape of the element, the corotated system does not change with time.

Three coordinate systems are used in the following formulation (see Fig. 5.5)

- **G**: Global coordinate system. Fixed and common for all the elements, the equations of motions are solved in this system.

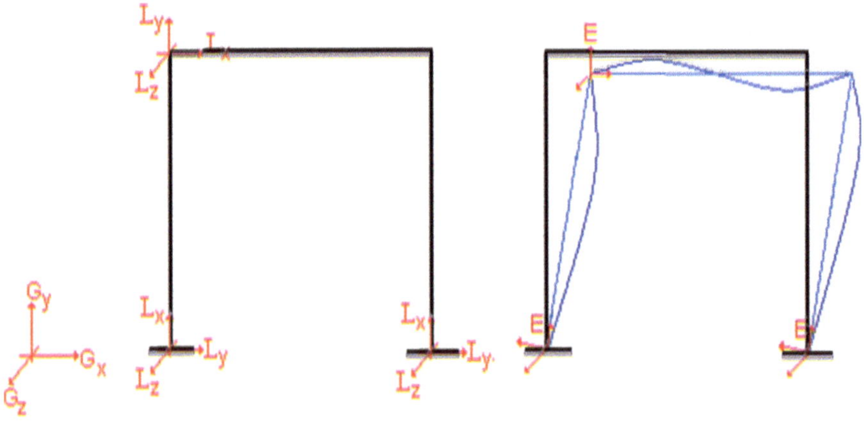

Fig. 5.5 Definition of coordinate system

- **L**: Local coordinate system. Defines the position of the element in its original unstressed position. This system serves as a reference for the calculation of nodal rotations.
- **E**: Corotated coordinate system. Defines the position of the chord of the element at each time step.

5.10 Degrees of Freedom (DOFs)

According to the defined coordinate systems, the DOFs can be written in global, local and corotated coordinates. In global and local coordinates, each node has 6 DOF: 3 displacements and 3 rotations.

$$\mathbf{u} = \left\{ \mathrm{u_x} \ \mathrm{u_y} \ \mathrm{u_z} \ \theta_\mathrm{x} \ \theta_\mathrm{y} \ \theta_\mathrm{z} \right\} \tag{5.1}$$

And its conjugated forces are:

$$\mathbf{F} = \left\{ \mathrm{F_x} \ \mathrm{F_y} \ \mathrm{F_z} \ \mathrm{M_x} \ \mathrm{M_y} \ \mathrm{M_z} \right\} \tag{5.2}$$

An element in its corotated coordinates has only a total of 6 DOF based on the following assumptions: the axial force and torsion are constant along the member and the moments can vary only linearly (Fig. 5.6).

$$\mathbf{q} = \left\{ \mathrm{q_1} \ \mathrm{q_2} \ \mathrm{q_3} \ \mathrm{q_4} \ \mathrm{q_5} \ \mathrm{q_6} \right\} \tag{5.3}$$

And its conjugated forces are:

$$\mathbf{Q} = \left\{ \mathrm{Q_x} \ \mathrm{Q_y} \ \mathrm{Q_z} \ \mathrm{Q_x} \ \mathrm{Q_y} \ \mathrm{Q_z} \right\} \tag{5.4}$$

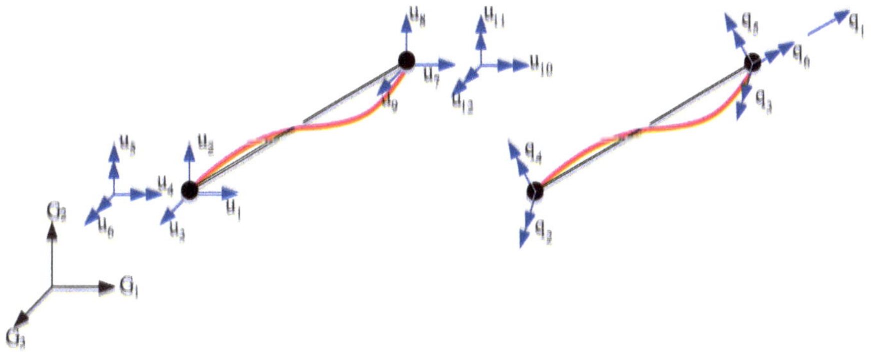

Fig. 5.6 Definition of DOFs in global coordinates (*left*) and corotated coordinates (*right*)

5.11 Global Variables

The global variables are: total displacements, total displacement rates and internal forces.

$$\begin{aligned} \mathbf{y}_1 &= \mathbf{u} \\ \mathbf{y}_2 &= \dot{\mathbf{u}} \\ \mathbf{y}_3 &= \mathbf{Q} = \mathbf{k}\mathbf{u} \end{aligned} \tag{5.5}$$

With these set of variables, the well known equation of motion:

$$\mathbf{m}\ddot{\mathbf{u}} + \mathbf{c}\dot{\mathbf{u}} + \mathbf{k}\mathbf{u} = \mathbf{P} \tag{5.6}$$

Becomes:

$$\begin{aligned} \mathbf{m}\dot{\mathbf{y}}_2 + \mathbf{c}\dot{\mathbf{y}}_1 + \mathbf{y}_3 &= \mathbf{P} \\ \mathbf{y}_2 &= \dot{\mathbf{y}}_1 \\ \dot{\mathbf{y}}_3 &= \mathbf{k}\dot{\mathbf{y}}_1 \end{aligned} \tag{5.7}$$

The first equation corresponds to the equation of motion, the second one gives the relationship between the total displacements y_1 and their rates y_2 and the third one establishes a relationship between the internal force Q and the total displacements y_1 in rate form through the element stiffness matrix k. Therefore the second order equation of motion has been written as a set of 3 first order differential equations.

5.12 Local Variables

The selection for local variables will depend of the type of analysis and constitutive laws (Simeonov 1999). In the present formulation, a flexibility based element and a constitutive law based on a stress decomposition model are chosen. Therefore, the total strains and curvatures are the independent local variable.

$$\boldsymbol{\varepsilon} = \left\{ \varepsilon \; k_x \; k_y \; k_z \right\} \tag{5.8}$$

5.13 Rigid Body Transformation Matrix

The transformation matrix **E** is particular to each element. Vector $\mathbf{E}_1$ defines the local corotated x coordinate and it is calculated as the vectorial difference of the nodes' actual positions **p** in global coordinates:

$$\mathbf{E}_1 = \frac{\mathbf{p}_i - \mathbf{p}_j}{\left\| \mathbf{p}_i - \mathbf{p}_j \right\|} \tag{5.9}$$

The variables i and j represent the end nodes of the element. The rotation matrix associated with the rotation of $\mathbf{G}_1$ into $\mathbf{E}_1$ is used to find the corotated y and z coordinates. The origin of the corotated coordinate system **E** is placed at one of the end nodes, say node i. This location simplifies latter calculations.

5.14 Evolution Equations

The evolution equations are the basic laws of analysis of structures: Equilibrium, Compatibility and Plasticity.

5.15 Equilibrium

Equilibrium equations constitute the evolution equations for global variables. Since the internal forces Q are state variables, their values are known at every time step. Therefore, equilibrium consists on transforming the local forces (written in corotated coordinates) into global forces (with respect to a global coordinate system). The transformation is done with a static transformation matrix T_{EC}. This matrix has two components: the first transforms the global coordinate system (G) into the corotated one (E) and the second has the static condensation of the degrees of freedom (DOFs) q:

$$\mathbf{q} = \begin{bmatrix} -1 & 0 & 0 & 0 & 0 & 0 & 1 & 0 & 0 & 0 & 0 & 0 \\ 0 & {}^{1}/_{\xi_L} & 0 & 0 & 0 & 1 & 0 & {}^{-1}/_{\xi_L} & 0 & 0 & 0 & 0 \\ 0 & {}^{1}/_{\xi_L} & 0 & 0 & 0 & 0 & 0 & {}^{-1}/_{\xi_L} & 0 & 0 & 0 & 1 \\ 0 & 0 & {}^{-1}/_{\xi_L} & 0 & 1 & 0 & 0 & 0 & {}^{1}/_{\xi_L} & 0 & 0 & 0 \\ 0 & 0 & {}^{-1}/_{\xi_L} & 0 & 0 & 0 & 0 & 0 & {}^{1}/_{\xi_L} & 0 & 1 & 0 \\ 0 & 0 & 0 & -1 & 0 & 0 & 0 & 0 & 0 & 1 & 0 & 0 \end{bmatrix} \begin{bmatrix} \mathbf{u}_i \\ \mathbf{u}_j \end{bmatrix} = \mathbf{T}_{EC} \cdot \mathbf{u}^E \tag{5.10}$$

Matrix **T** transforms global displacements **u** into Corotated coordinates **q** and its transpose transforms Corotated forces **Q** into Global coordinates **F**:

$$\mathbf{F} = \mathbf{T}^t \cdot \mathbf{Q} \tag{5.11}$$

5.16 Compatibility

Consider an element and its Corotated coordinate system E. Let ε be the axial deformation of the centerline and let P any point in the centerline (Fig. 5.7).

P has originally coordinates (x, 0, 0) respect to E. At the deformed configuration, its new coordinates are (ξ, η, χ). Since large displacements/large strains are assumed at this level, in order to characterize the complete movement of the section, a rotation matrix R(x) is defined at each point P.R(x) describes the rotation of the section in respect to the chord.

$$t = \mathbf{R} \cdot \mathbf{E} \tag{5.12}$$

The variations of the coordinate components with the x axis are:

$$\frac{d\xi}{dx} = \frac{d\xi}{ds}\frac{ds}{dx} = \frac{d\xi}{ds}(1+\varepsilon) \quad \frac{d\eta}{dx} = \frac{d\eta}{ds}\frac{ds}{dx} = \frac{d\eta}{ds}(1+\varepsilon) \quad \frac{d\chi}{dx} = \frac{d\chi}{ds}\frac{ds}{dx} = \frac{d\chi}{ds}(1+\varepsilon) \tag{5.13}$$

where, s is the arc-length parameter tracking the deformed configuration, thus ds/dx represents the variation of the axial deformation in respect to the original length of the element and $d\xi/ds$, $d\eta/ds$ and $d\chi/ds$ represent the components of the vector $\mathbf{t_1}$ in the corotated coordinates $\mathbf{E}$, i.e. the change in the orientation of the coordinate system along the length of the element. ε, the axial strain, is by definition:

$$\varepsilon = \frac{\delta L}{\delta L_o} - 1 = \frac{ds}{dx} - 1 \tag{5.14}$$

Considering a small variation about this deformed position, the incremental compatibility is:

$$\frac{d\dot{\xi}}{dx} = \left(\mathbf{E}_1^t \cdot \dot{\mathbf{t}}_1\right)(1+\varepsilon) + \left(\mathbf{E}_1^t \cdot \mathbf{t}_1\right)\dot{\varepsilon} \quad \frac{d\dot{\eta}}{dx} = \left(\mathbf{E}_2^t \cdot \dot{\mathbf{t}}_1\right)(1+\varepsilon) + \left(\mathbf{E}_2^t \cdot \mathbf{t}_1\right)\dot{\varepsilon}$$
$$\frac{d\dot{\chi}}{dx} = \left(\mathbf{E}_3^t \cdot \dot{\mathbf{t}}_1\right)(1+\varepsilon) + \left(\mathbf{E}_3^t \cdot \mathbf{t}_1\right)\dot{\varepsilon} \tag{5.15}$$

The variations of the vector $\mathbf{t_1}$ with respect to time are:

$$\dot{\mathbf{t}}_1 = \dot{\mathbf{R}}\mathbf{E}_1 = \dot{\boldsymbol{\Omega}} \cdot \mathbf{R}\mathbf{E}_1 = \boldsymbol{\Omega} \cdot \mathbf{t}_1 \tag{5.16}$$

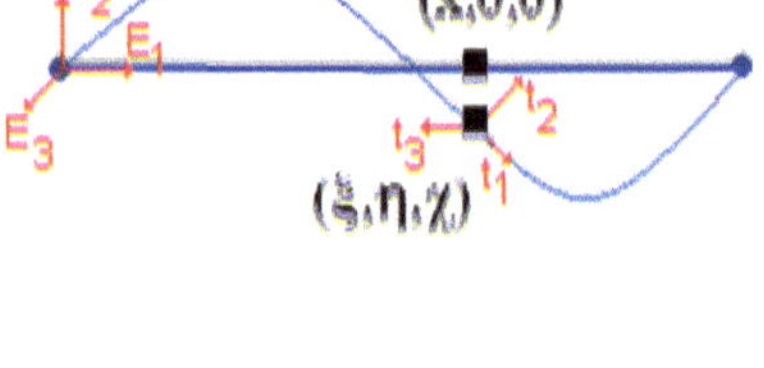

Fig. 5.7 Compatibility formulation

Where, $\boldsymbol{\Omega}$ is a skew-symmetric matrix representing an instantaneous infinitesimal rotation or spin and $\boldsymbol{\omega}$ will be its correspondent rotational vector. $\boldsymbol{\Omega}$ represents the time variations of the rotations over the **t** coordinate system.

It is important to note that, at the deformed level, the variations of the CR system **E** with time are irrelevant because the strains and stresses are not affected by rigid body motion. Only an observer located at the Global coordinate **G** will notice that the CR system **E** moves with time.

The compatibility equations for rotations are derived from the relationship between rotation and curvature, in 3D:

$$\frac{d\mathbf{R}}{ds} = \mathbf{KR} \tag{5.17}$$

Where **K** is a skew-symmetric matrix containing the local corotated curvatures. Taking variations on the Eq. 5.17 and after some mathematical manipulations the following relationship is found:

$$\frac{d\omega}{dx} = \left(\dot{\mathbf{k}} - \omega x \mathbf{k}\right)(1+\varepsilon) + \mathbf{k}\dot{\varepsilon} \tag{5.18}$$

The first term of the right hand side of Eq. 5.18 gives the rate of change of the curvature with respect to an observer which moves with the frame **t**. The material representation takes the form:

$$\frac{d\omega^{m}}{dx} = \dot{\mathbf{k}}^{m}(1+\varepsilon) + \mathbf{k}^{\mathbf{m}}\dot{\varepsilon} \tag{5.19}$$

Integrating the compatibility equations and integrating over the length of the element, it can be shown that the following relationship is obtained:

$$\mathbf{T} \cdot \dot{\mathbf{u}} = \dot{\mathbf{q}} = \int_{0}^{L} \mathbf{B}^{*} \begin{bmatrix} \dot{\varepsilon} \\ \dot{\mathbf{k}} \end{bmatrix} dx \tag{5.20}$$

Where B* is:

$$\mathbf{B}^{*} = \begin{bmatrix} t_{11} & R_{31}\eta - R_{21}\chi & R_{32}\eta - R_{22}\chi & R_{33}\eta - R_{23}\chi \\ \frac{-t_{12}}{\xi_L} & \left(\frac{\xi}{\xi_L}-1\right)R_{31} - R_{11}\frac{\chi}{\xi_L} & \left(\frac{\xi}{\xi_L}-1\right)R_{32} - R_{12}\frac{\chi}{\xi_L} & \left(\frac{\xi}{\xi_L}-1\right)R_{33} - R_{13}\frac{\chi}{\xi_L} \\ \frac{-t_{12}}{\xi_L} & \frac{\xi R_{31} - R_{11}\chi}{\xi_L} & \frac{\xi R_{32} - R_{12}\chi}{\xi_L} & \frac{\xi R_{33} - R_{13}\chi}{\xi_L} \\ \frac{t_{13}}{\xi_L} & \left(\frac{\xi}{\xi_L}-1\right)R_{21} - R_{11}\frac{\eta}{\xi_L} & \left(\frac{\xi}{\xi_L}-1\right)R_{22} - R_{12}\frac{\eta}{\xi_L} & \left(\frac{\xi}{\xi_L}-1\right)R_{23} - R_{13}\frac{\eta}{\xi_L} \\ \frac{t_{13}}{\xi_L} & \frac{\xi R_{21} - R_{11}\eta}{\xi_L} & \frac{\xi R_{22} - R_{12}\eta}{\xi_L} & \frac{\xi R_{23} - R_{13}\eta}{\xi_L} \\ 0 & R_{11} & R_{12} & R_{13} \end{bmatrix} \tag{5.21}$$

B* matrix is called "Force interpolation matrix" because the transpose interpolates the nodal forces Q to get the internal forces P(x) at any point x along the length of the element.

$$\mathbf{P}(x) = \mathbf{B}^{t}(x)\mathbf{Q} \tag{5.22}$$

5.17 Plasticity

The plasticity equations represent the evolution equations for the local variables. Any constitutive law can be used as a plasticity equation. In the present formulation, a stress decomposition model developed by Simeonov (1999) and Sivaselvan (2003) is adopted:

$$\dot{F} = K_e\dot{\varepsilon} + (K_0 - K_e) \times \left[\dot{\varepsilon} - \frac{\left(\frac{\partial\phi}{\partial F_h}\right)^{t} (K_0 - K_e)\,\dot{\varepsilon}}{\left(\frac{\partial\phi}{\partial F_h}\right)^{t} (K_0 - K_e) \left(\frac{\partial\phi}{\partial F_h}\right)} \frac{\partial\phi}{\partial F_h} \left[(\phi + 1)^{n}\right] \left[\eta_1 + \eta_2 \mathrm{sgn}\left(\frac{\partial\phi}{\partial F_h} \cdot \dot{F}_h \right) \right] \right] \tag{5.23}$$

Where F(x) is the flexibility matrix at the internal point x. Equation 5.23 represents the relationship between the rate of strains and the rate of forces i.e. the stiffness matrix of the section. K_e is the elastic stiffness representing the hardening of the material, K_0 is the initial total elastic stiffness and ϕ is the plastic yield surface.

5.18 Solution of the System of Equations

The total system of equations is composed by:

- Set of differential and/or algebraic Equilibrium equations.
- Set of differential Compatibility equations.
- Set of differential Plasticity equations.

The equilibrium equations are differential only if there is mass present in the DOF. Otherwise they are algebraic (Q = P). Thus, in general, the system is a set of Differential Algebraic Equations (DAE). The software IDA developed by the Lawrence Livermore National Laboratory is used to solve the system. The total number of equations to solve depends on the number of elements and DOFs involved. Because the Compatibility equation is in integral form, a numerical solution involves the selection of an integration method. In this case the Lobatto formula is chosen because it considers the nodal values as integration points. The integral is replaced by the values of the formula evaluated a t the selected integration points time a weight. The position of the integration points as well as the weights are given by the Lobatto rule.

5.19 Zipper Frame Verification

Only the first story of the zipper frame was modeled. The objective of the modeling is to test the capability of the new formulation to capture the behavior of the braces and to check the sensitivity to modeling details. Although an analysis of a full three stories model could reduce the number of modeling unknowns, the one story model was chosen in order to reduce the number of free parameters. The analytical model is presented in Fig. 5.8. Each one of the braces is divided into eight elements, while each of the beams is composed of a single element. The location of the nodes in the braces is the same as the location of the Krypton probes. The variables used in the analytical model are: the unknown material properties of the gusset plates and the equivalent stiff-ness of the vertical spring that models the zipper strut and the upper floors.

After comparing the performance of the model versus the testing (see Schachter and Reinhorn 2011), the sensitivity of the analytical performance to the variations in the modeling parameters is performed. The model is subjected, at the beam

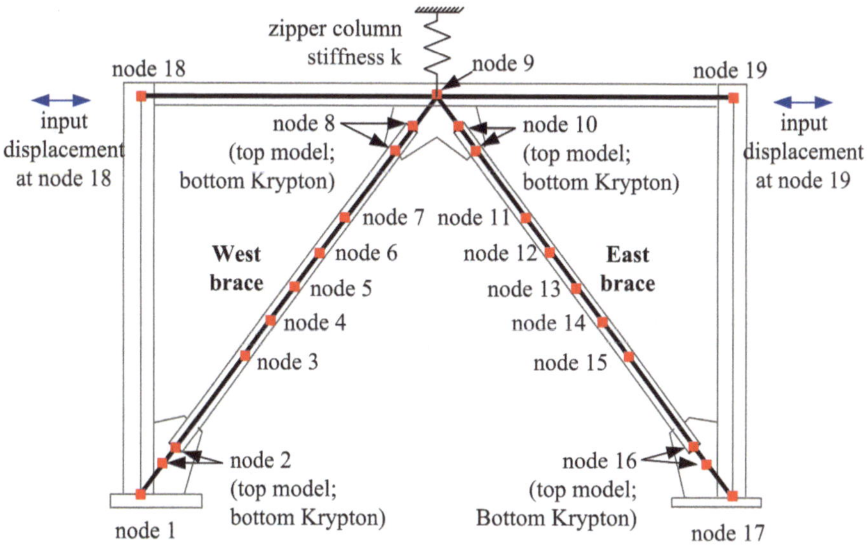

Fig. 5.8 Zipper frame model

elevation, to a lateral displacement history, which was recorded in the shake table experimental study. The model is selected as a base of comparison for this analysis with the following assumed properties:

(i) The stiffness of the equivalent "zipper column" is calculated as the axial stiffness of the element k = (EA/L) zipper column.
(ii) The out of plane inertia of the gusset plates is assumed to be the same as the out of plane inertia of the braces (I).
(iii) The yield strength of the gusset plates is assumed to be the same as the yield strength of the braces (σy).
(iv) The yield function for the gusset plates is assumed to be the same as the yield function of the braces (HSS section).

The sensitivity was evaluated varying the properties of the assumed values as follows:

1. The influence of the stiffness of the vertical spring was tested by changing its value to a very small amount ($0.03 \times k$) and to a very large amount ($2.61 \times k$) while leaving all the other variables unchanged;
2. The influence of the out of plane inertia of the gusset plates was tested by changing its value to an amount ten times smaller (I/10) and ten times larger ($10 \times I$), while leaving all the other variables unchanged;
3. The influence of the yield strength of the gusset plates was evaluated changing it to $0.6 \times \sigma y$ and to $1.67 \times \sigma y$, while leaving the other variables unchanged.

Normalized results for the in plane and out of plane displacements of the middle node of the left brace (node 13) are presented in Fig. 5.9 and the normalized results for the hysteretic curve of the left brace (axial displacement vs. axial force) are presented in Fig. 5.10. In plane and out of plane displacements are normalized to zmax = 66.55 mm (2.62 in), axial displacements are normalized to dy = 2.54 mm (0.1 in) and axial forces are normalized to Py = 22.86 ton (50.4 kip). Similar results are obtained for the right brace which are not included here for brevity but can be found in (2011). Test results are included for comparison purposes in Fig. 5.3.

From the figures it can be noted that: (a) The vertical stiffness of the equivalent zipper strut controls the vertical displacement of the brace to beam connection (node 9 in Fig. 5.9), thus it controls the axial displacement of the braces (see Fig. 5.10a left). A model using a softer vertical stiffness has larger vertical displacements, thus increasing the deformation to the compression brace and reducing the axial deformation of the brace in tension (which does not yield). (b) The inertia of the gusset plate has an important influence in the out of plane displacement of the brace. The model with a flexible gusset (see Fig. 5.9b left) has 25 % more out of plane displacements and the shape of the curve is very smooth, like the one obtained from the test data (Fig. 5.3a). A model with a stiffer gusset plate (Fig. 5.9b right) has displacements of the same order of magnitude as those measured during testing, but the shape is not as smooth. (c) When the yield strength of the gusset plates is very small, the braces do not buckle and the gusset plates yield in both tension and compression (Fig. 5.10c left). In this case, the gusset plates are weaker than the brace and attract

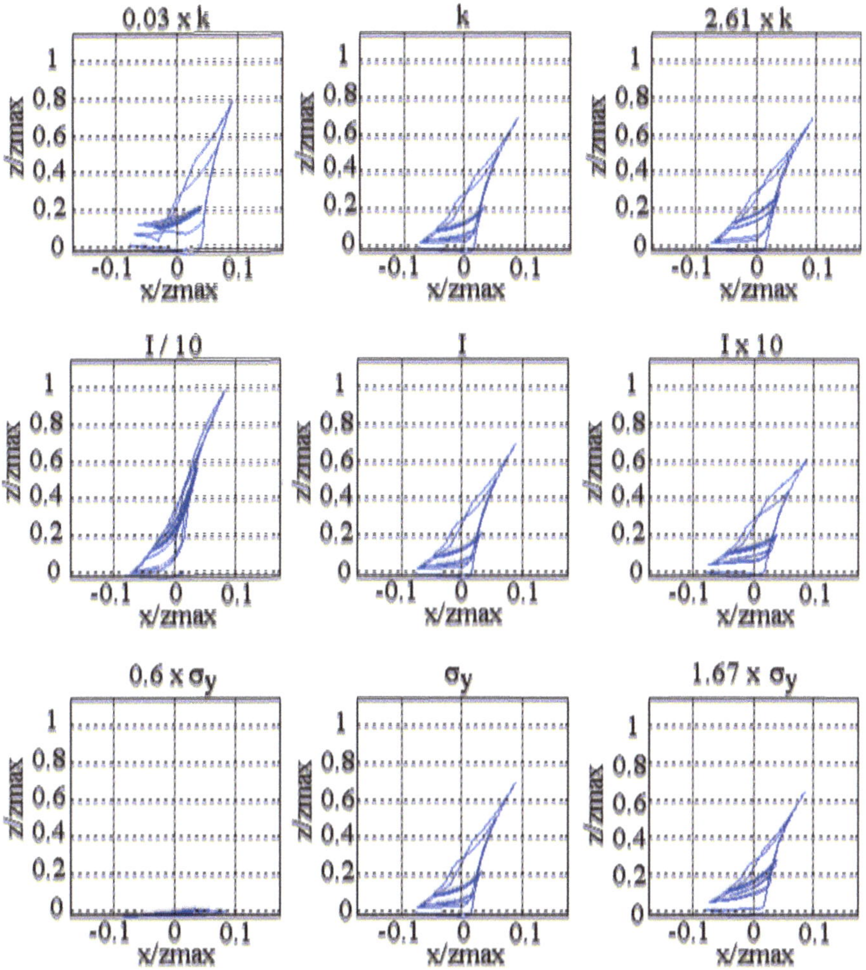

Fig. 5.9 Normalized in-plane versus out-of-plane displacements for the right brace

all plastic deformations. On the other hand, when the gusset plates have high yield strength (Fig. 5.9c right), this forces the braces to buckle and yield. Thus, this stiffness controls the distribution of plasticity between the brace and the gusset plates.

From the results of the bounding analysis it can be concluded that the new formulation is sensitive to the modeling parameters and can reproduce the inelastic buckling observed during testing. Moreover, all the lower and upper bound predictions are in agreement with the expected behavior obtained from structural mechanics theory. The differences between the analytical and test results are due to the incomplete modeling of the frame and the lack of detailed knowledge about the gusset plates properties. The "real" behavior of the structure is bounded between the extreme values for each of the variables analyzed.

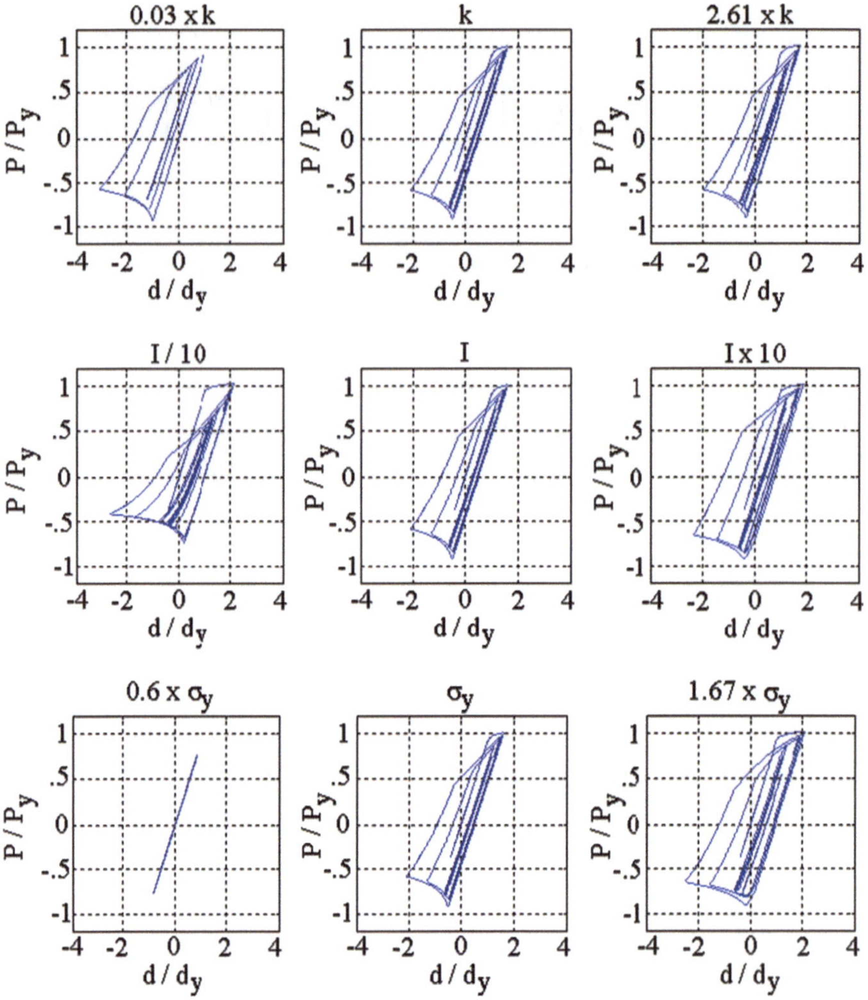

Fig. 5.10 Normalized force- displacement curves left brace

5.20 Conclusions

Three shaking table tests of an incomplete one-third scale suspended zipper frame were completed. It was proven that the expected zipper mechanism develops. Multiple stories were engaged in the resistance of lateral force, however test #3 shows that damage still can concentrate on the first floor. The brace to beam gusset plate connection is a key factor on the success of the frame. It has to be ensured that the plate is flexible enough to deform. If not possible, the beam has to be prevented to rotate through lateral restraints.

A new formulation was developed to model geometric nonlinearities in general, and the zipper test results in particular. Simple examples show very good results and convergence with both elastic and nonlinear materials, with a minimum number of elements. Working is under way to model the full 3D movement of the braces as seen in testing.

References

The American Institute of Steel Construction (2002) ANSI/AISC 341-10 seismic provisions for structural steel buildings. The American Institute of Steel Construction, Chicago

Bruneau M, Uang C-M, Whittaker A (1998) Ductile design of steel structures. McGraw-Hill, New York

Crisfield MA (1990) A consistent co-rotational formulation for non-linear, three dimensional, beam-elements. Comput Methods Appl Mech Eng 21:131–150

Goel SC (1992) Stability and ductility of steel structures under cyclic loading. In: Earthquake resistant design of ductile braced steel structures. CRC Press, Boca Raton, pp 297–308

Khatib IF, Mahin SA, Pister KS (1988) Seismic behavior of concentrically braced steel frames. Report no. UCB/EERC-88/01. Earthquake Engineering Research Center, University of California, Berkeley

Leon RT, Yang, CS (2003) Special inverted-V-braced frames with suspended zipper struts. In: International workshop on steel and concrete composite construction. IWSCCC, National Center for Research on Earthquake, Taipei

Schachter M, Reinhorn A (2011) Dynamic analysis of three dimensional frames with material and geometric nonlinearities. J Struct Eng 137(2):207–219

Simeonov V (1999) Three dimensional inelastic dynamic structural analysis of frame systems. PhD Thesis, University at Buffalo, SUNY

Simo JC, Vu-Quoc L (1986) A three dimensional finite strain rod model. Part II computational aspects. Comput Methods Appl Mech Eng 58:79–116

Sivaselvan MV (2003) Nonlinear analysis towards collapse simulation – a dynamical systems approach. PhD thesis, University at Buffalo, SUNY

Chapter 6
Simplified Seismic Evaluation of Structures Using Adaptive Pushover Analysis

Joseph M. Bracci

Abstract A simplified procedure for evaluating the seismic performance and retrofit of existing low to mid-rise reinforced concrete buildings was first presented by Bracci et al. (ASCE J Struct Eng 123(1):3–10, 1997). The procedure is derived from the well-known Capacity Spectrum Method (Freeman, Prediction of response of concrete buildings to severe earthquake motion. In: Douglas McHenry international symposium on concrete and concrete structures, SP-55. ACI, Detroit, pp 589–605, 1978) and is intended to provide practicing engineers with a relatively simple methodology for estimating the margin of safety against structural failure. The procedure consists of constructing a series of seismic story demand curves from modal superposition analyses wherein changes in the dynamic characteristics of the structure (natural frequencies, mode shapes and linear viscous damping) at various response phases ranging from elastic to full failure mechanism are considered. Then these demands are compared to the lateral story capacities as determined from an independent adaptive pushover analysis, where the distribution of lateral forces is based on stiffness dependent story shear demands. The adaptive pushover analysis forms a critical aspect of the methodology which relies on a reasonable estimate of the failure modes of the structure. The proposed technique is applied to a one-third scale three story reinforced concrete frame model building that was subjected to repeated shaking table excitations, and later retrofitted and tested again at the same intensities. This study shows that the procedure can provide reliable estimates of story demands vs. capacities for use in seismic performance and retrofit evaluation of structures.

J.M. Bracci (✉)
Zachry Department of Civil Engineering, Texas A&M University,
College Station, TX 77843-3136, USA
e-mail: j-bracci@tamu.edu

G.P. Cimellaro et al. (eds.), *Computational Methods, Seismic Protection, Hybrid Testing and Resilience in Earthquake Engineering*, Geotechnical, Geological and Earthquake Engineering 33, DOI 10.1007/978-3-319-06394-2_6,
© Springer International Publishing Switzerland 2015

6.1 Introduction

In regions of moderate to high seismic risk, there is a significant need for evaluating the seismic adequacy of existing structures given the damage and collapse of numerous structures during earthquakes in recent history throughout the world. In particular, the seismic rehabilitation of concrete buildings in regions of moderate to high seismicity without properly seismic detailing and appropriate lateral force capability is a continued matter of concern. Structures that may be vulnerable to damage should be identified and an acceptable level of seismic upgrade might be considered if the societal risk of seismic damage is high. The task of making such an assessment and establishing the level of seismic upgrading rests with the structural engineer. Fortunately over the past 20 years or so, simplified analytical tools and sophisticated analytical platforms have advanced and code-based performance guidelines have been developed to assist the engineer to carry out the task.

A simple analytical technique that engineers have utilized for a quick estimate to evaluate the performance and survivability of structures during earthquakes is the Capacity Spectrum Method (Freeman 1978, 1987; ATC 1982). The Capacity Spectrum Method was also incorporated in the U.S. Army, Navy and Air Force Technical Manual on Seismic Design Guidelines for Essential Buildings (Army 1986). In its original form, the Capacity Spectrum Method compares the strength capacity curve expressed in the standard spectral acceleration vs. period format with the assumed site-specific elastic response spectrum earthquake demands. The strength capacity curve was typically established from a pushover analysis by the code-based inverted triangular loading distribution or a distribution proportional to the fundamental mode of the structure. The method was advanced by Mahaney et al. (1993) by expressing the strength capacity curve in Acceleration-Displacement Response Spectrum (ADRS) format, which had the advantage of representing structural capacity and demand in terms of story forces and displacements. Up until this time, the Capacity Spectrum Method was used for evaluation of low to mid-rise structures in which the response was primarily characterized by the fundamental mode of vibration, since the formulation used a predetermined lateral load distribution for the pushover.

To consider the impact of high modes of vibrations and resulting mid-story failure mechanisms, the Capacity Spectrum Method was extended by Bracci et al. (1997) by using an adaptive pushover analysis and a story-by-story performance evaluation using modal superposition. This adaptive pushover procedure was first used by Reinhorn and Vladescu (Vladescu 1995) and uses a stiffness-dependent lateral force distribution derived from story shear demands. Seismic demand curves are established at varying levels of inelasticity: from initial elastic response to final failure mechanism. The resulting range of demands are compared with the inelastic force-deformation capacities for each story level, which can provide an estimate of the margin of safety against collapse for the selected seismic demand level. This procedure was validated using dynamic fundamental properties and experimental response of a one-third scale three story reinforced concrete frame model building that was tested, before and after retrofit, at varying levels of seismic excitation on a shaking table (Bracci et al. 1995a, b). In this paper, the procedure and a case study using the shaking table tests are summarized. In addition, an extension of the procedure is introduced by considering the equivalent dynamics characteristics at varying levels of ductility demands to formulate a single expected performance.

6.2 Methodology

In the paper Bracci et al. (1997), the basis of the method was to develop a range of site-specific demand curves and compare them to the computed adaptive pushover capacities at each story level of the structure. The primary differences between the proposed scheme and the Capacity Spectrum Method were: (i) the treatment of the applied lateral load during the pushover analysis; (ii) the use of story level demands and capacities as opposed to overall base shear vs. top story displacement; and (iii) the consideration of different levels of inelasticity to generate a range of seismic demands curves.

The step-by-step approach used in the evaluation process is described below:

1. Identify the site-specific ground spectral demand (design response spectra) and the equivalent viscous damping, typically taken as 5 % for concrete structures but can vary from about 2 to 10 % depending on structural type, structural material, element connectivity, and soil conditions to name a few.
2. Conduct an adaptive pushover analysis of the structure as described in the next section. Plots of the story shear vs. drift are established beyond yielding and prior to failure of the structure. Failure can be defined in terms of an acceptable story drift limit following the formation of a failure mechanism in the structure or other criteria depending upon the type of structure. Four distinct response phases during the pushover analysis can be identified for computing the story demands as follows:

    (i) Initial elastic stage of linear structural response;
    (ii) First yield stage, identified by the first major transition from elastic to inelastic response;
    (iii) Incipient plastic mechanism stage when a complete mechanism develops; and
    (iv) Final failure mechanism or specified ultimate drift limit stage.

3. Conduct a complete eigenvalue analysis of the system at each of the above response phases to identify the respective mode shapes and frequencies.
4. Determine the equivalent elastic story shear and drift demands using modal superposition, as described later, at each of the four response phases identified in Step 2.
5. Superimpose the story response demand curves from Step 4 with the pushover analysis story capacities from Step 2 and identify approximate damage potential.

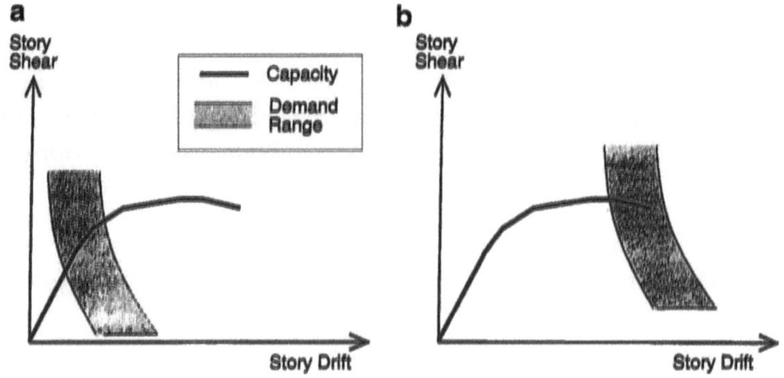

Fig. 6.1 Typical seismic story demand vs. capacity. (**a**) Safe design, (**b**) unsafe design (Bracci et al. 1997)

A qualitative representation of a story response demand vs. capacity is shown in Fig. 6.1. If the demand curves intersect the capacity envelope near the elastic region or near a region with a significant amount of strength and deformation reserve capacity (Fig. 6.1a), then the structure should be capable of resisting both the imposed seismic story force and drift demands. If a demand curve intersects the force-deformation capacity envelope with little story strength and deformation reserve capacity (Fig. 6.1b), then it can be concluded that the structure would not perform in a sufficient manner during the imposed seismic excitation and that the structure may be in need of seismic retrofit to avoid major damage or collapse.

The two essential evaluation tasks outlined in the step-by-step process above are described in detail in the subsequent sections.

6.3 Capacity Evaluation Using Adaptive Pushover Analysis

The computation of system capacity using pushover analysis requires a rational consideration of the potential failure modes of the structure during earthquake loading. The typical inverted triangular loading distribution or fundamental mode loading distribution may not be consistent with the internal resistance of the structure especially if local column sidesway mechanisms (story mechanisms) develop under the imposed loading. Vladescu (1995) proposed an adaptive pushover method which uses a self-correcting load distribution based on the internal story resistance of the system at the prior analysis step as follows:

$$\Delta F_i^{j+1} = V^j\left(\frac{F_i^j}{V^j} - \frac{F_i^{j-1}}{V^{j-1}}\right) + \Delta P^{j+1}\left(\frac{F_i^j}{V^j}\right) \tag{6.1}$$

where i = story number; j = analysis step; $\Delta F_i{}^{j+1}$ = incremental ith story force at step $j+1$; V^j = base shear at step j; ΔP^{j+1} = incremental base shear applied at step $j+1$; and $F_i{}^j$ = ith story force at step j.

The initial step of the pushover analysis can begin with any assumed lateral force distribution. For each subsequent step, the story level force to be applied is obtained from Eq. (6.1). The state of stress in each member of the structure is evaluated at the end of each step and the structural stiffness matrix is updated if any element changes state during that step. The load is continually applied in small increments until the inter-story drift at any level in the structure exceeds a specified drift limit or a P-delta collapse is imminent, as indicated by a negative story stiffness.

The four phases at which an eigenvalue analysis of the system is required are determined from the pushover force-displacement curves. The elastic (initial) phase identified as the tangent stiffness of the linear response and is straightforward. The second phase corresponds to the first major deviation from the elastic behavior, which usually corresponds to first yielding of a structural element. The third phase represents yielding of a story level where a story mechanism has just formed and the system has softened considerably. The final phase corresponds to system failure (or the end of the analysis) as defined in the previous paragraph.

6.4 Equivalent Elastic Story Response Spectral Demands

The story shear force and inter-story drift demands of a structure during an earthquake excitation can be estimated using elastic response spectrum analysis and modal superposition. By uncoupling the dynamic equations of motions, the maximum relative story displacements for a MDOF system produced by the n-th mode of vibration ($v_{n,max}$) can be represented directly from the dynamic characteristics of the structure at a particular response phase and the elastic spectral acceleration demands for a SDOF system (Clough and Penzien 1993):

$$v_{n,\max} = \phi_n \frac{\Gamma_n}{M_n \omega_n^2} S_{pa}\left(\zeta_n, \mathrm{T}_n\right) \tag{6.2}$$

where ϕ_n is the nth mode shape; Γ_n is the nth modal participation factor ($\phi_n{}^T m$, where m is the story mass vector); M_n is the nth modal mass ($\phi_n{}^T m \phi_n$); ω_n is the nth natural frequency; $S_{pa}(\zeta_n, \mathrm{T}_n)$ is the elastic spectral acceleration demand for damping ζ_n and natural period T_n.

The maximum total story displacement response vector ($v_{\max}$) can be estimated using the square root of the sum of the squares (SRSS) of the significant maximum modal responses from Eq. (6.2):

$$v_{\max} = \sqrt{\left(v_{1,\max}\right)^2 + \left(v_{2,\max}\right)^2 + \ldots + \left(v_{N,\max}\right)^2} \tag{6.3}$$

where N is the total number of significant modes.

Similarly, the maximum elastic force vector for the nth mode of vibration ($f_{n,max}$) can be determined by:

$$f_{n,\max} = m\phi_n \frac{\Gamma_n}{M_n} S_{pa}\left(\zeta_n, \mathrm{T}_n\right) \tag{6.4}$$

The corresponding maximum story shear force vector for the n-th mode of vibration ($S_{\mathrm{n,max}}$) can be determined by summing the maximum elastic forces from Eq. (6.4) above the story of interest. The maximum total story shear force vector ($S_{\max}$) can then be estimated using the square root of the sum of the squares (SRSS) of the maximum modal story shear responses:

$$S_{\max} = \sqrt{\left(S_{1,\max}\right)^2 + \left(S_{2,\max}\right)^2 + \ldots..\left(S_{N,\max}\right)^2} \tag{6.5}$$

To facilitate the development of the four spectral demands for each story of the structural system during a ground motion excitation (represented in terms of equivalent story shear forces and displacements), two simplifying assumptions must be made about the dynamic characteristics of the structure throughout the modal superposition analyses: (i) the ratios of natural frequencies of the structure at each response phase, identified from an eigenvalue analysis, are assumed to remain constant, i.e. $\omega_1 : \omega_2 : \ldots \omega_N$; and (ii) the mode shapes of the structure at each response stage, also identified from an eigenvalue analysis, remain constant during the analysis. To develop a demand curve for each of the phases identified in Step 2, story shears and drifts are determined using Eqs. (6.1), (6.2), (6.3), (6.4) and (6.5) by sweeping the ratio of natural frequencies across the design response spectra and using the associated mode shapes. This process yields a story shear force-drift demand curve for each phase.

6.5 Validation of Evaluation Methodology

The analytical developments required to carry out the proposed evaluation, the adaptive pushover analysis, and a reduced-order eigenvalue analysis where the global degrees-of-freedom (DOF) are condensed to lateral story DOF were implemented in the inelastic damage analysis program IDARC (Kunnath et al. 1992). A case study of a one-third scale, three story reinforced concrete frame model building was used to validate the effectiveness of the proposed seismic evaluation procedure. A description of both the original model building, and the subsequent retrofit of the same building after severe seismic damage, is first presented to develop an understanding of the dynamic response of both structural systems. Then, the imposed shaking table excitation and the corresponding response spectrum acceleration demands are identified for calculating the story demands. The pushover analyses are carried out using IDARC, and story response demands are determined using the formulations presented earlier. The obtained demand and capacity curves are compared to the experimentally measured responses.

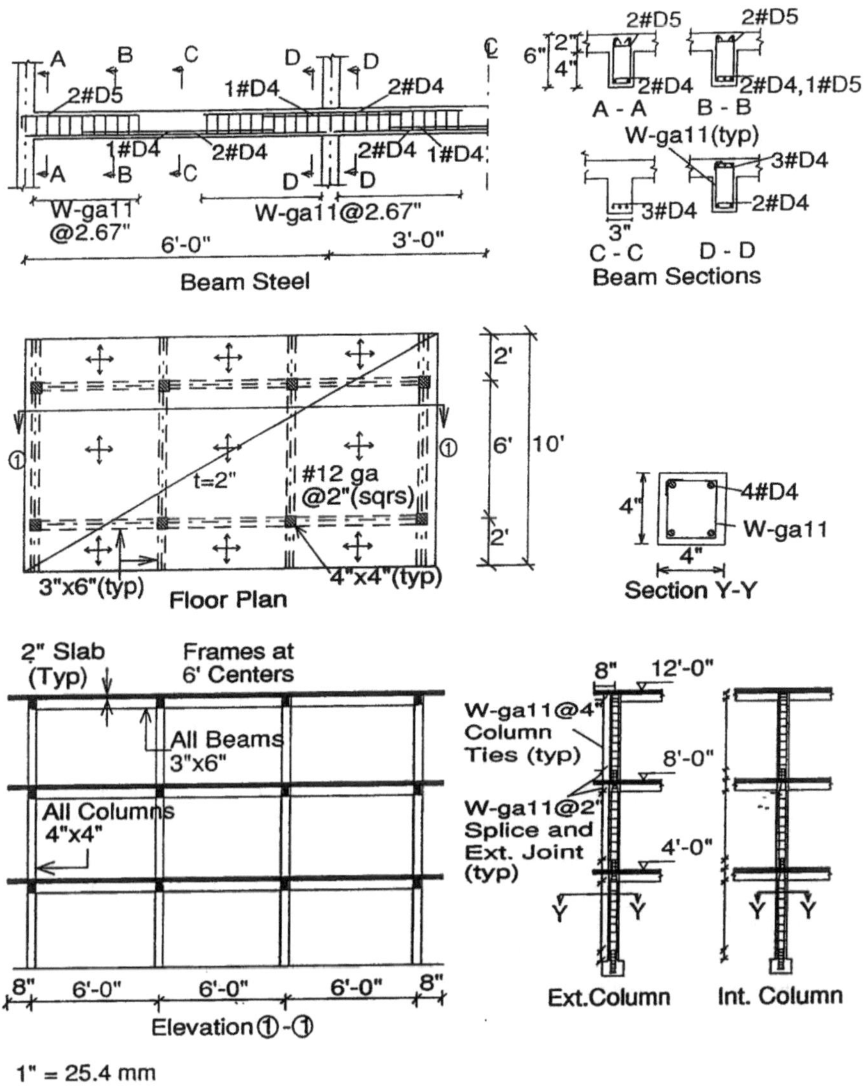

Fig. 6.2 Plan, elevation and reinforcing details of original model (Bracci et al. 1995a)

6.5.1 Description of Original Frame Model Building

Past design practice in the United States for low-rise reinforced concrete frame buildings constructed in low to moderate seismic risk areas have generally been for only gravity loads (gravity load design, GLD) according to the non-seismic detailing provisions of the American Concrete Institute ACI 318 code. The seismic response of such structures was evaluated in a dual analytical-experimental shaking table study of a one-third scale model building (see Fig. 6.2)

(Bracci et al. 1995a). The seismic deficiencies incorporated in that study, all of which violated the seismic provisions of Chapter 21 in ACI 318, included: (i) weak-column strong-beam behavior creating a structure prone to a soft-story collapse mechanism; (ii) inadequate transverse reinforcement in columns and joints for shear and confinement; (iii) column lap splices located in potential hinge zones; and (iv) discontinuous positive (bottom) beam reinforcement in beam-column joints. The imposed shaking table motions on the building simulated the 1952 Taft N21E earthquake component with normalized peak ground accelerations (PGA) of 0.05, 0.20, and 0.30 g, where g is the gravitational acceleration constant. The strong motion tests indicated that column end yielding developed, particularly on the second floor level with the initiation of a damaging soft-story mechanism.

6.5.2 Description of Retrofitted Frame Model Building

Following the series of damaging shaking table excitations on the original model structure, several seismic retrofit alternatives for GLD R/C frame structures were investigated for improving their local and global response performance during low to moderate type earthquakes by comparing structural response behavior from non-linear dynamic analyses (Bracci et al. 1995b). These retrofit schemes were designed to distribute damage throughout the building with acceptable control of story deformations by averting a catastrophic soft-story failure and enforcing a more ductile beam-sidesway mechanism. A minimum seismic retrofit was performed on the interior columns of the previously damaged one-third scale model building using a prestressed concrete jacketing technique (see Fig. 6.3) by: (i) encasing interior columns in a concrete jacket with additional longitudinal and transverse reinforcement; (ii) providing a reinforced concrete fillet around the unreinforced beam-column joints; and (iii) post-tensioning the added longitudinal column reinforcement. By strengthening only the interior columns of the model building, the retrofit scheme was considered to provide a minimum seismic resistance acceptable for low to moderate seismicity zones and considered potentially inadequate for higher seismic risk zones. The retrofitted model building was then experimentally tested on the shaking table with the same excitations as in the original model tests with PGAs of 0.20 and 0.30 g.

When compared to pre-retrofit performance, the experimental results of the retrofitted model structure showed that the overall behavior and damage could be more effectively controlled using only minimum column and joint strengthening techniques, which may be suitable for structures in low to moderate seismic risk zones. The retrofit scheme was successful in changing the failure mode from an undesirable column sidesway mechanism to a more ductile beam sidesway mechanism.

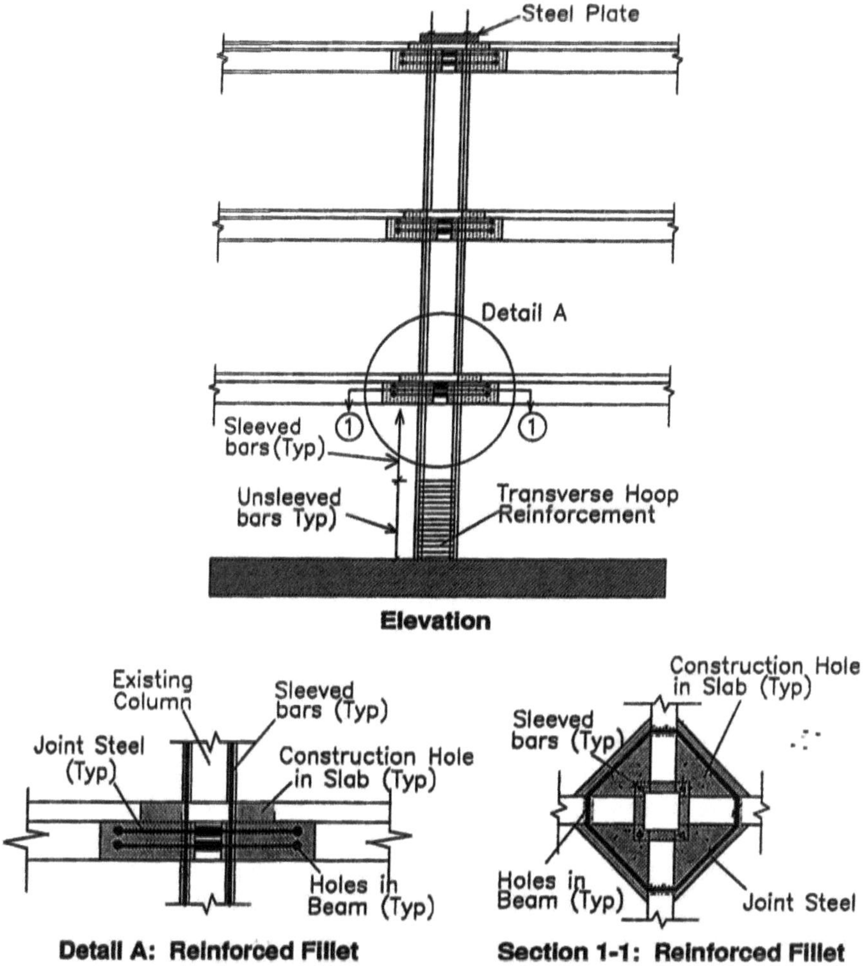

Fig. 6.3 Prestressed concrete jacketing retrofit (Bracci et al. 1995b)

6.5.3 Shaking Table Excitation and Acceleration Response Spectrum

The simulated shaking table excitations used in the experimental studies for both the original and retrofitted model structures were derived from the Taft N21E accelerogram with appropriate normalizing of the peak ground acceleration (PGAs of 0.05, 0.20 and 0.30 g) and required time scaling for the one-third scale model

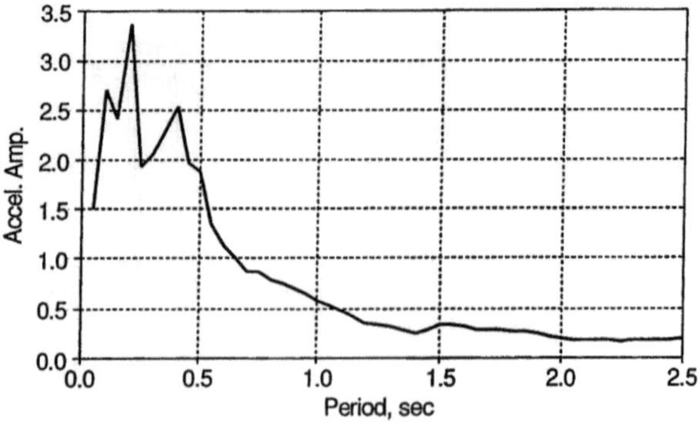

Fig. 6.4 Normalized response spectrum – Taft N21E component, 5 % damping

studies. The elastic response spectrum acceleration demand for this ground motion with 5 % equivalent viscous damping is shown in Fig. 6.4, after being normalized with respect to the peak ground acceleration.

6.5.4 Adaptive Pushover Analysis and Eigenvalue Analyses

Adaptive nonlinear pushover analyses were performed on both the original and retrofitted model buildings. The corresponding force-deformation story capacities are shown in Figs. 6.5, 6.6, 6.7, 6.8 and 6.9, respectively, for the original and retrofitted buildings. The four response phases, corresponding to elastic, first yield, incipient failure mechanism, and full failure mechanism, are also identified for each structure. Eigenvalue analyses were then performed at each of the four response phases and the resulting dynamic characteristics are shown in Tables 6.1 and 6.2, which correlated well with the experimentally measured characteristics. It is important to note that significant changes developed in the mode shapes, especially the first mode of vibration (see column [4]), throughout the various response phases of the pushover analysis. In the original building, the change in first mode shape identifies a possible development of a soft story mechanism on the second floor. In the retrofitted building, it can be observed that the change in the first mode shape has a significant influence on the first floor. It is important to point out that the observed changes in mode shapes are facilitated by the adaptive pushover analysis.

6.5.5 Response Spectral Demands and Performance Evaluation

Using the dynamic characteristics at the four response phases of the pushover analysis (see Tables 6.1 and 6.2) and the input base acceleration response spectrum from Fig. 6.4, the four corresponding force-deformation story demands for the

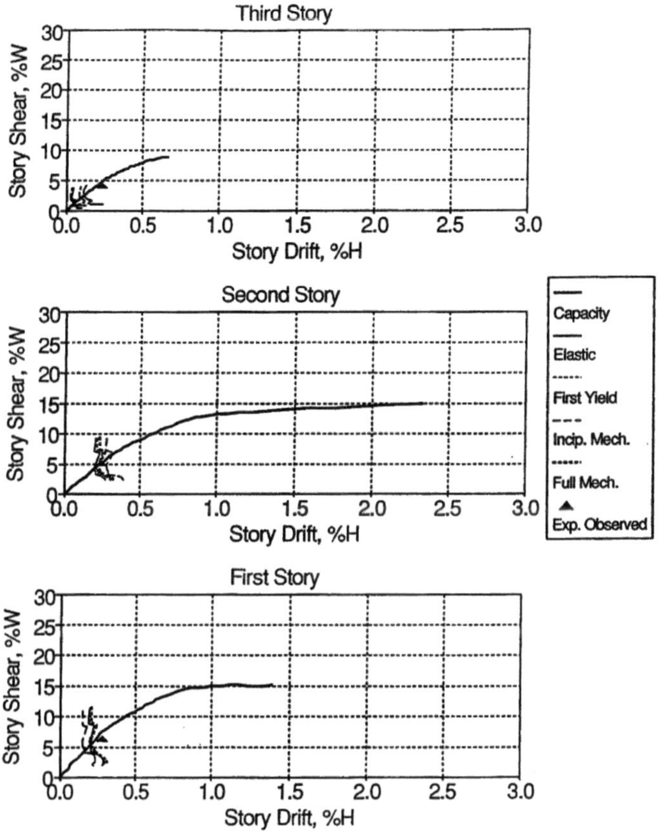

Fig. 6.5 Seismic demand vs. capacity, original building, PGA 0.05 g (Bracci et al. 1997)

various shaking table excitations are calculated using Eqs. (6.1), (6.2), (6.3), (6.4) and (6.5) for both the original and retrofitted model buildings. The resulting story demands and pushover capacities of the original model building are shown in Figs. 6.5, 6.6 and 6.7 for the 0.05, 0.20, and 0.30 g PGA shaking table excitations, respectively. The resulting story demands and pushover capacities of the retrofitted model building are shown in Figs. 6.8 and 6.9 for the 0.20 and 0.30 g PGA shaking table excitations, respectively. It can be observed that the maximum story response predictions for the original building are similar to the experimentally measured maximum story response. For the 0.05 g PGA excitation, the demand region intersects the story capacity envelopes near the elastic portion of the response, implying elastic behavior. For the 0.20 and 0.30 g PGA excitations, the demand curves intersect the capacity envelope in the region near incipient and full failure mechanisms, respectively. In particular, note that the soft-story response on the second floor is captured with reasonable accuracy using the appropriate inelastic characteristics of the building (incipient mechanism phase). From the standpoint of

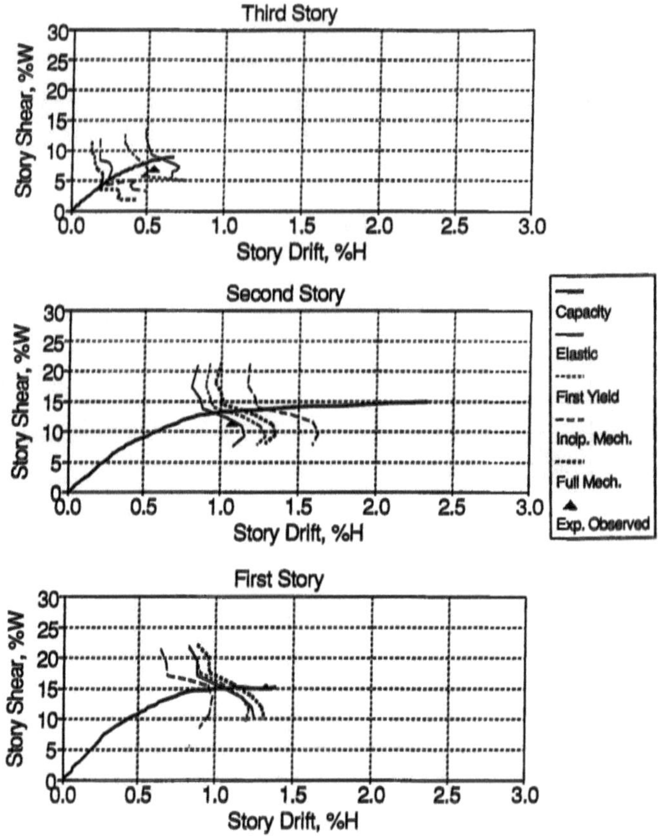

Fig. 6.6 Seismic demand vs. capacity, original building, PGA 0.20 g (Bracci et al. 1997)

estimating the margin of safety of the structure against the imposed loading, the following observations can be made from Figs. 6.5, 6.6, and 6.7 for the original building: the structure performed in the elastic range during the 0.05 g shaking; for the 0.20 g intensity shaking, the structure responded in the inelastic range with likely yielding in some members, but had a sufficient margin of safety against structural collapse as seen from the strength and deformation capacity reserve beyond the intersection of the demand and capacity lines; however, for the 0.30 g shaking, the demands intersect the capacities with little strength and deformation reserve. Therefore, it can be concluded that the margin of safety against collapse for the original building may be inadequate at this intensity of shaking.

For the retrofitted model building under the same earthquake ground motions (0.20 and 0.30 g), it can be observed from Figs. 6.8 and 6.9 that most of the inelastic response occurred on the first and second stories. The predicted maximum response for each story correlates well with the experimentally observed maximum behavior. From the standpoint of evaluating the adequacy of the retrofit behavior, it can be

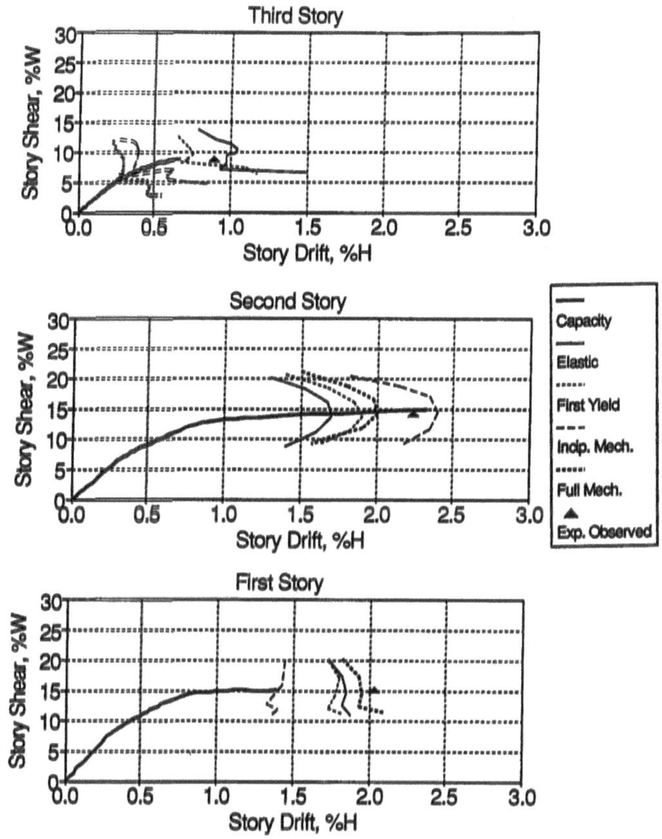

Fig. 6.7 Seismic demand vs. capacity, original building, PGA 0.30 g (Bracci et al. 1997)

observed that the story demands intersect the capacity envelopes with sufficient strength and displacement reserves. Although inelastic response is evident during these base motions, collapse of the retrofitted structure is not imminent for these levels of ground motion excitation.

6.6 Extended Methodology for a Single Demand Estimation

O'Boyle (1996) presented an extended methodology to estimate a single seismic demand curve that utilizes the relevant equivalent dynamic characteristics of the structure at the appropriate level of deformation ductility demand. These dynamic characteristics include the natural periods, mode shapes, and equivalent linear viscous damping of the structure at the varying levels of ductility as determined from the adaptive pushover analysis as presented above. Figures 6.10 and 6.11 show

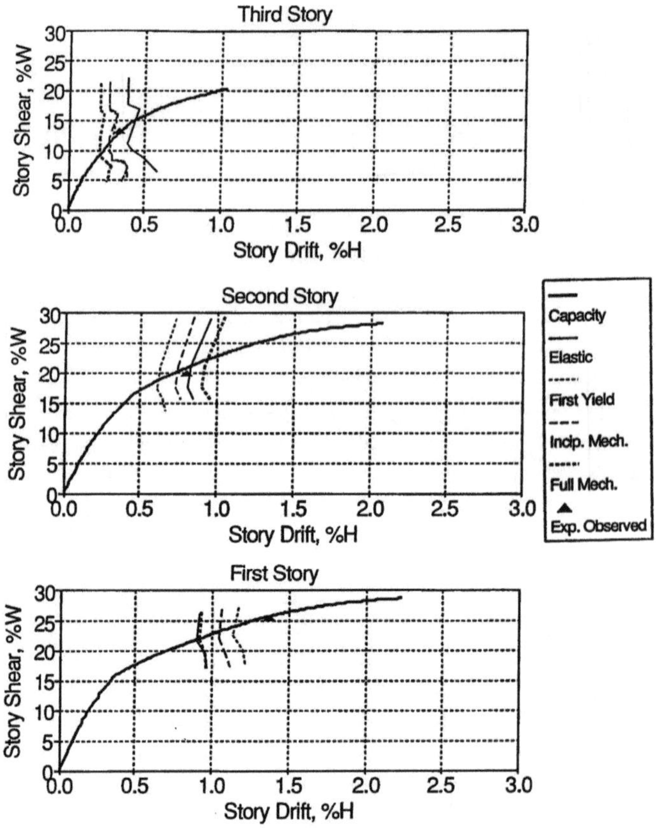

Fig. 6.8 Seismic demand vs. capacity, retrofitted building, PGA 0.20 g (Bracci et al. 1997)

the pushover response of the original and retrofitted buildings, respectively, and the unloading at varying levels of structural ductility where an eigenvalue analysis is performed within a modified version of the IDARC program. Using the updated structural stiffness characteristics for that level of ductility demand, equivalent natural periods and mode shapes can be identified.

In addition for structural deformations beyond yielding, the equivalent linear viscous damping that combines the effect of all damping mechanisms in the structural behavior will increase due to hysteretic response in the yielded regions of the structure. O'Boyle (1996) proposed a method for estimating the level of equivalent linear viscous damping (ζ_{eq}) as a function of the level of deformation ductility demand (μ) as follows:

$$\zeta_{eq} = \frac{2\,(\mu - 1)\,\beta}{\pi\ \mu} \tag{6.6}$$

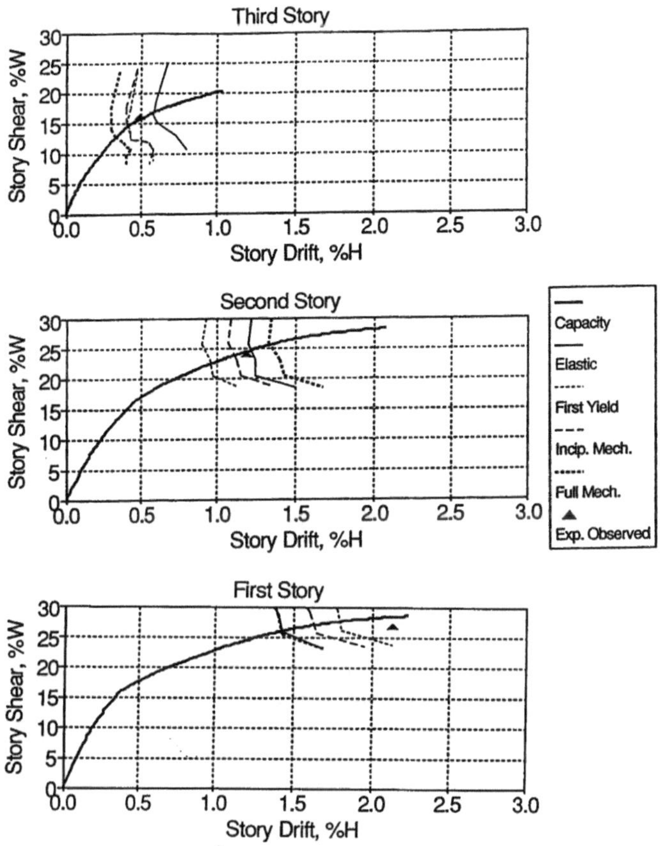

Fig. 6.9 Seismic demand vs. capacity, retrofitted building, PGA 0.30 g (Bracci et al. 1997)

where β is the energy dissipation effectiveness factor that is the ratio of the actual energy dissipation of the structure to the energy approximated by the elastic-perfectly plastic parallelogram behavior as shown in Fig. 6.12. Using nonlinear seismic analyses within IDARC, the energy dissipation effective factor for the original and retrofitted reinforced concrete buildings was determined to be about 0.25.

The resulting dynamic characteristics of the original and retrofitted buildings showed softened natural periods, changing mode shapes, and increased levels of equivalent viscous damping ranging from an assumed 2 % for elastic behavior up to about 10 % for larger ductility levels (see O'Boyle 1996 for more details). The resulting story demands and pushover capacities for the original and retrofitted buildings are selectively shown in Figs. 6.13, 6.14, and 6.15 for the Taft 0.3 g PGA shaking table excitations, respectively. For the original building, the demand curve which accounts for the dynamic characteristics at varying levels of ductility intersects the capacity curve with little strength and deformation capacity remaining.

Table 6.1 Dynamic characteristics of original building

Response phase (1)	Ratio of frequencies (2)	Mode shapes (3)	Approx. 1st mode (4)
Elastic (initial)	1.00: 3.00: 4.74	$\begin{Bmatrix} 1.000 & -0.823 & -0.422 \\ 0.772 & -0.577 & -1.000 \\ 0.378 & -1.000 & -0.925 \end{Bmatrix}$	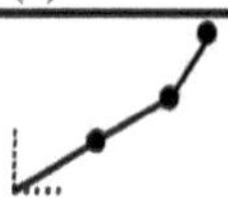
First yield ■	1.00: 3.50: 5.65	$\begin{Bmatrix} 1.000 & -0.697 & -0.540 \\ 0.832 & -0.366 & 1.000 \\ 0.392 & 1.000 & -0.743 \end{Bmatrix}$	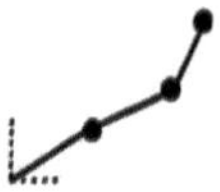
Incipient mechanism □	1.00: 4.70: 8.17	$\begin{Bmatrix} 1.000 & -0.392 & -0.786 \\ 0.909 & 0.077 & 1.000 \\ 0.321 & 1.000 & -0.385 \end{Bmatrix}$	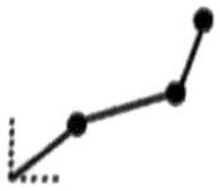
Full mechanism ▲	1.00: 3.30: 6.10	$\begin{Bmatrix} 1.000 & -0.359 & -0.835 \\ 0.930 & -0.080 & 1.000 \\ 0.433 & 1.000 & -0.220 \end{Bmatrix}$	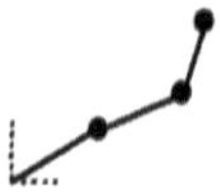

Table 6.2 Dynamic characteristics of retrofitted building

Response phase (1)	Ratio of frequencies (2)	Mode shapes (3)	Approx. 1st mode (4)
Elastic (initial)	1.00: 3.61: 6.53	$\begin{Bmatrix} 1.000 & -0.847 & -0.436 \\ 0.813 & 0.520 & 1.000 \\ 0.425 & 1.000 & -0.889 \end{Bmatrix}$	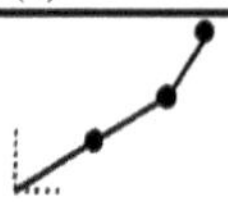
First yield ■	1.00: 4.22: 9.25	$\begin{Bmatrix} 1.000 & -0.737 & -0.532 \\ 0.865 & 0.203 & 1.000 \\ 0.561 & 1.000 & -0.595 \end{Bmatrix}$	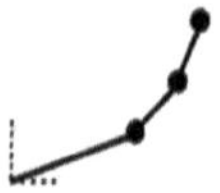
Incipient mechanism □	1.00: 4.65: 9.37	$\begin{Bmatrix} 1.000 & -0.779 & -0.508 \\ 0.868 & 0.310 & 1.000 \\ 0.510 & 1.000 & -0.706 \end{Bmatrix}$	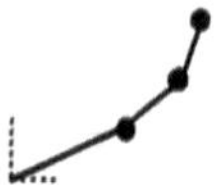
Full mechanism ▲	1.00: 5.68: 10.77	$\begin{Bmatrix} 1.000 & -0.882 & -0.490 \\ 0.901 & 0.477 & 1.000 \\ 0.452 & 1.000 & -0.910 \end{Bmatrix}$	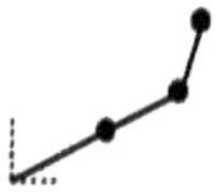

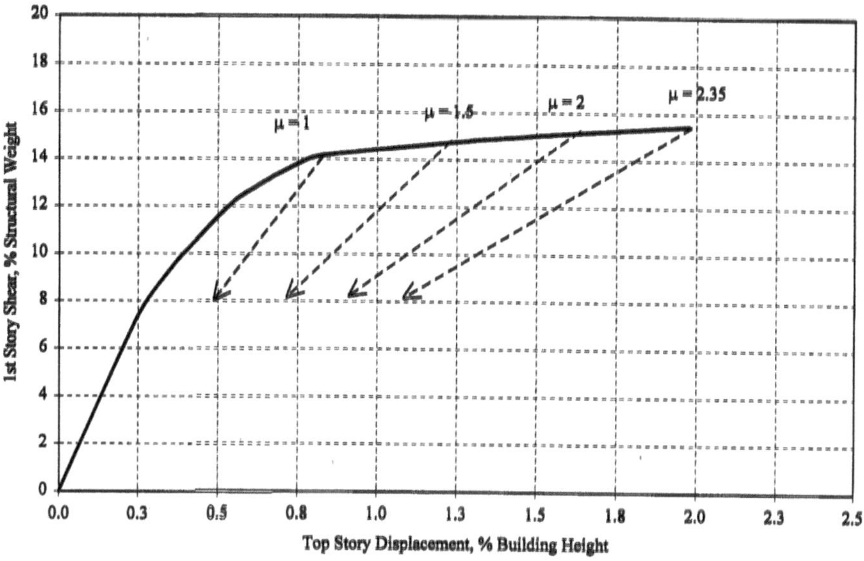

Fig. 6.10 Dynamic characteristics of the original building at varying ductility

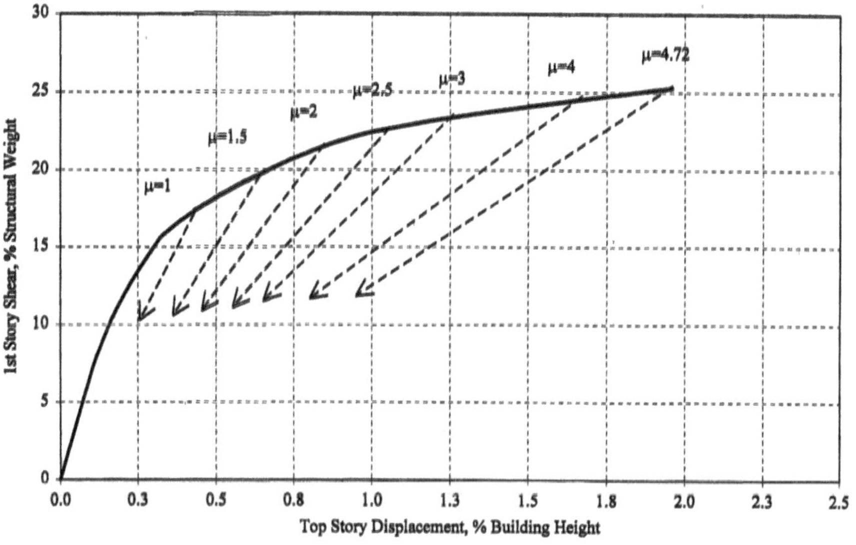

Fig. 6.11 Dynamic characteristics of the retrofitted building at varying ductility

This implies that the original building has an inadequate margin of safety against collapse for this level of excitation. For the retrofitted building, the demand curve intersects the capacity curve at a maximum ductility of 3.2; however, there is still sufficient strength and deformation capability. Therefore the retrofitted building would not be expected to collapse, but some level of damage would be evident follow the earthquake.

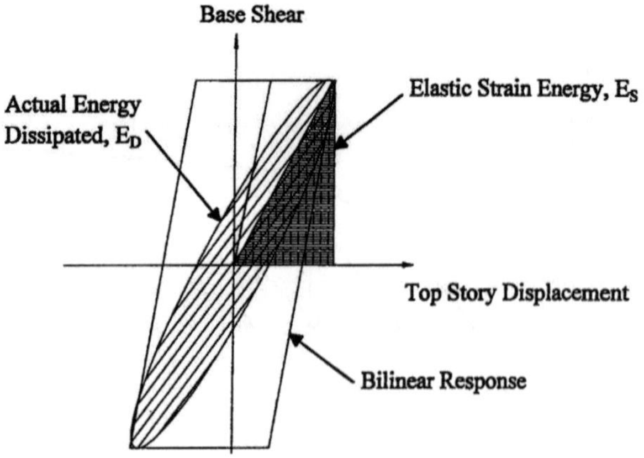

Fig. 6.12 Energy dissipation effectiveness

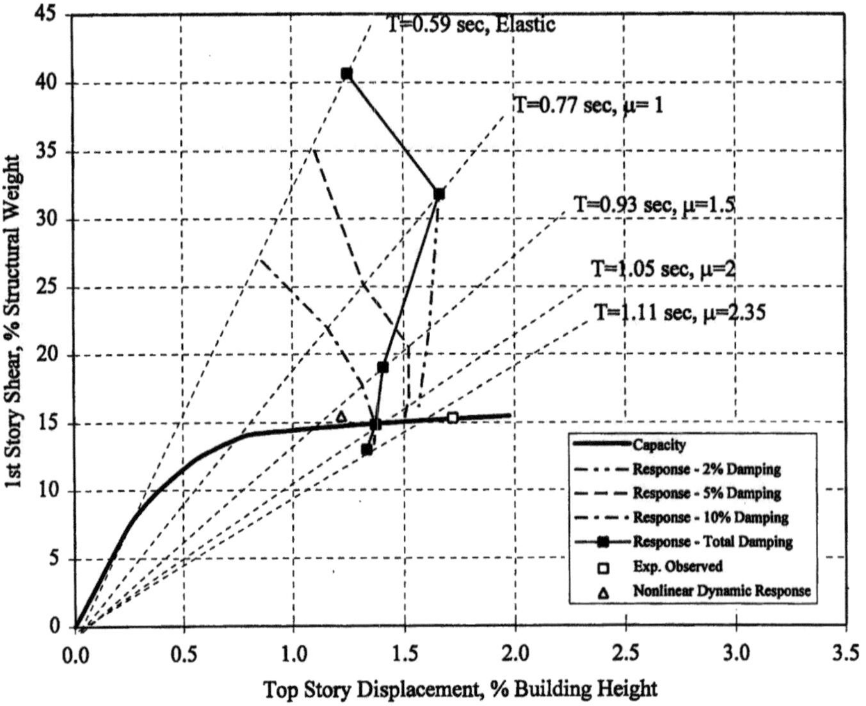

Fig. 6.13 Original building, Taft 0.3 g, 1st story shear vs. drift

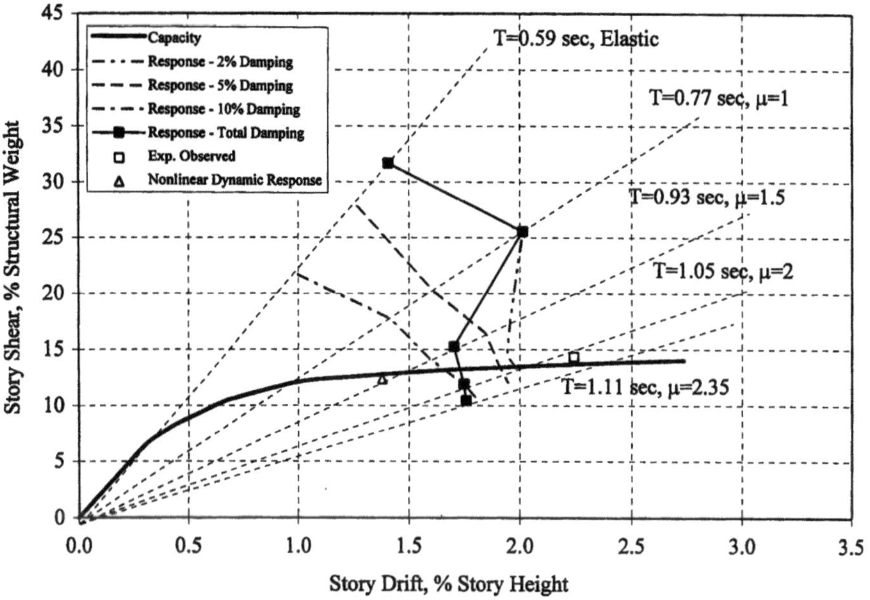

Fig. 6.14 Original building, Taft 0.3 g, 2nd story shear vs. drift

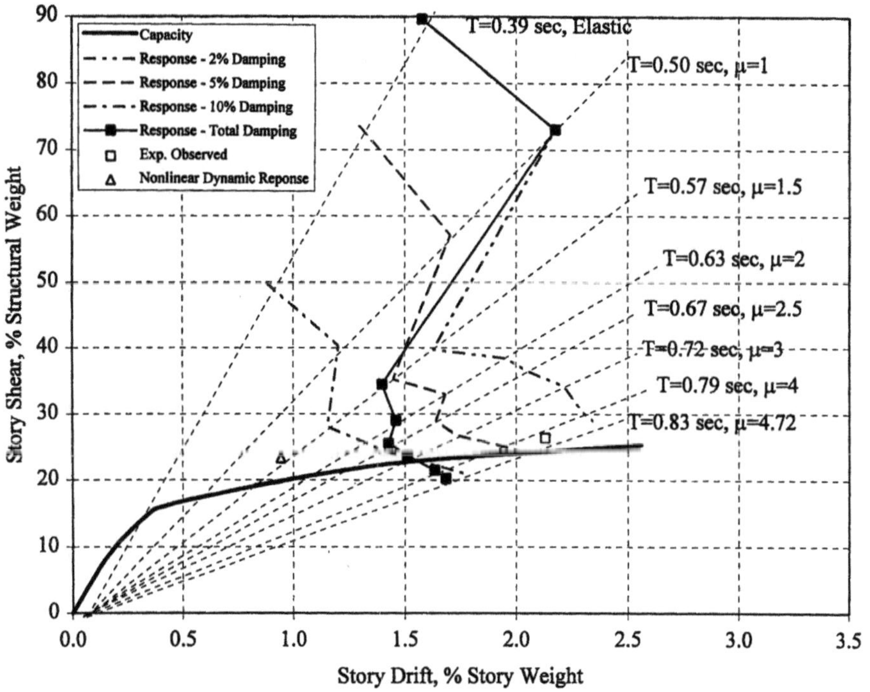

Fig. 6.15 Retrofitted building, Taft 0.3 g, 1st story shear vs. drift

6.7 Conclusions

An evaluation procedure to estimate the maximum seismic story demands and capacities using an enhanced capacity-spectrum based approach was first proposed by Bracci et al. (1997). The procedure utilizes the adaptive pushover analysis to provide a more rational basis for applying the lateral forces which in turn yields a response that is more consistent with the expected system behavior under seismic loading. The paper also documents an extended method to determine a single demand representation utilizing the equivalent dynamic characteristics of the structure at varying levels of imposed deformation ductility. The original and extended methodologies were able to successfully evaluate the experimental seismic response of a low-rise reinforced concrete frame model building and the subsequent response of the same building after retrofit.

References

Army (1986) Seismic design guidelines for essential buildings. Department of the Army (TM-809-10-1), Navy (NAVFAC P355.1), and Air Force (AFM 88-3, Chapter 13, Section A), Washington, DC

ATC (1982) An investigation of the correlation between earthquake ground motion and building performance (ATC-10). Applied Technology Council, Redwood City

Bracci JM, Reinhorn AM, Mander JB (1995a) Seismic resistance of reinforced concrete frame structures designed for gravity loads: performance of structural system. ACI Struct J 92(5):597–609

Bracci JM, Reinhorn AM, Mander JB (1995b) Seismic retrofit of reinforced concrete buildings designed for gravity loads: performance of structural model. ACI Struct J 92(6):711–723

Bracci JM, Kunnath SK, Reinhorn AM (1997) Seismic performance and retrofit evaluation of RC structures. ASCE J Struct Eng 123(1):3–10

Clough RW, Penzien J (1993) Dynamics of structures, 2nd edn. McGraw-Hill, Inc., New York

Freeman SA (1978) Prediction of response of concrete buildings to severe earthquake motion. In: Douglas McHenry international symposium on concrete and concrete structures, SP-55, ACI, Detroit, pp 589–605

Freeman SA (1987) Code designed steel frame performance characteristics. In: Proceedings of the dynamics of structures congress, structural division, ASCE, Orlando, pp 383–396

Kunnath SK, Reinhorn AM, Lobo RF (1992) IDARC version 3.0 – a program for inelastic damage analysis of reinforced concrete structures. National Center for Earthquake Engineering Research, State University of New York at Buffalo, Technical report NCEER-92-0022

Mahaney JA, Paret TF, Kehoe BE, Freeman SA (1993) The capacity spectrum method for evaluating structural response during the Loma Prieta earthquake. In: Proceedings of the national earthquake conference, vol 2, Memphis, pp 501–510

O'Boyle M (1996) Evaluation methodology for structures subjected to seismic loading. MS thesis, Texas A&M University

Vladescu A (1995) Push-over analyses of inelastic structures. MS thesis, State University of New York at Buffalo

Chapter 7
Evaluation of the Seismic Capacity of Nonstructural Components

Gennaro Magliulo, Crescenzo Petrone, Giuseppe Maddaloni, Pauline Lopez, and Gaetano Manfredi

Abstract Nonstructural components (NSC) economic impact and the extensive damages due to NSC after an earthquake motivate the research studies conducted in the past few years at the Department of Structures for Engineering and Architecture, University of Naples Federico II on this topic. The seismic qualification of continuous ceiling systems, plasterboard and brick internal partitions via shake table tests is described in the paper. The test campaign on continuous ceiling systems highlights the low fragility of the tested specimen, primarily caused by: (a) the continuous nature of the ceiling, (b) the dense suspen-sion grid, and (c) the large number of hangers being used. In order to test the internal partitions, which are mainly displacement-sensitive components, an appropriate steel test structure is designed. This structure simulates the behavior of a generic floor in a structure that exhibits an interstorey drift equal to 0.5 % for a frequent earthquake, according to Eurocode 8 prescriptions. Three possible damage states are considered in the study and correlated to an engineering demand parameter, i.e. the interstorey drift ratio, through the use of a damage scheme. Extensive tests show an excellent seismic performance of the plasterboard partition walls, which are characterized by innovative anti-seismic details. In fact, they show minor damage when subjected to interstorey drifts even larger than 1 %. The shake table tests performed at different

G. Magliulo (✉) • C. Petrone • G. Manfredi
Department of Structures for Engineering and Architecture, University of Naples Federico II, Via Claudio 21, 80125 Naples, Italy
e-mail: gmagliul@unina.it

G. Maddaloni
Department of Engineering, University of Naples Parthenope, Centro Direzionale Isola C4, 80143 Naples, Italy

P. Lopez
Direction Innovation and R&D, Siniat, 500 Rue Marcel Demonque - Zone Agroparc, 84915 Avignon Cedex 9, France

G.P. Cimellaro et al. (eds.), *Computational Methods, Seismic Protection, Hybrid Testing and Resilience in Earthquake Engineering*, Geotechnical, Geological and Earthquake Engineering 33, DOI 10.1007/978-3-319-06394-2_7,
© Springer International Publishing Switzerland 2015

intensity levels on hollow brick partitions, widespread in the European zone, denote significant damage in the tested specimen for 0.3 % interstorey drift and extensive damage for drift close to 1 %.

Keywords Nonstructural components • Seismic assessment • Shake-table test • Internal partition • Plasterboard ceiling

7.1 Introduction

In the last years, the scientific community research effort is moving towards the investigation of the seismic behavior of nonstructural components. Nonstructural elements are those systems and components attached to the floors, roof and walls of a building or industrial facility that are not part of the main load-bearing structural system, but may also be subjected to large seismic actions (Villaverde 1997).

Studies available in bibliography (Taghavi and Miranda 2003) point out that the damage of the nonstructural components gives the largest contribution to the economic loss due to an earthquake. Within the nonstructural components category, the interior construction, which the partition and ceiling systems belong to, represents the largest source of the building cost. The economic impact is much more severe if losses of inventory and downtime cost are considered. Moreover, the failure of nonstructural components may also represent a threat to life safety. A partition or infill overturning may easily result in injuries or casualties.

Nonstructural components category behavior is nowadays recognized to be a relevant issue within the framework of Performance-Based Earthquake Engineering (Bertero and Bertero 2002). The nonstructural components economic impact and their extensive damages due to an earthquake require much more research effort on the seismic performance assessment of such components. The research studies conducted in the past few years at the Department of Structures for Engineering and Architecture at the University of Naples Federico II tend to meet this objective.

The research is focused on the seismic qualification of plasterboard ceiling systems and plasterboard and brick internal partitions through shake table tests. The use of the shake table system plays an important role in the development of the research project.

7.2 Seismic Capacity of Plasterboard Continuous Ceiling Systems

7.2.1 Description of the Experimental Tests

Two typologies of plasterboard continuous suspended ceilings are tested: single frame ceiling (SFC) and double frame ceiling (DFC) systems. A schematic representation of the two used systems is shown in Fig. 7.1a, b respectively.

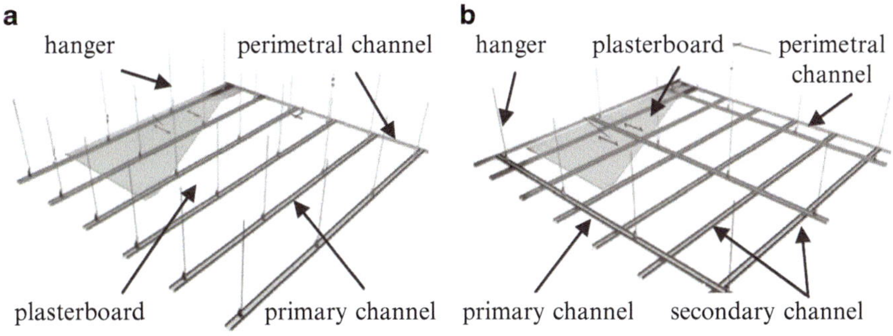

Fig. 7.1 Suspended plasterboard continuous ceilings: (**a**) single frame ceiling (SFC); (**b**) double frame ceiling (DFC)

Fig. 7.2 Test frame installed on the shake table

The seismic qualification of suspended ceilings is carried out by the earthquake simulator system available at the laboratory of Department of Structures for Engineering and Architecture at the University of Naples Federico II. The system consists of two 3 m × 3 m square shake tables; only one table is used in this experimental campaign. With the purpose of simulating the seismic effects on the ceilings, a steel test frame is properly designed and built (Fig. 7.2). A very stiff test frame is required in order to control the acceleration on the specimen and to avoid shake table – test frame – ceiling system double-resonance issues. Two U-section profiles (UPN100) are welded around the perimeter of the test frame, at a distance of 20 and 50 cm from the roof; a 40 mm × 100 mm timber ledger is inserted in the U-section profile in order to easily laterally restrain the ceiling system. Indeed, the ceiling system plasterboards are connected along the perimeter of the frame to the timber ledger. Consequently, the light mass and the large stiffness

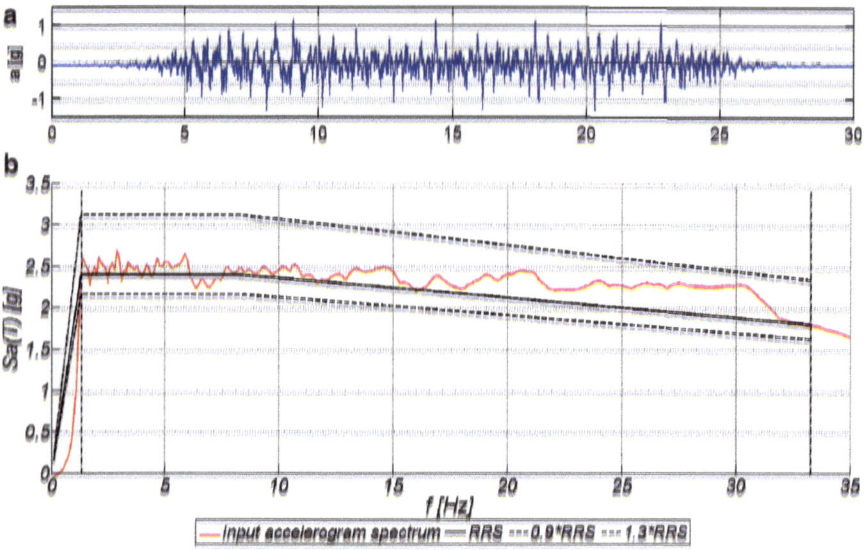

Fig. 7.3 Earthquake time history and spectra for a level of shaking corresponding to SDS equal to 1.50g: (**a**) acceleration time-history; (**b**) input accelerogram spectrum, RRS (*bold line*), upper and lower matching limits (*dashed line*)

of the timber-channel profiles system represent the typical boundary conditions of a ceiling on structural elements. The total weight of the test frame is equal to 19.2 kN.

The tested specimen is composed by three plasterboards connected one another via stucco, both for SFC and DFC. The total dimension of the specimen is 2.20 m × 2.20 m.

Accelerometers and strain gauges are used to monitor the response of the test frame and plasterboard ceilings in both ceiling system configurations.

A set of 5 accelerograms, used as input for the shakings, are opportunely selected matching a target response spectrum or required response spectrum (RRS), provided by the ICBO-AC156 code "Acceptance criteria for seismic qualification testing of nonstructural components" (International Conference of Building Officials (ICBO) 2000).

Two artificial acceleration time histories are defined so as their response spectra, i.e. test response spectra, envelope the target spectrum over the frequency range from 1.3 to 33.3 Hz. The test response spectrum ordinates do not have to be lower than 0.9 times RRS and larger than 1.3 times RRS (according to EC8 (Maddaloni et al. 2012) and AC156 rules respectively). The selection procedure of the accelerograms is performed for a RRS corresponding to $S_{DS} = 1.50$ g; the so obtained record is then scaled to match other four levels of the target spectrum (corresponding to SDS 0.30 g, 0.60 g, 0.90 g and 1.20 g).

The input motion for $S_{DS} = 1.50$ g is shown in Fig. 7.3. Additional information on testing specifications is present in Magliulo et al. (2012a).

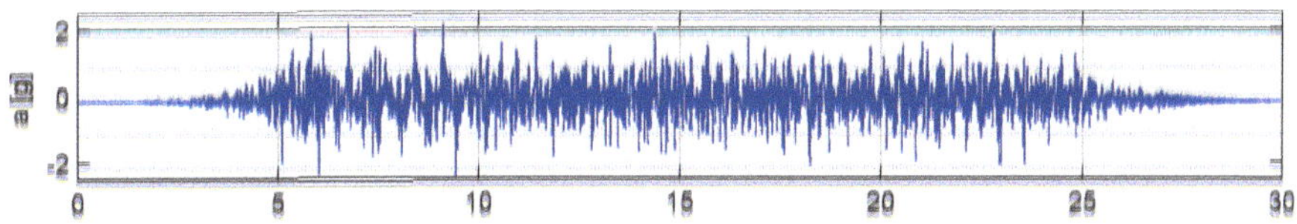

Fig. 7.4 Acceleration time histories recorded on the single frame ceiling (SFC) for the intensity levels corresponding to SDS equal to 1.50 g

7.2.2 Description of the Experimental Tests

Using the selected drive motions, five unidirectional shaking tests are performed for both ceiling systems (Fig. 7.4). The tested ceilings shows negligible damage at all intensity levels (the acceleration on the ceiling ranges from 0.40 to 2.36g), resulting in a low fragility both for single and for double frame ceilings.

An interesting comparison with a previous vulnerability study performed by Badillo-Almaraz et al. (2006) on typical U.S. ceiling system with discrete tiles is made. The tests were performed at the Structural Engineering and Earthquake Simulation Laboratory (SEESL) at the University of Buffalo.

In order to make a comparison, the fragility curve for ceilings with undersized tiles (Badillo-Almaraz et al. 2007) in terms of PFA is considered. This fragility curve, at the acceleration equal to 1.35 g, which is the maximum acceleration of the shaking table attained during the tests performed in Naples, gives 100 % probability of exceeding minor and moderate damage states and 29 % probability of exceeding major damage state. The ceilings tested in Naples, instead, show no damage at all intensity levels of the tests, resulting in a lower fragility with respect to the ceiling systems tested in Buffalo. Three main reasons may be the cause of this different vulnerability: (a) the continuous nature of the tested ceiling, that improves the seismic behavior with respect to the ceilings with tiles; (b) the dense steel channel grid (the "primary channel" span is 500 mm and 1,000 mm for SFC and DFC systems respectively, the "secondary channel" span is 500 mm for DFC system), that connects one another the plasterboards in a unique horizontal element, ensuring high in plane stiffness and strength; (c) the large number of hangers that connects the ceiling system to the roof, ensuring an adequate out of plane stiffness and strength, avoiding any ceiling vertical movement; (d) the smaller dimensions of the specimen tested in Naples with respect to the specimen tested in Buffalo (2.20 m × 2.20 m vs 4.88 m × 4.88 m), considering that very recent studies seem to show that specimen dimensions can affect the ceiling seismic response.

7.3 Seismic Capacity of Plasterboard and Brick Internal Partitions

7.3.1 Description of the Experimental Tests

In Fig. 7.5 the test setups for the two different test campaigns are shown. They consist of: (a) a shake table simulator; (b) a 3D steel test frame, defined in Magliulo et al. (2013), able to transfer the seismic demand to the partitions; (c) the specimen, i.e. plasterboard partitions or hollow brick partitions.

A steel test frame is properly designed and built with the purpose of simulating the seismic effects on the partitions. The test frame is designed so as to simulate the behavior of a generic story of a building, in which the partitions are installed. The geometry of the test frame is defined taking into account three requirements: (i) realistic value of mass; (ii) realistic interstorey height h, assumed equal to 2.74 m; (iii) realistic interstorey displacement d_r, assumed equal to $0.005\,h$ for a Damage Limit State earthquake with 50 year return period. The final result is a 2.50 m (X direction) × 2.00 m (Y direction) × 2.89 m (Z direction) inverted pendulum test fixture. A reinforced concrete slab is placed on the roof of the structure. The dynamic identification procedure on the bare test frame is described in Magliulo et al. (2012b).

The plasterboard partition is a plain drywall partition with dimensions of 2.68 m (height) by 2.27 m (width) with 12.5 cm thickness. It is installed between two columns, in particular along the longest side (2.50 m) of test fixture, connected by the perimeter U-section runners to a timber ledger covering the steel elements.

a b

Fig. 7.5 Test setup in (**a**) plasterboard partition and (**b**) hollow brick partition tests

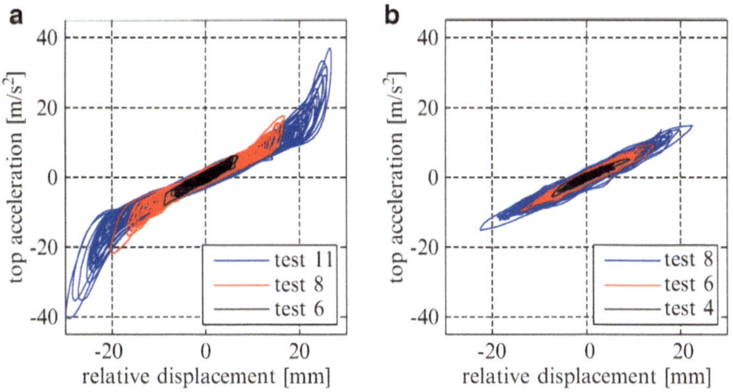

Fig. 7.6 Top acceleration vs relative displacement plot for the different seismic tests in (**a**) X direction (**b**) Y direction

Between the plasterboards and the wooden supports, a 0.8–1 cm gap is left. Such partitions are designed in order to not interfere with the hosting structure up to moderate level of interstorey drifts (0.5 %). The detailed mounting phase is included in Magliulo et al. (2013).

The hollow brick partition specimen is doubly symmetric and presents a 150 cm wide partition along the reference Y direction and two smaller 80 cm wide partitions in the orthogonal direction. The partitions are 2.6 m high; the connection between the specimen and the shake table is ensured by an "I" shape RC slab. The walls are made with hollows bricks 250 × 250 × 80 mm connected together and plastered with mortar (Petrone et al. 2014).

The table seismic input is similar to the one defined for the ceiling systems; however, in this case bidirectional input motions are provided.

7.3.2 Plasterboard Partitions Test Results

In order to analyze the test frame–partition interaction, the top acceleration, representative of the total inertia force, is plotted versus the relative displacement for different intensity levels (Fig. 7.6).

From the hysteretic curve in X direction, i.e. partition in-plane direction, some comments can be stated:

- the trend of the hysteretic curve is linear until a 0.5 % drift (14 mm) is reached, denoting no interaction between the partitions and the hosting structure; the slope of the initial linear envelope confirms the numerical model of the bare test frame;
- the contribution of the partitions is initially frictional, resulting in an increase of damping; for displacement close to twice the clearance between the partitions and the test frame, the energy dissipated increases due to the damage recorded within the partitions.

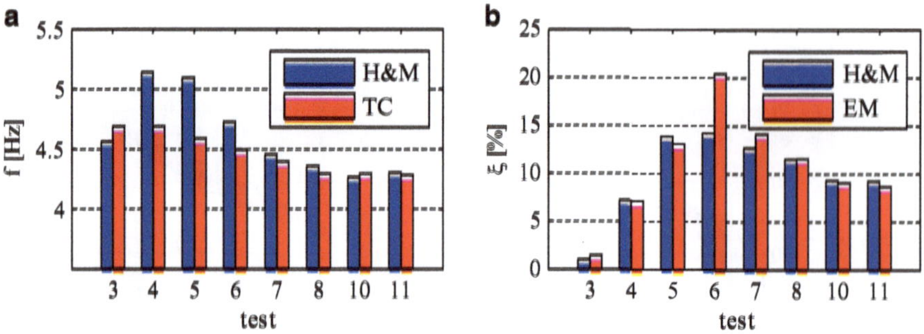

Fig. 7.7 (**a**) Natural frequency evaluation according to Hashemi and Mosalam (H&M) procedure (Hashemi and Mosalam 2006) and transfer curve method (TC) for the different seismic tests; (**b**) damping ratio evaluation according to Hashemi and Mosalam (H&M) procedure (Hashemi and Mosalam 2006) and the energetic method (EM) for the different seismic tests

The hysteretic curve in Y direction, i.e. partition out of plane direction, clearly shows that the partitions do not contribute to the lateral stiffness and strength in this direction, since the linear slope of the hysteresis loops corresponds to the bare test frame natural frequency.

Both bidirectional and unidirectional tests show a limited damage up to 1.08 % drift, including:

- acrylic silicone, inserted in the separation between partition and wooden vertical support, detachment for moderate intensity tests;
- gypsum dust fall from moderate intensity tests, with increasing intensity as the demand increases;
- cracking of the vertical joints between plasterboards for the highest intensity tests;
- crushing of the corners of the plasterboards for the highest intensity tests.

The natural frequency trend for the seismic tests is investigated in order to correlate the damages to the dynamic characteristics of the setup.

The transfer function, estimated as the ratio between the top and the base acceleration in the frequency domain, is applied with respect to the time histories recorded during the different seismic tests. This method allows following the change of the natural frequency during the tests, as shown in Fig. 7.7a. The procedure proposed by Hashemi and Mosalam (2006), which allows evaluating the average values of stiffness k from the dynamic equilibrium, is also implemented (Fig. 7.7a).

Assuming an exclusively viscous dissipation, the damping ratio ξ is proportional to the ratio between the dissipated energy per cycle, W_D (area enclosed within each hysteresis cycle), and the elastic energy E (Chopra 1995). In Fig. 7.7b the median values of damping coefficient are plotted for each test. The damping ratio ξ is also evaluated from the procedure proposed by Hashemi e Mosalam (Hashemi and Mosalam 2006) and plotted in Fig. 7.7b.

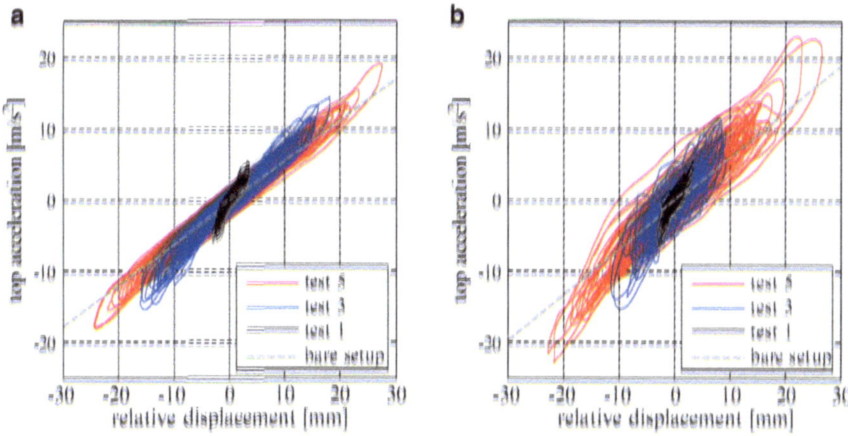

Fig. 7.8 Top acceleration vs relative displacement plot for different seismic tests in (**a**) X direction and (**b**) Y direction

The trend of both the natural frequency and the damping ratio confirms the recorded damage. The presence of the acrylic silicone in the clearance between the test frame and the partition lightly contributes to the lateral stiffness in the first tests. The decrease of the frequency in tests 6 and 7 denotes the failure of the acrylic silicone. Consequently the natural frequency tends to the bare test frame natural frequency.

Similarly, from test 3 to test 6 an increase in the damping ratio and then a following decrease is exhibited, essentially due to the silicone progressive damaging. In test 6 a very high value of damping is recorded, mainly owed to two factors: the presence of the acrylic silicone which is a material characterized by high damping and the friction developed by the plasterboards that slide with respect to the top runner. Once the silicone is detached, the damping ratio decreases, since the damping due to the silicone vanishes.

7.3.3 Hollow Brick Partitions Test Results

Using the selected drive motions, five bidirectional shaking tests are performed. In order to analyze the partition behavior and its contribution to the global behavior of the test setup, the top acceleration, representative of the total inertia force, is plotted versus the relative displacement for different intensity levels (Fig. 7.8). A dotted line denotes the behavior of the bare test frame based on its natural frequency and assuming to be in absence of damping.

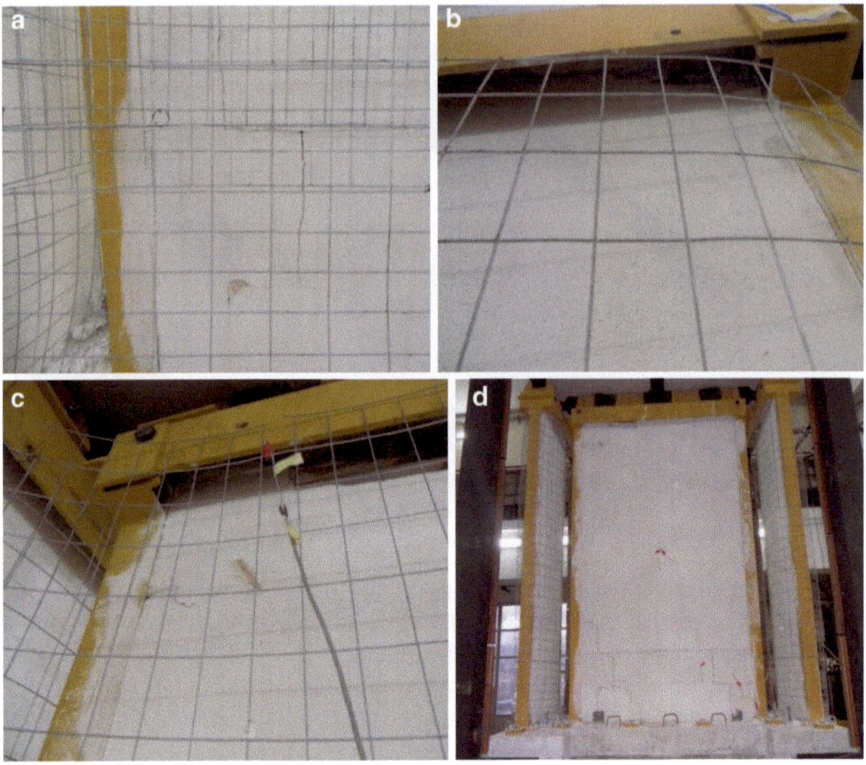

Fig. 7.9 Recorded damage after different shaking tests

From the analysis of the so-obtained hysteretic curve it can be noted that:

- a significant interaction between the partitions and the hosting structure is exhibited during the first tests; during the fifth test, the hysteretic behaviour is very close to the bare frame response;
- the secant stiffness, evaluated at the maximum displacement of each test, decreases as the relative displacement increases;

Bidirectional tests show a slight damage already up to 0.35 % drift in X direction and 0.20 % in Y direction. The damage level increases according to the shaking test intensity and the following damages are noticed (Fig. 7.9):

- cracks along the perimeter of the specimen due to the partitions slip from the surrounding frame for low intensity tests;
- fall of plaster and pieces of brick from the top of the specimen from moderate intensity tests with increasing intensity as the demand increases;
- horizontal cracks, wider than 0.3 mm, in the lower part of the walls for moderate intensity tests;

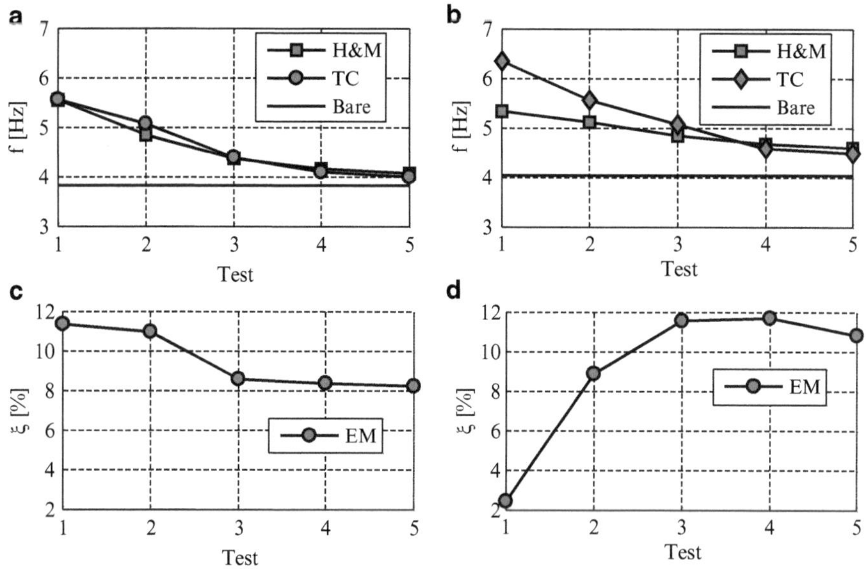

Fig. 7.10 Test frame natural frequency evaluation according to the Transfer Curve method (TC) and to the Hashemi and Mosalam (H&M) procedure (Hashemi and Mosalam 2006) and compared to the bare frame natural frequency (Bare) for the different seismic tests in (**a**) X and (**b**) Y directions; damping ratio evaluation according to the Energetic Method (EM) for the different seismic tests in (**c**) X and (**d**) Y directions

- wide sliding cracks in mortar, crushing of mortar at the corner of the specimen and collapse of a brick in the top of the partition for high intensity tests (Fig. 7.9b, c);
- deep extended horizontal cracks in mortar in the lower part of the walls, that let the part above the crack move as a rigid block with a rocking behavior with respect to the surrounding frame for high intensity tests (Fig. 7.9d, a). At this damage level, the specimen do not offer any resistance against lateral displacement since it rigidly moves and rotates within the surrounding frame that restraints it in the out of plane direction.

In the following the change of the natural frequency during the seismic tests is investigated in order to correlate the damage to the dynamic characteristics of the specimen, adopting the same techniques used for the tests on plasterboard partitions (see Sect. 7.3.2). The assessment is performed in both the horizontal directions (Fig. 7.10).

The trend of both the natural frequency and the damping ratio confirms the recorded damage.

Natural frequency shows a great reduction during the first three tests due to the damage suffered by the specimen during the first tests and the consequent decrease of the lateral stiffness. Subsequently natural frequency values converge towards the bare frame natural frequency.

Similarly, from test 1 to test 3, in the Y direction, an increase in the damping ratio is exhibited, essentially due to the wall progressive damaging, while in the X direction the damping is almost constant for the different tests.

7.4 Conclusions

Nonstructural components category behavior is nowadays recognized to be a relevant issue within the framework of Performance-Based Earthquake Engineering. The nonstructural components economic impact and their extensive damages due to an earthquake require much more research effort on the seismic performance assessment of such components. The research studies conducted in the past few years at the Department of Structures for Engineering and Architecture at the University of Naples Federico II tend to meet this objective.

The research is focused on the seismic qualification of plasterboard ceiling systems and plasterboard and brick internal partitions through shake table tests. The use of the shake table system plays an important role in the development of the research project.

The test campaign of continuous ceiling systems highlights the low fragility of the tested specimen, primarily caused by: (a) the continuous nature of the ceiling, (b) the dense suspension grid, and (c) the large number of hangers being used.

In order to test the internal partitions, which are mainly displacement-sensitive components, an appropriate steel test space frame is designed. This frame simulates the behavior of a generic floor in a structure that exhibits an interstorey drift equal to 0.5 % for an earthquake characterized by a 50 years return period.

Extensive tests show an excellent seismic performance of the plasterboard partition wall, characterized by innovative anti-seismic details. In fact, they show minor damage when subjected to interstorey drifts even larger than 1 %.

The dynamic tests performed at different intensity levels on hollow brick partitions, instead, highlight an extensive damage for moderate interstorey drift demand that may compromise the integrity of the specimen.

Acknowledgments This research study has been funded both by Italian Department of Civil Protection in the frame of the national project DPC – ReLUIS 2010–2013 and by Siniat that also provided the plasterboard specimens for the testing program.

The support provided by Eng. Giuseppe Campanella in the execution of the different test campaigns is gratefully acknowledged.

References

Badillo-Almaraz H, Whittaker AS, Reinhorn A, Cimellaro GP (2006) Seismic fragility of suspended ceiling systems, Report MCEER-06-0001. MCEER/SUNY, Buffalo

Badillo-Almaraz H, Whittaker AS, Reinhorn AM (2007) Seismic fragility of suspended ceiling systems. Earthq Spectra 23(1):21–40. doi:10.1193/1.2357626

Bertero RD, Bertero VV (2002) Performance-based seismic engineering: the need for a reliable conceptual comprehensive approach. Earthq Eng Struct Dyn 31(3):627–652. doi:10.1002/Eqe.146
Chopra AK (1995) Dynamics of structures: theory and applications to earthquake engineering. Prentice Hall, Englewood Cliffs
Hashemi A, Mosalam KM (2006) Shake-table experiment on reinforced concrete structure containing masonry infill wall. Earthq Eng Struct Dyn 35(14):1827–1852. doi:10.1002/Eqe.612
International Conference of Building Officials (ICBO) (2000) AC 156 acceptance criteria for the seismic qualification of nonstructural components. ICBO Evaluation Service, Inc., Whittier
Maddaloni G, Magliulo G, Cosenza E (2012) Effect of the seismic input on non-linear response of R/C building structures. Adv Struct Eng 15(10):1861–1877
Magliulo G, Pentangelo V, Maddaloni G, Capozzi V, Petrone C, Lopez P, Talamonti R, Manfredi G (2012a) Shake table tests for seismic assessment of suspended continuous ceilings. Bull Earthq Eng 10(6):1819–1832. doi:10.1007/s10518-012-9383-6
Magliulo G, Petrone C, Capozzi V, Maddaloni G, Lopez P, Talamonti R, Manfredi G (2012b) Shake table tests on infill plasterboard partitions. Open Constr Build Technol J 6(Suppl 1-M10):155–163. doi:10.2174/1874836801206010155
Magliulo G, Petrone C, Capozzi V, Maddaloni G, Lopez P, Manfredi G (2013) Seismic performance evaluation of plasterboard partitions via shake table tests. Bull Earthq Eng. doi:10.1007/s10518-013-9567-8
Petrone C, Magliulo G, Manfredi G (2014) Shake table tests for the seismic assessment of hollow brick internal partitions. Eng Struct 72(1):203–214
Taghavi S, Miranda E (2003) Response assessment of nonstructural building elements, PEER report 2003/05. College of Engineering University of California, Berkeley
Villaverde R (1997) Seismic design of secondary structures: state of the art. J Struct Eng-ASCE 123(8):1011–1019. doi:10.1061/(Asce)0733-9445(1997)123:8(1011)

Chapter 8
Reference Quantities and Values for a Possible Interpretation of the Data Acquired from Monitoring System of Historical Buildings

Tomaso Trombetti, Stefano Silvestri, Giada Gasparini, Michele Palermo, and Simonetta Baraccani

Abstract Preserving historic buildings is essential in the safeguard of the cultural heritage of any country. The need to carry out structural analyses by means of non-destructive methods has made structural monitoring ever more widespread in the diagnosis and control of historic buildings. The aim of this study is to introduce a standardized approach for the analysis of the data acquired from a monitoring system of historic buildings. This approach is based on the definition of specific reference quantities (extrapolated from the recorded time series) able to characterize the main features of the structural response and the preliminary identification of the order of magnitudes of these quantities. It is assumed that the recorded time series can be decomposed into two components: a periodical one depending on with the natural forces acting on the building and an unknown one related to the variation of the state of the structure. Exploiting the properties of periodic functions, one may identify these reference quantities, which are based on the year and the day variability and allow to monitor the evolution of the phenomena under observation. These reference quantities may be collected in a database and may become fundamental for comparing the response of similar buildings. This approach has been applied to the data obtained from the monitoring systems of two important Italian monuments: the Cathedral of Modena and the "Two Towers" of Bologna.

T. Trombetti (✉) • S. Silvestri • G. Gasparini • M. Palermo • S. Baraccani
Department DICAM, University of Bologna, Viale Risorgimento 2, 40136 Bologna, Italy
e-mail: tomaso.trombetti@unibo.it; stefano.silvestri@unibo.it; giada.gasparini4@unibo.it; michele.palermo7@unibo.it; simonetta.baraccani2@unibo.it

G.P. Cimellaro et al. (eds.), *Computational Methods, Seismic Protection, Hybrid Testing and Resilience in Earthquake Engineering*, Geotechnical, Geological and Earthquake Engineering 33, DOI 10.1007/978-3-319-06394-2_8,
© Springer International Publishing Switzerland 2015

8.1 Introduction

In recent years, structural monitoring has acquired an ever increasing importance in the diagnosis and control of buildings, in particular for historical buildings.

The main objective leading to the installation of a monitoring system is the evaluation of the building "structural health". The engineering discipline dealing with this topic is called "Structural Health Monitoring" (SHM) (Sohn et al. 2004).

Data acquired from a SHM system may be considered reliable only if related to a "sufficiently long" time period, provided that their are highly correlated to external factors, such as temperature, wind and others, which significantly vary over the seasons.

The data acquired from SHM systems may be used to detect potential criticalities and/or incoming structural "damages". The evolution of damages can be slow as a consequence of thermal stresses, fatigue, corrosion, subsidence phenomena or fast as it happens as consequence of extreme events such as earthquakes, hurricanes and explosions. Therefore, structural monitoring may represent a valid tool for a non-destructive assessment of the structural health of the building.

The SHM system is composed of several devices that permit to monitor the time evolution of parameters concerning the structural behavior of buildings – like the strain and stress state of structural elements, the development of specific existing cracks, the inclination in specific critical points of the structure ....

The amount of data acquired need to be analyzed in order to gather some useful information on the condition of the building under observation. Analyses carried out on these data can have multiple purposes.

The present paper is aimed at introducing a standardized approach for the SHM data analysis with specific reference to historical buildings.

The approach is applied to two important Italian monuments: the Cathedral of Modena and the "Two Towers" of Bologna.

8.2 The Approach for the Data Analysis

Theoretically speaking, once a certain structure can be idealized by a reliable mathematical model, the evaluation of its structural response under a given set of actions could be straightforward. However, in most cases, and in particular with reference to historical buildings, the assesment of the structural behaviour, and therefore its idealization (the structural model), may be characterized by high uncertainties.

As a consequence, the direct approach consisting in computing the response of the structure by means of a mathematical model cannot be feasible. In these cases, SHM may provide information on the evolution of the structural response. A simplified standardized approach is presented hereafter for the interpretation of the monitored data.

8.2.1 The Objective

The approach that is here proposed for the analysis of SHM data is aimed at:

1. introducing reference quantities based on the identification of the main physical components characterizing the recorded data;
2. providing order of magnitudes of these physical quantities.

Then the approach is applied to two different structural typologies such as a Romanesque cathedral and two medieval masonry towers.

8.2.2 The Main Components of the Recorded Data

SHM systems return discrete time series which can be generally referred to as x(t), where x represents the monitored parameter (e.g. displacement, strain, angle inclination, crack width, temperature, ...) and t represents a specific instant of time. Each time series can be seen as a function of two main factors:

$$x(t) = f[F(t), S(t)] \tag{8.1a}$$

where F(t) are the (time-dependent) forces acting on the structure, and S(t) represent the time evolution of the "state" of the structure, i.e. the condition of the structure due to its geometrical configuration, the materials mechanical properties, its boundary conditions, etc. The state of the structure can be stationary or variable with time. The variation of the state may be due to different factors such as material degradation, incoming coactive states, soil-structure interaction, etc.

The actions F(t) on the building can be classified into three main groups:

- Dead loads D(t): the permanent forces acting on a structure such as the self-weight of the structure;
- Live loads L(t): the non-permanent forces acting on the structure. In detail, these encompass the forces that depend on the weather effects, which are herein referred to as natural forces, N(t), such as wind, temperature, precipitations, etc.
- Accidental loads A(t): the forces depending on rare events, such as earthquakes, hurricanes, explosions, etc.;

Thus:

$$F(t) = D(t) + L(t) + A(t) \tag{8.1b}$$

The state of the structure may be seen as:

$$S((t)) = S((t_0)) + \Delta S((t_0, t)) = S_0 + \Delta S((t_0, t)) \tag{8.2}$$

where

- S_0 is the state of the structure at time t_0, which is assumed as known;
- $\Delta S(t_0,t)$ represents the variation of the state over the time. If $\Delta S(t_0,t) = 0$ and $S(t) = S(t_0)$, then the state is stationary. On the contrary, if $\Delta S(t_0,t)$ differs from zero, then it clearly indicates a variation in the state.

8.2.3 *The Assumptions*

If the dead and live loads are within their usual ranges, it is expected that: (i) the state of the structure will not substantially change, (ii) the recorded data may be characterized by predominant periodical components due to seasonal and daily temperature excursions.

In the absence of rare events inducing accidental actions $A(t) = 0$, and assuming that the live loads are predominately due to the natural forces $L(t) = N(t)$, $F(t)$ can be expressed as the sum of the two components $D(t)$ and $N(t)$. Moreover, assuming that the dead loads may be taken as constant, $D(t) = \overline{D}$, Eq. 8.1b can be specified as follows:

$$F((t)) = \overline{D} + N(t) \tag{8.3}$$

Substituting Eqs. 8.1b, 8.2 and 8.3 in Eq. 8.1a leads to:

$$x(t) = f\left[\overline{D}, N(t), S_0, \Delta S(t_0, t)\right] \tag{8.4}$$

The natural forces are assumed to be periodic functions with two fundamental contributions:

$$N(t) = N_1(T_1, t) + N_2((T_2, t)) \tag{8.5}$$

a contribution N_1 with period T_1 equal to 365 days (due to the revolution of the earth round the sun) leading to the annual oscillations and a contribution N_2 with a period T_2 equal to 1 day (relative to the earth rotation around its axis) leading to the daily oscillations. The actions on the structure due to the earth motion of revolution are considered periodic functions with constant amplitude. The actions on the structure due to the Earth rotation are considered periodic functions with variable amplitude depending on the season.

Based on the above considerations, it is assumed that the time series $x(t)$ are made of two components:

$$x(t) = x_1(t) + x_2(t) \tag{8.6}$$

Table 8.1 The reference quantities for the analysis of the recorded data

Reference quantity	Definition	
Daily amplitude	$\delta_{day,j} = [max\ x(t_i) - min\ x(t_i)] = \sigma$	$\forall_{ti} \in_{day\ j}$
Mean daily value	$\mu_{day,j} = \frac{1}{n_j}\sum_{i=1}^{n_j} X(t_i) = m$	$\forall_{ti} \in_{day\ j}$
Annual amplitude	$\delta_{year,j} = [max\ x(t_i) - min\ x(t_i)] = \Sigma$	$\forall_{ti} \in_{year\ j}$
Mean annual value	$\mu_{year,j} = \frac{1}{n_j}\sum_{i=1}^{n_i} X(t_i) = M$	$\forall_{ti} \in_{year\ j}$
Daily residual on σ	$r_{\delta day,j(k-1)} = \delta_{day,j}\vert_{year\ k} - \delta_{day,j}\vert_{year\ l} = r_\sigma$	
Daily residual on m	$r_{\mu day,j(k-1)} = \mu_{day,j}\vert_{year\ k} - \mu_{day,j}\vert_{year\ l} = r_m$	
Annual residual on Σ	$r_{\delta year,j(k-1)} = \delta_{year,j}\vert_{year\ k} - \delta_{year,j}\vert_{year\ l} = R_\Sigma$	
Annual residual on M	$r_{\mu year,j(k-1)} = \mu_{year,j}\vert_{year\ k} - \mu_{year,j}\vert_{year\ l} = R_M$	
Drop	Δ	

$x_1(t)$ is the periodic component of x(t) depending on N(t) and $\overline{D}$ actions; $x_2(t)$ is the component of x(t) depending on the state S(t). The component $x_2(t)$ could have some periodicity whose characterization generally requires several years of monitoring.

8.2.4 *Analysis of the Data*

Based on the assumed functional form of the recorded time series x(t), the data analysis is conducted introducing the quantities summarized in Table 8.1.

With reference to the j-th generic day, the quantities of interest are the daily amplitude and the mean daily value. With reference to the j-th year, the quantities of interest are the annual amplitude and mean annual value. It is also of interest to compute the daily residuals at the j-th day of the amplitude ($r_{\delta day,j(K-1)}$) and of the mean value between two consecutive years k and l ($r_{\mu day,j(K-1)}$). These quantities are useful to obtain information on the evolution of the state S(t). Also, the annual residual of the amplitude ($r_{\delta year,j(K-1)}$) and of the mean value($r_{\mu year,j(K-1)}$) contribute to the knowledge of the evolution of the state S(t).

As illustrative example, a time series x(t) simulating generic data from a SHM system has been generated and displayed in Fig. 8.1. For the sake of clearness, Fig. 8.2 illustrates some reference quantities: daily amplitude (σ), mean daily value (m), annual amplitude (Σ), mean annual value (M), and daily residual of the mean daily value (r_m).

From the graph of the residual (i.e. Fig. 8.2c), it is possible to evaluate:

- linear regression of r_σ and r_m (trend of the residuals)
- cumulative of the residuals (stability of the measured event).

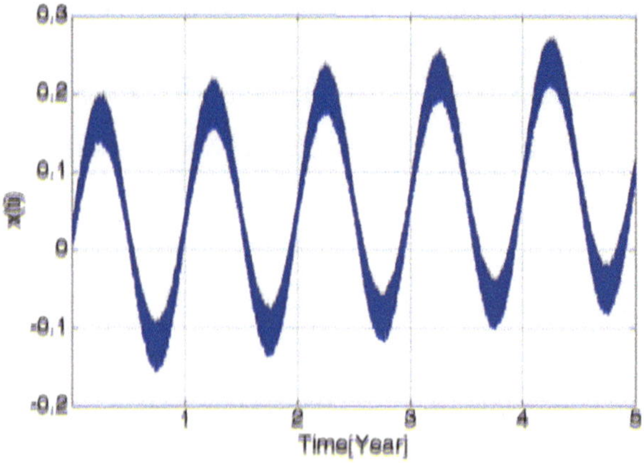

Fig. 8.1 Time series x(t)

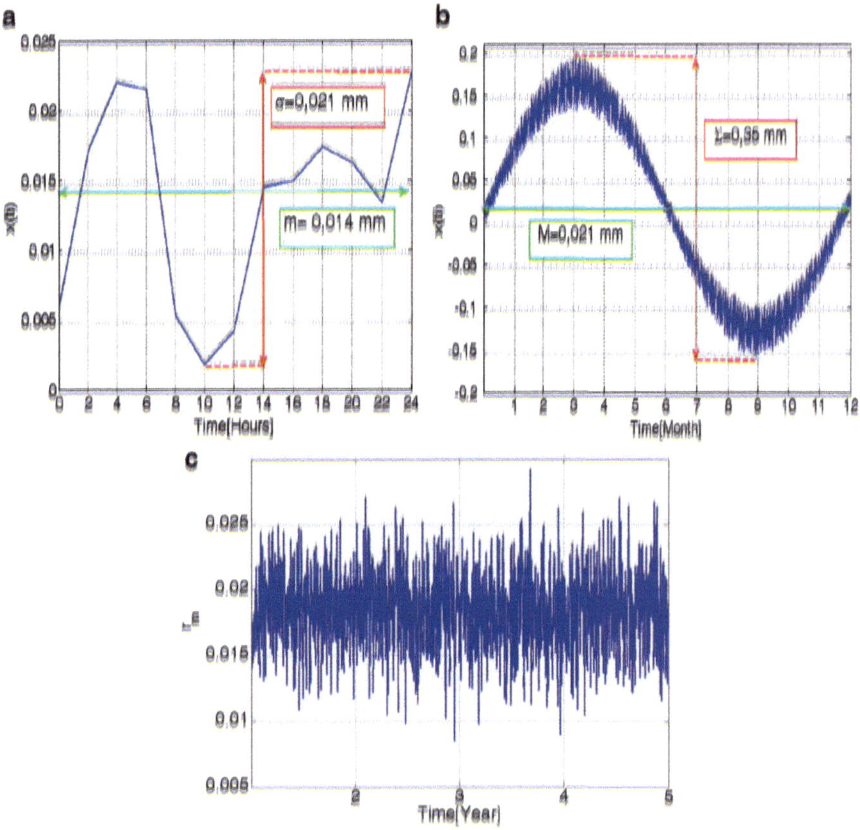

Fig. 8.2 (**a**) Daily amplitude and mean daily value, (**b**) annual amplitude and mean annual value, (**c**) daily residual of the mean daily value

In addition to the above described quantities, also sudden drops (Δ) may be present in the recorded time series x(t). These sudden drops may be related either to an instrument malfunctioning, to external factors or to accidental events (earthquakes, hurricanes, ...).

The introduction of the reference quantities summarized in Table 8.1 can be useful in order to collect data in a systematic way and to compare them with those of similar buildings. In the following sections, some selected data from SHM related to two important Italian monuments systems are analyzed and preliminary reference values are obtained for the defined quantities.

8.3 Applicative Examples

8.3.1 The Cathedral of Modena

The Cathedral of Modena, whose construction took place between 1099 and 1184 through the instrument of architect Lanfranco, is one of the most important examples of the Romanesque art in Italy (Fig. 8.3). In 1997 it has been declared "UNESCO Cultural Heritage" site.

The Cathedral has a basilica plan with three naves culminating in three apses. The cathedral is connected to the contiguous Ghirlandina Tower (a tower of about 86 m height) through two masonry arches. During the centuries the monument experienced various interventions and transformations (Armandi 1984; Sandonni 1983). Recently, a conservation project which is still under development has been planned with the purpose of strengthening the main walls, the vaults and of providing the structure with a box behavior. In the context of this retrofit project a SHM system was also installed.

Fig. 8.3 The Cathedral of Modena

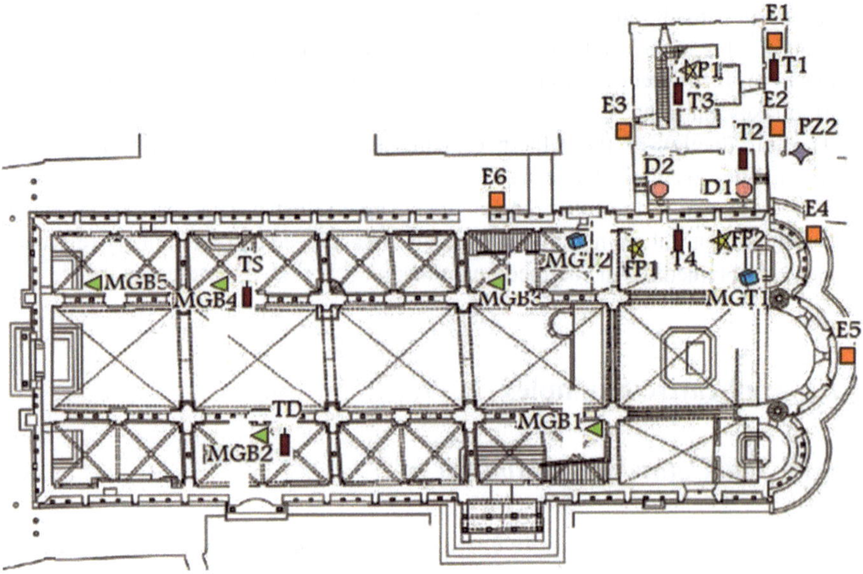

Fig. 8.4 Location of the sensors

8.3.1.1 The Monitoring System

The monitoring system allows to monitor the main cracks across the walls and vaults, the relative displacements between the cathedral and the tower, the inclination of the external longitudinal walls, the foundation settlements, the water table level and the temperature. Most of the instruments were installed in 2003, while others (such as the inclinometers) were installed in 2010. Data are acquired at time intervals of 30 min. Figure 8.4 shows the location of the sensors. The symbols indicate: D: deformometer, E: extensimeter, MGB: biaxial joint meter, MGT: triaxial joint meter, FP: inclinometer, T: thermometer.

8.3.1.2 The Data Obtained from the SHM

The values of the reference quantities defined in Sect. 8.2 are reported in Table 8.2. Each value refers to a single specific device except for the biaxial joint meter whose value is obtained computing the average of all instruments of that type.

As illustrative example, a more detailed data analysis following the approach introduced in the previous section is conducted with reference to the biaxial joint meter MGB1 placed on a crack on the south longitudinal wall (Figs. 8.5, 8.6, 8.7 and 8.8).

Figure 8.5a displays the complete time series over years 2004–2012. With the exception of few spikes, this series clearly shows the daily and annual periodicity with an almost constant average value, thus not showing a significant evolution

Table 8.2 Reference values for each type instrumental

Device			σ [mm]	Σ [mm]	r_m [mm]	R_M [mm]
Deformometers			0.02	0.70	0.10	0.30
Extensimeter	E4		0.16	4.7	−0.25	−1.55
	E5		0.02	1.05	−0.07	−0.04
Biaxial joint meters		Comp.X	0.02	0.35	−0.012	−0.03
		Comp.Y	0.006	0.10	−0.005	−0.01
Triaxial joint meters	MGT1	Comp.X	0.03	0.60	0.02	0.04
		Comp.Y	0.01	0.15	−0.004	−0.006
		Comp.Z	0.01	0.20	0.014	0.020
	MGT2	Comp.X	0.026	0.70	0.034	−0.11
		Comp.Y	0.010	0.50	0.013	0.011
		Comp.Z	0.006	0.22	−0.06	−0.08
Inclinometer	FP1	Comp.X	0.007	1.5	−0.37	−0.70
		Comp.Y	0.008	0.84	−0.50	−1.0
	FP2	Comp.X	0.012	0.90	−0.58	−1.10
		Comp.Y	0.008	0.30	−0.16	−0.31

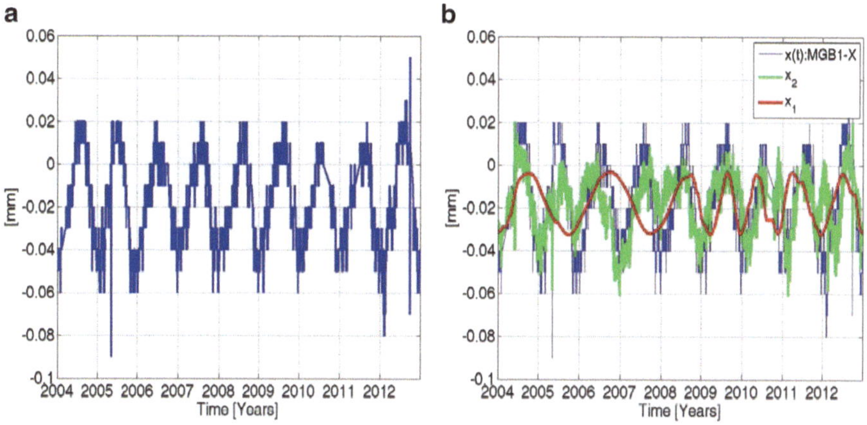

Fig. 8.5 (**a**) Overall trend MGB1-X, (**b**) reconstructed signal with only the harmonics corresponding to T1 = 365 day, T2 = 1 day, and signal less harmonics corresponding to T1 and T2

of the state S(t). A more in depth inspection of the time series is provided by the Fig. 8.5b which provides the two main components of the signal x_1 and x_2. For this specific case, the periodic contribution is essentially given by the harmonic with period $T_1 = 365$ days, whilst the amplitude of the harmonic $T_2 = 1$ day is order of magnitudes lower. Also, the signal component x_2 is characterized by a significant periodicity whose fundamental harmonic has a period T of 300 days.

Figure 8.6a displays the daily amplitude (σ about 1/100 mm) and the mean daily value related to a generic day. Figure 8.6b represents the annual amplitude (Σ about 1/10 mm) and the mean annual value related to year 2006. Note that the ratio σ/Σ seems to be equal to 1/10.

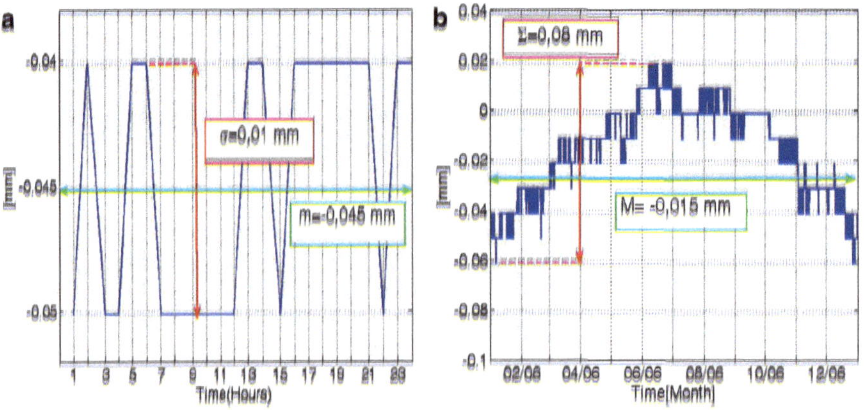

Fig. 8.6 (**a**) Daily amplitude and middle value, (**b**) annual amplitude and middle value

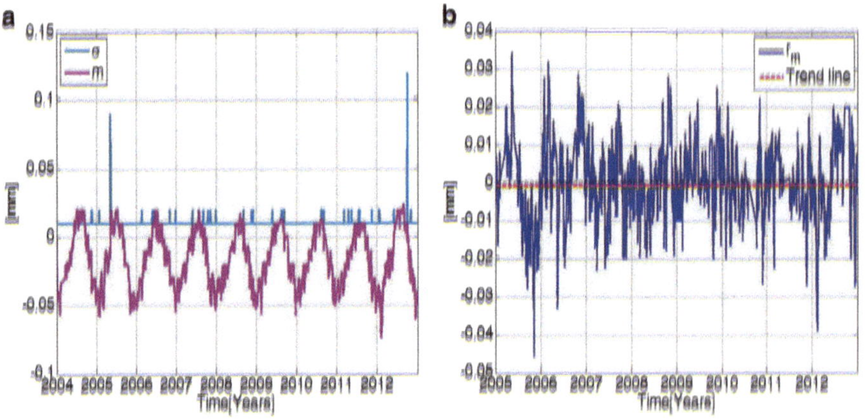

Fig. 8.7 (**a**) Trend in the year of daily amplitude and middle value, (**b**) daily residual on m and linear regression line

Figure 8.7a plots the daily amplitude (σ) and the daily mean value (m) over the entire observation period (9 years). It can be noted that m oscillates around an almost constant value, indicating a stable behavior of the monitored crack (no evolution in time), whilst σ, despite isolated spikes, is small and bounded within 0.02 mm. Figure 8.7b plots the daily residual of the mean daily value r_m. These graphs further confirm that no significant evolution of the state S(t) is observed: the trend regression line is practically horizontal.

Figure 8.8a, b represents the histograms of the annual residuals R_Σ and R_M, and their cumulative curves. The cumulative curves indicate a quite stable behavior, with compensations along the years (except for year 2012 in R_Σ).

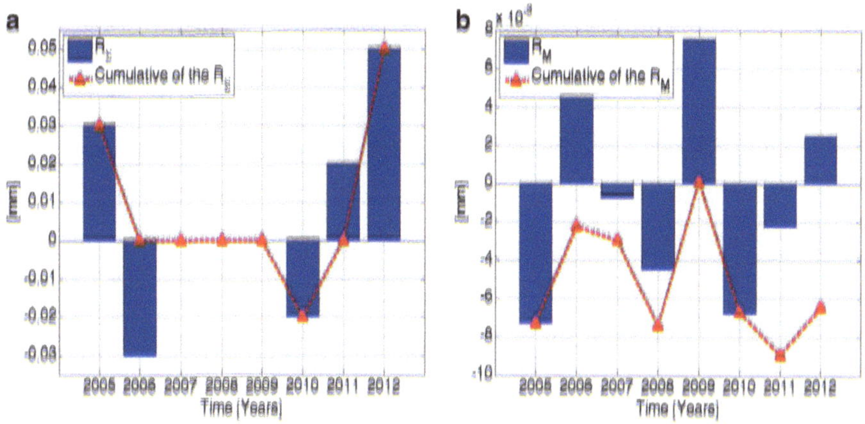

Fig. 8.8 (**a**) Annual residual on $\sum$ and cumulative curve, (**b**) annual residual on M and cumulative curve

Fig. 8.9 The Asinelli and Garisenda Towers of Bologna

8.3.2 *The Asinelli and Garisenda Towers of Bologna*

The city of Bologna preserves traces of past civilizations and the mark of medieval splendor. During the twelfth and thirteenth centuries, a large number of towers were erected by the richest families (Costa 2011). Among these, the towers of Garisenda and Asinelli, both pending, arose as a symbol of the city (Fig. 8.9).

The Garisenda, the oldest of the two, can be dated around the last two decades of the eleventh century. During the construction phases the foundation soil underwent subsidence phenomena which caused a consistent tilt of the tower (Ceccoli et al. 2001). The Tower is 48 m high and has a slope of 3.22 m towards North-East.

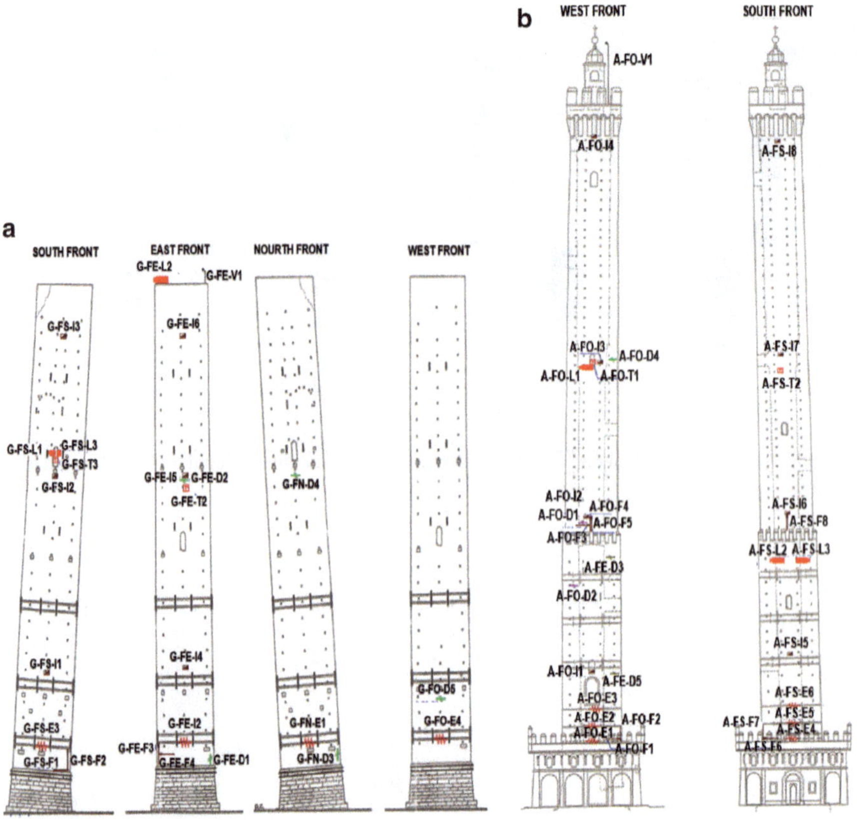

Fig. 8.10 (**a**) Monitoring system of Garisenda Tower, (**b**) monitoring system of Asinelli tower

The Asinelli tower was most probably completed in 1119 reaching an height of almost 100 m thanks to the introduction of advanced construction methods for the time (Cavani 1912). During the Second World War, the Asinelli Tower was used as a watchtower. It tilts toward West of 2.23 m.

8.3.2.1 The Monitoring System

Several strengthening interventions were performed in the last decade (1998–2008) on both towers. After the end of the works, at the beginning of year 2011, a monitoring system was installed in both the Towers in order to monitor the opening of the main cracks, the variation of the stress level of the steel ties, the masonry compression at critical locations, the variation in the inclination of the Towers, and the trend of meaningful environmental parameters (temperature, wind direction and wind speed). Figure 8.10 shows the location of the sensors installed into the Two Towers. The symbols indicate: G: Garisenda, A: Asinelli, FN: North front, FS: South

Table 8.3 Reference values for each typology instrumental installed in the Asinelli tower

Asinelli tower					
Device		σ	Σ	r_m	R_M
Long base deformometer		0.02 [mm]	0.26 [mm]	0.04 [mm]	0.005 [mm]
Deformometer		0.009 [mm]	0.09 [mm]	0.01 [mm]	0.009 [mm]
Extensimeter		46 [με]	173 [με]	−4.6 [με]	−1.9 [με]
Laser displacement sensor		0.013 [m]	0.053 [m]	−0.068 [m]	−0.068 [m]
Inclinometer	Comp.X	0.009 [°]	0.09 [°]	0.007 [°]	0.006 [°]
	Comp.Y	0.008 [°]	0.15 [°]	0.016 [°]	0.028 [°]

Table 8.4 Reference values for each typology instrumental installed in the Garisenda tower

Garisenda tower					
Device		σ	Σ	r_m	R_M
Long base deformometer		0.06 [mm]	0.45 [mm]	−0.02 [mm]	−0.03 [mm]
Deformometer		0.01 [mm]	0.08 [mm]	0.02 [mm]	0.02 [mm]
Extensimeter		22 [με]	90 [με]	−11 [με]	−11 [με]
Laser displacement sensor		0.010 [m]	0.044 [m]	0.008 [m]	0.005 [m]
Inclinometer	Comp.X	0.006 [°]	0.06 [°]	0.003 [°]	0.003 [°]
	Comp.Y	0.007 [°]	0.06 [°]	0.003 [°]	0.003 [°]

front, FE: East front, FO: West front, E: extensimeter, D: deformometer, F: long base deformometer, I: Inclinometer, L: laser displacement sensor, T: thermometer, V: gonioanemometer sensor.

8.3.2.2 The Data Obtained from the SHM

The values of the reference quantities defined in Sect. 8.2 are reported in Tables 8.3 and 8.4. Each value is related to a device typology and is obtained computing the average over all instruments of that type.

As illustrative example, a more detailed data analysis following the approach here introduced is conducted with reference of the deformometer A-FS-F1, installed at the south-west corner base of the Asinelli tower measuring the horizontal masonry deformation (displacement between two points).

Figure 8.11a shows the complete time series. No data are recorded from July to August due to an interruption of the system. Data oscillates around an almost constant average value, thus not showing a significant evolution of the state S(t). Figure 8.11b shows the two main components of the signal x_1 and x_2. For this specific case, the periodic contribution is essentially given by the harmonic with period $T_1 = 365$ days, whilst the amplitude of the harmonic with period $T_2 = 1$ day is order of magnitudes lower. Component x_2 oscillates with a not clear periodicity and is characterized by a small amplitude.

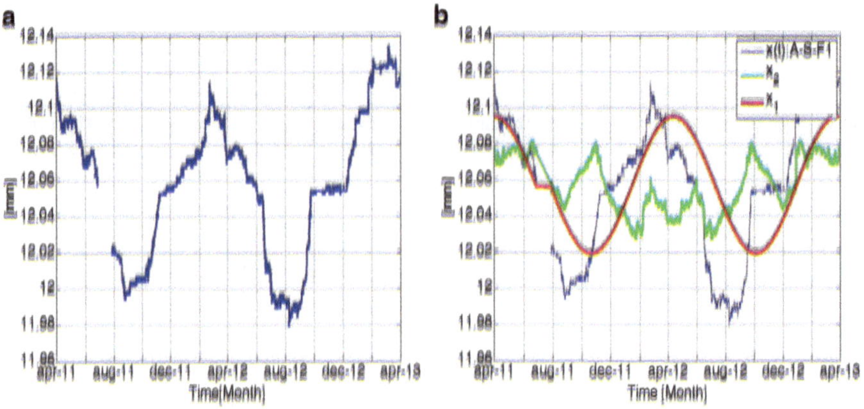

Fig. 8.11 (**a**) Overall trend F1, (**b**) reconstructed signal with only the harmonics corresponding to T1 = 365 day, T2 = 1 day, and signal less harmonics corresponding to T1 and T2

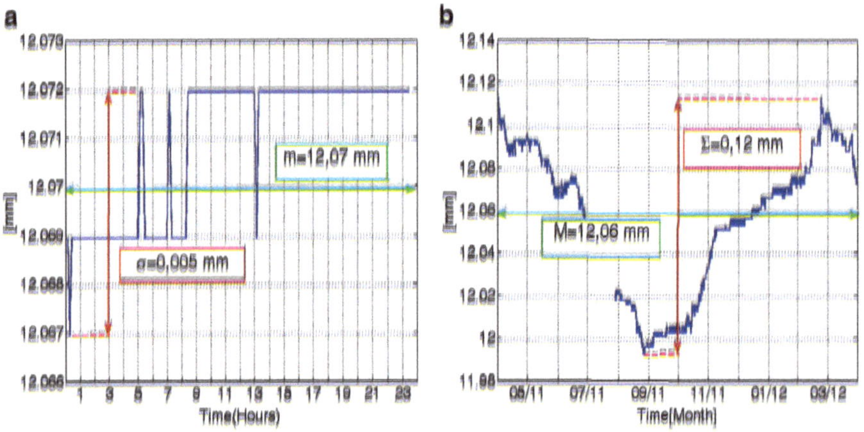

Fig. 8.12 (**a**) Daily amplitude and middle value, (**b**) annual amplitude and middle value

Figure 8.12a displays the daily amplitude (σ about 0.5/100 mm) and the mean daily value related to a generic day. Figure 8.12b represents the annual amplitude (Σ about 1.2/10 mm) and the mean annual value related to year 2011. It can be noted that the daily amplitude is around 4 ÷ 5 % of the annual amplitude. Note that the ratio σ/Σ seems to be roughly equal to 1/10.

Figure 8.13a plots the daily amplitude (σ) and the daily mean value (m) over the entire observation period (2 years). It can be noted that the daily amplitude σ is quite variable and no clear trend is observed: most values (except for isolated spikes) are between 0.002 and 0.005 mm. Figure 8.13b plots the daily residual of the mean daily value r_m. An increasing trend is observed. The trend should be related with the temperature. Nonetheless, in order to give a more reliable interpretation of the observed trends, a more extended observation period appears necessary.

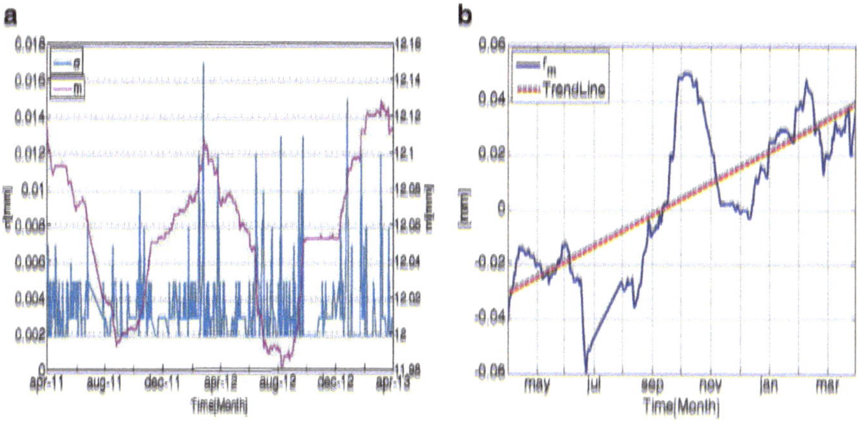

Fig. 8.13 (**a**) Trend in the year of daily amplitude and middle value. (**b**) Daily residual on m and linear regression line

8.4 Conclusions

This paper presents a simple approach for a standardized analysis of the data recorded by a Structural Health Monitoring system installed in an historical building. It is assumed that the recorded time series may be decomposed into two fundamental components: the first one related with the natural actions and characterized, in absence of extreme events (such as explosion earthquake, hurricanes, ...), by a substantial periodic behavior, the second one related to the other factors such as the evolution of the state of the structure due to material degradation, soil settlements and others.

The main objective is to introduce significant reference quantities from the analysis of recorded data in order to detect possible anomalies from the usual structural behavior. This information may be collected in a database and adopted as reference quantities for the analyses of data of similar buildings in order to have reference values leading to a more sound interpretation of the SHM data. The approach has been applied to the analysis of the data as obtained from the SHM systems of two important monuments: the Cathedral of Modena and the Two Towers of Bologna.

Based on the data recorded by the two SHM systems, the following reference values for deformometers monitoring the amplitudes of existing cracks are obtained: values of the daily amplitude σ about $0.2 \div 1/100$ mm and values of the annual amplitude Σ about $1 \div 1.2/10$ mm. These values may represent a preliminary indication of possible typical orders of magnitudes for such historical masonry buildings. Also, the ratio σ/Σ seems to be roughly equal to 1/10.

Acknowledgments This paper was inspired by an informal conversation between Prof. Andrei Reinhorn and two of the authors (Stefano Silvestri and Michele Palermo) during lunchtime in the sixth European Workshop on the Seismic Behaviour of Irregular and Complex Structures

(6EWICS), Haifa, Israel, September 2011. In that occasion, Prof. Reinhorn praised Italy for its cultural heritage and underlined the importance of taking care of such unique monumental buildings, also by adopting controlling and monitoring systems.

References

Armandi M (1984) Lanfranco e Wiligelmo. II Duomo di Modena

Cavani F (1912) Sulla pendenza e sulla stabilità della torre degli Asinelli di Bologna. University of Bologna, Bologna (in Italian)

Ceccoli C, Diotallevi P, Pozzati P, Sanpaolesi L, Dallavalle G (2001) Indagini inerenti le strutture murarie e fondali e consolidamento delle parti in elevazione della Torre Garisenda. Gruppo di studio della Torre Garisenda Bologna

Costa T (2011) Il grande libro delle Torri Bolognesi. Studio Costa

Sandonni T (1983) Cronaca dei restauri del Duomo di Modena (1897–1925). Baracchi Giovanardi, O., Modena

Sohn H, Farrar CR, Hemez FM, Shunk DD, Stinemates DW, Nadler BR, Czarnecki JJ (2004) A review of structural health monitoring literature: 1996–2001. Los Alamos National Laboratories, Los Alamos. Retrieved 2010-07-10 (2004)

Part III
Hybrid Testing and Control

Chapter 9
A Versatile Hybrid Testing System and Its Application in Developing Hybrid Simulation Methods for NEESR Projects

Xiaoyun Shao, Adam Mueller, Chelsea Griffith, and Griffin Enyart

Abstract Hybrid simulation method in earthquake engineering, which combines physical testing and numerical simulation, was developed to evaluate seismic performance of civil structural systems. Thus, instead of constructing a full sized structural specimen, hybrid simulation allows researchers to build a complex experimental substructure tested experimentally while the relatively simple part of the structure is numerically simulated to economically obtain the full structural responses. Recently a versatile hybrid testing system was built at the Laboratory of Earthquake and Structural Simulation (LESS) at Western Michigan University. The major equipment consists of a seismic simulator (often called shake table), an actuator/reaction system and an advanced hybrid testing controller. Such testing system is capable of conducting various hybrid simulation experiments such as displacement-based pseudodynamic substructure testing as well as force-based real time dynamic hybrid testing. The benchmark scale testing system at LESS is particularly suitable for development of hybrid simulation techniques and earthquake engineering education and outreach activities. The development of this testing system including both hardware and software integration is presented. Example hybrid simulation methods that can be conducted using the developed testing system as well as its applications in the hybrid simulation method development of two NEESR projects are discussed.

X. Shao (✉) • A. Mueller
Western Michigan University, 1903 W Michigan Ave, Kalamazoo, MI 49008, USA
e-mail: Xiaoyun.shao@wmich.edu

C. Griffith
Department of the Navy, Naval Sea Systems Command, 1333 Isaac Hull Ave SE, Washington, DC (202) 781-1869, USA
e-mail: Chelsea.Griffith@navy.mil

G. Enyart
City of Decatur, #1 Gary K. Anderson Plaza, Decatur, IL 62523, USA
e-mail: griffinenyart@gmail.com

G.P. Cimellaro et al. (eds.), *Computational Methods, Seismic Protection, Hybrid Testing and Resilience in Earthquake Engineering*, Geotechnical, Geological and Earthquake Engineering 33, DOI 10.1007/978-3-319-06394-2_9,
© Springer International Publishing Switzerland 2015

9.1 Introduction

To achieve safe and economical structural design under severe earthquake shaking, earthquake engineering studies structural performance when subject to seismic loading. Though numerical simulation has been continuously advancing in the last 30 years, experimental investigation remains indispensable. It provides crucial information of structural seismic behavior and provides data used to calibrate computer models in numerical simulation. More recently, hybrid simulation was proposed and developed as an economic alternative experimental method that combines numerical simulation and physical testing. Part of the structure, called *experimental substructure*, is physically loaded using hydraulic equipment, such as shake tables and/or actuators/reaction wall, while the rest of the structure, named *numerical substructure*, is numerically simulated using an analytical model. During a hybrid simulation, the numerical simulation and the physical experimentation are executed in parallel with designated data transferred in between through a step-by-step manner. With advancing of time steps, the response of the whole structural system subjected to a seismic excitation would be obtained at the end of a hybrid simulation.

A versatile benchmark scale hybrid testing system was designed and constructed in the Laboratory of Earthquake and Structural Simulation (LESS) at Western Michigan University. The major equipment consists of a uniaxial shake table, an actuator/reaction system and a real-time hybrid testing controller. The main objective of building such a testing system is to facilitate efficient and economical development of hybrid testing techniques through a flexible and versatile testing system, with both hardware and software integrated. Several common testing methods, including quasi-static testing (QST), shake table testing (STT), effective force testing (EFT), pseudodynamic (PSD) testing and real-time dynamic hybrid simulation (RTDHS) can be performed using the developed hybrid testing system.

This paper first describes the testing system development including hardware and software integrations. Then four examples of hybrid simulation that can be conducted using the developed testing system are presented followed by discussions of its application to two US National Science Foundation funded Network for Earthquake Engineering Simulation Research (NEESR) projects to develop their respective hybrid simulation methods.

9.2 Hybrid Testing System

To provide an efficient, flexible and economic testing platform that can be used in continuous improvement of various hybrid simulation techniques, a versatile hybrid testing system was developed. The major components of the testing system with their respective specifications are listed in Table 9.1. Figure 9.1 schematically illustrates these components and their connections to form the hybrid testing system, which are described in detail in this section.

Table 9.1 Major components in the developed hybrid testing system

		Components	Specifications	
Hardware		Uniaxial shake table (built in house)	Table size	3 ft × 3 ft (915 mm × 915 mm)
			Maximum specimen mass	500 lbf (228 kg)
			Frequency of operation	0–20 Hz
			Maximum acceleration	4 g (with 500 lbf specimen)
			Maximum displacement	±3 in. (±76.2 mm)
		Hydraulic actuators Shore Western model #910D-1.08-6(0)-4-1348	Force	±3,240 lb (±13.3 kN) at 3,000 psi
			Stroke	6 in. (±2.5 in plus ±0.5 in.)
			Swivel base and end rod	±90° swivel, ±7 tilt
			Servo valve	2.5 Gpm at 1,000 psi
		Steel reaction frame	Height	15 ft (4.6 m)
			Maximum capacity	15 lb·ft (20.4 N·m) overturning moment
		Hydraulic power supply (Shore Western 110.11S model)	20 hp TEFC motor	
			380–480 V, 3 phase, 50/60 Hz power	
			Local and remote control, 24 VDC control voltage	
			High/low pressure controls	
	Chassis (PXI-1050)	Embedded (real-time) processor (PXI-8108)	2.53 GHz Intel Core 2 Duo T9400 dual-core processor	
			1 GB 800 MHz DDR2 RAM standard	
			10/100/1000BASE-TX (gigabit) Ethernet	
			Integrated hard drive, GPIB, serial, and peripheral I/O	
		Data acquisition (PXI-6229, PXI-6221)	Integrated chassis with 4 signal conditioning module slots	
			16-Bit, 48 analog inputs, 6 analog outputs, 72 digital I/O	
		Instrumentations: Linear variable differential transducer (LVDT)	Stroke	10 in. (254 mm)
			Frequency response	200 Hz
			Power supply converter	115 V AC → 15 V DC
		Accelerometers	Peak value	4 g
			Direction	3 axis: x, y, and z-axis

(continued)

Table 9.1 (continued)

	Components	Specifications	
Software	Generic hydraulic controller (Shore Western SC6000)	Two servo amplifiers	
		Two servo valve drivers	
		Two internal function generators	
		Either voltage or amperage valve drive signals	
		Four transducer amplifiers per controller card	
		Integrated data acquisition	
	Hybrid testing controller	Hybrid testing model	Programmed in Matlab/Simulink
			Deployed using NI Veristand

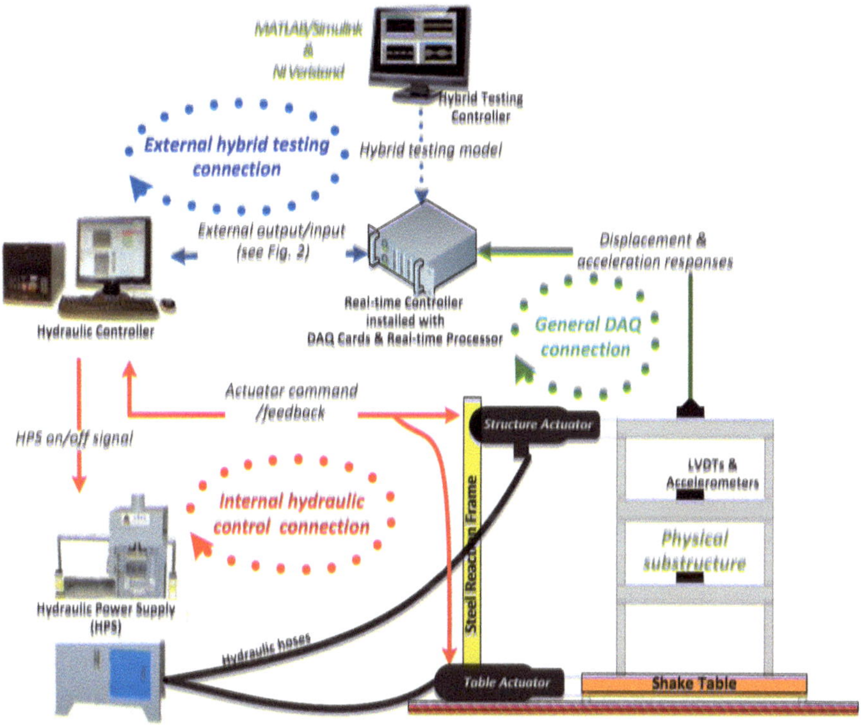

Fig. 9.1 Schematic diagram of the hybrid testing system (Reproduced from Shao and Enyart (2012))

9.2.1 *Testing System Hardware Integration*

Various types of cables were used to connect the hardware components, including the sensors as shown in Fig. 9.1. The connections between the hydraulic controller, two actuators (i.e. the Table actuator and the Structure actuator) and the hydraulic power supply (HPS) are provided by the manufacturer of the equipment as shown in red arrows. The hydraulic controller sends command signals to the servo-valve and receives feedback from the embedded LVDTs and load cells within the actuators. The HPS can be turned on/off by the hydraulic controller through the cable connection. With these connections, the hydraulic hardware components form an *internal closed loop hydraulic control* that is conventionally utilized in structural experiments to apply a predefined loading history to the test specimen (i.e. quasi-static testing and shake table testing). These experiments are considered open-loop tests during which no feedback from the test specimen is necessary to determine actuators' commands.

On the other hand, during a hybrid simulation, an online numerical computation is necessary to generate the loading commands of the actuators based on the feedback from the test specimen and/or actuators. The numerical computation is conducted

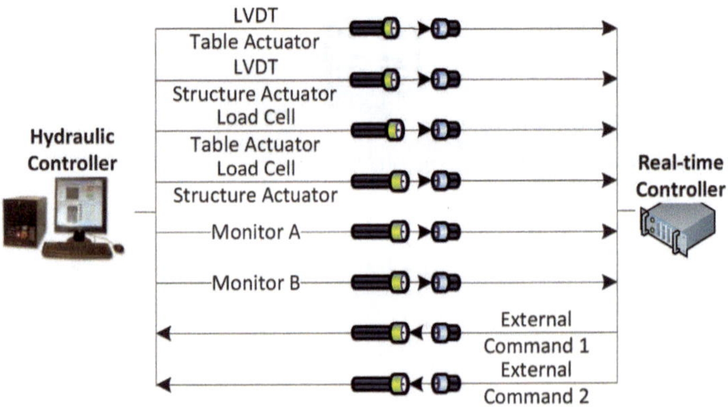

Fig. 9.2 External BNC connections

in the real-time controller and the feedback is collected using both the general DAQ system and the DAQ embedded in the hydraulic controller. Therefore, a *general DAQ connection* (green arrows) and an *external hybrid testing connection* (blue arrows) were created for hybrid simulation purposes. The general DAQ connection is a standard one way connection that transfers measured structural responses from the sensors (i.e. LVDTs and accelerometers) to the general DAQ system. Utilizing a single chassis (i.e. NI-PXI 1050) housing both the real-time processor and the DAQ system, the structural response data is immediately available for the hybrid testing controller once they are collected from the sensors.

The external hybrid testing connection connects the hydraulic controller, the real-time controller, and the hybrid testing controller as shown in Fig. 9.1. The two actuators' positions and forces are fed back to the general DAQ system through the standard Bayonet Neill–Concelman (BNC) connectors (see Fig. 9.2). Along with the four actuators' feedback signals, two monitor signals (Monitor A and B) are available for parallel real-time simulation that can be used to check any critical point within the hydraulic control loop. In the opposite direction, two external command signals can be sent to the hydraulic controller from the real-time processor. These two external commands are used to control the table actuator and the structure actuator respectively during hybrid simulations. The BNC connections shown in Fig. 9.2 are essential for hybrid simulation since they enable data transfer between physical experiments conducted using the hydraulic equipment and numerical simulation running in the real-time processor. The connection between the real-time controller and the hybrid testing controller is realized through an internet (Ethernet) cable (blue dashed arrow). A hybrid testing model defining a numerical simulation, and/or a control algorithm developed in Matlab/Simulink, is deployed (downloaded) through this connection to the real-time controller prior to testing, using a software named NI Veristand that will be discussed next.

9.2.2 Software Integration

To perform various hybrid simulations, the hybrid testing system was designed to connect the numerical simulation and the physical tests by sending interface loading commands between the substructures to the hydraulic controller, which further drives actuators to impose this interface loading to the experimental substructure. Meanwhile required sensor data is fed back to the hybrid testing model to calculate the next step's structural response and/or actuator's commands.

The hybrid testing controller runs two software programs. Matlab/Simulink is used to develop the hybrid testing model. The hybrid testing model can be a numerical substructure simulation utilizing specimen's response and defining the interface loadings, or it can be an advanced control compensation algorithm that uses actuators' feedback and generates compensated driving commands. This hybrid testing model is then deployed using NI Veristand to the real-time controller that will be running the model in real-time during a hybrid simulation. NI Veristand is a testing software tool that allows developing control systems and performing real-time testing using hardware input/output and numerical simulation models. The user interface of the hydraulic controller software has a function to receive external command from the real-time controller while setting the internal command to zero. This function is activated during a hybrid simulation.

To facilitate fast model development in Simulink, a software platform was created as shown in Fig. 9.3 which integrates all the currently available input and output. The input to the hybrid testing model consists of structural response data (i.e. displacement/acceleration responses) collected by the general DAQ system and actuators' response data obtained from the embedded actuators' sensors. All of these feedback data are available for numerical simulation and control algorithm programmed in the hybrid testing model (the middle box). The output on the right has two parts. In addition to command signals of the actuators to apply interface loadings, general output from the numerical substructure model can be recorded and output as data files that can be combined with the physical testing results for complete structural response analysis after each test.

9.3 Examples of Hybrid Simulation

The versatility of the developed hybrid testing system allows several different hybrid simulations to be conducted simply by changing the configuration of the testing system and developing a corresponding hybrid testing model. Examples of hybrid simulation that the developed system is capable of include: pseudodynamic (PSD) simulation with or without substructure, shake table substructure testing, effective force substructure testing and real-time dynamic substructure simulation. In this section, four hybrid simulation examples are discussed using a three story shear frame as the prototype structure. Detailed formula discussion of these examples can

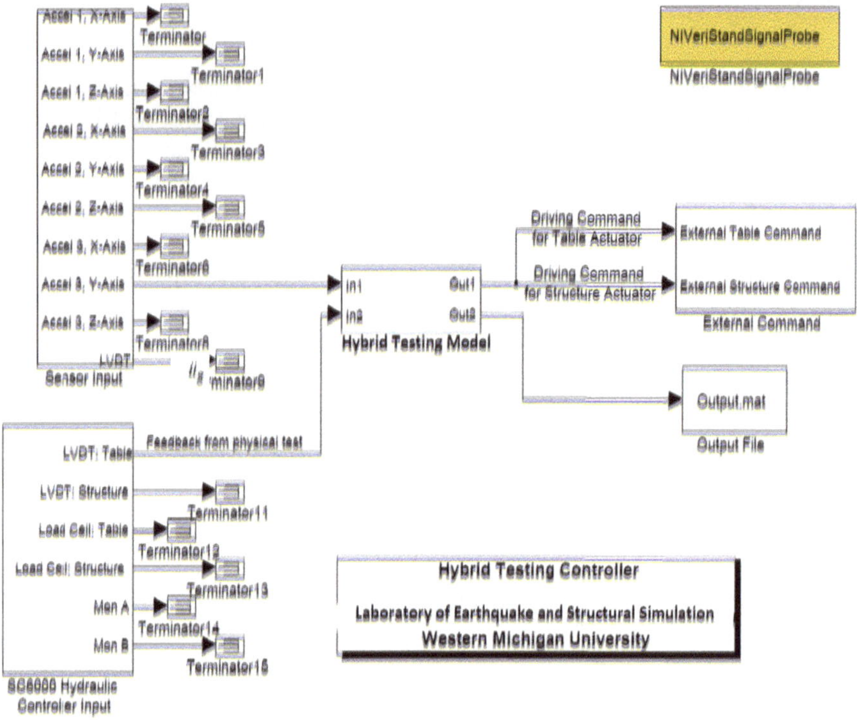

Fig. 9.3 Hybrid testing controller model

be found in Shao and Enyart (2012). Figure 9.4 shows the schematic diagrams of these examples. Please note that the developed testing system is also capable of conducting conventional open-loop tests as mentioned in Sect. 9.2.1.

9.3.1 Pseudodynamic Hybrid Simulation

In this example, the experimental substructure is the first story installed with base isolation. The mass and damping terms of this first story are modeled numerically, while the resisting force is captured experimentally to accurately consider the base isolated nonlinear response in the hybrid simulation. The numerical substructure is the top two stories which are assumed to be linear. To obtain the full response contributed from both substructures, the hybrid testing controller will perform two tasks. One is to solve the numerical substructural responses (top two stories) based on the physical response (first story). The other task is to solve the displacement response of the experimental substructure (first story) based on the measured restoring force from the physical testing. To perform this test, the structure actuator mounted on the reaction frame is used to load the experimental substructure with

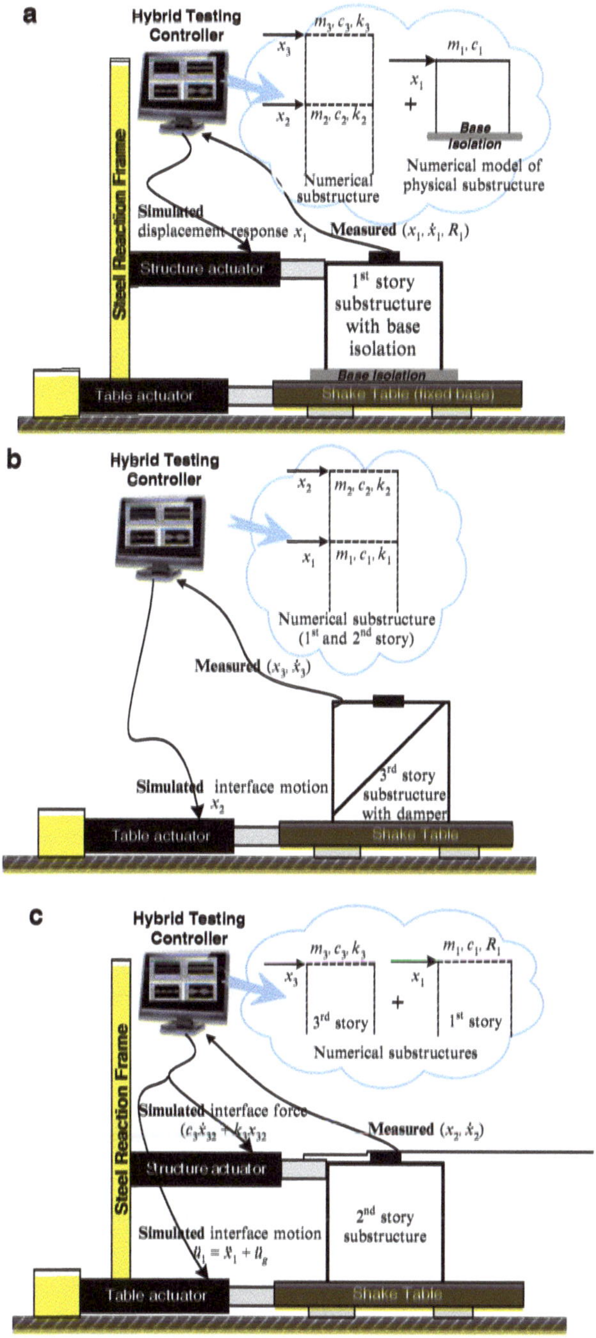

Fig. 9.4 Hybrid simulation examples: (**a**) PSD hybrid simulation, (**b**) Shake table substructure testing, (**c**) Real-time dynamic substructure simulation (Reproduced from Shao and Enyart (2012))

the first story simulated displacement response, while the resulting resisting force and structural responses are fed back to the hybrid testing controller to calculate the next step's displacement command (see Fig. 9.4a). The shake table is not used in this example but serves as a fixed base for the installation of the specimen.

9.3.2 Shake Table Substructure Testing

In a shake table substructure test, the experimental substructure is the third story installed with additional dampers for structural response mitigation, while the numerical substructure is the bottom two stories. Therefore, the third story experimental substructural specimen consists of physical mass, internal damping, stiffness and the supplemental viscous damping. The hybrid testing model solves the numerical responses (bottom two stories) due to the ground seismic excitation and the interface force based on the measured response of the third story. Meanwhile, the displacement response of the second story is applied by the shake table as the interface base motion to the top experimental substructure. Only the table actuator will be utilized in this hybrid simulation example as shown in Fig. 9.4b.

9.3.3 Effective Force Substructure Testing

The same substructural configuration as the first example utilizing only the structure actuator can also be used in an effective force substructure test with the developed hybrid testing system. The hybrid testing model computes the total force, i.e. the effective earthquake force plus the interface force, based on the measured and the simulated responses. This total force is then applied by the actuator to the physical specimen at its story level. A force control scheme for hydraulic actuators, such as the one proposed by Sivaselvan et al. (2008), is required herein to convert a generally displacement controlled actuator to a force controlled actuator. Similar to other hybrid simulations, the responses of the experimental substructure are fed back to the hybrid testing model to solve the numerical responses so that the whole structural responses will be obtained at the end of the test.

9.3.4 Real-Time Dynamic Hybrid Simulation

In real-time dynamic hybrid simulation (Shao and Reinhorn 2011), the experimental substructure is the second story which is loaded using both the shake table and the force controlled structure actuator from the bottom and top respectively (see Fig. 9.4c). The hybrid testing controller solves the numerical responses of the first and the third stories and utilizes them to calculate the top interface force and the base interface motion. The interface force is applied by the structure actuator while the shake table imposes the base motion so that the measured responses of the second story experimental substructure are equivalent to the responses that would occur if it were part of the whole structure.

9.4 Application to NEESR Projects

The developed versatile hybrid testing system has been utilized in the development of hybrid simulation methods in two NEESR projects, namely NEES-Soft: Seismic Risk Reduction for Soft-Story Woodframe Buildings and Near-Collapse Performance of Existing Reinforced Concrete Frame buildings. These two projects are referred to as NEES-Soft and NEES-RCCollapse hereafter.

9.4.1 NEES-Soft Project Overview

As early as 1970, the structural engineering and building safety community recognized that a large number of multi-story woodframe buildings designed with the first floor used either for parking or commercial space were built with readily identifiable structural system deficiencies, referred to as a "soft story". Thus, many older multi-story woodframe buildings (built prior to 1970s) are susceptible to collapse at the first story during earthquakes. The NEES-Soft project is a collaboration among Colorado State University (CSU), Clemson University (CU), Western Michigan University (WMU), Rensselaer Polytechnic Institute (RPI) and California State Polytechnic University, Pomona that aims to address the aforementioned structural system deficiencies with two main objectives: (1) to design and experimentally validate performance based seismic retrofit options for soft story woodframe buildings; and (2) to provide a fundamental understanding of the collapse mechanisms in woodframe buildings through hybrid simulation techniques. Retrofits have been designed and numerically modeled using a performance based seismic design approach developed by Bahmani and van de Lindt at CSU (2012) and utilizing energy dissipation devices proposed by Tian and Symans at RPI (2012). Numerical models of 3-D collapse mechanisms are being developed at CU (Pang and Ziaei 2012). Hybrid simulations are conducted to study multiple seismic retrofit options using the NEES facility at University at Buffalo (NEES@UB) as the most economic and efficient method. The development of the hybrid simulation controller is led by WMU (Griffith 2013).

The NEES-soft project adopts a displacement-based slow pseudodynamic (PSD) hybrid simulation method during which the inertia and damping effects of the whole structure and the hysteresis response of the numerical substructure are simulated computationally, while the restoring force of the experimental substructure is fed back to the numerical simulation to determine displacements through a step-by-step

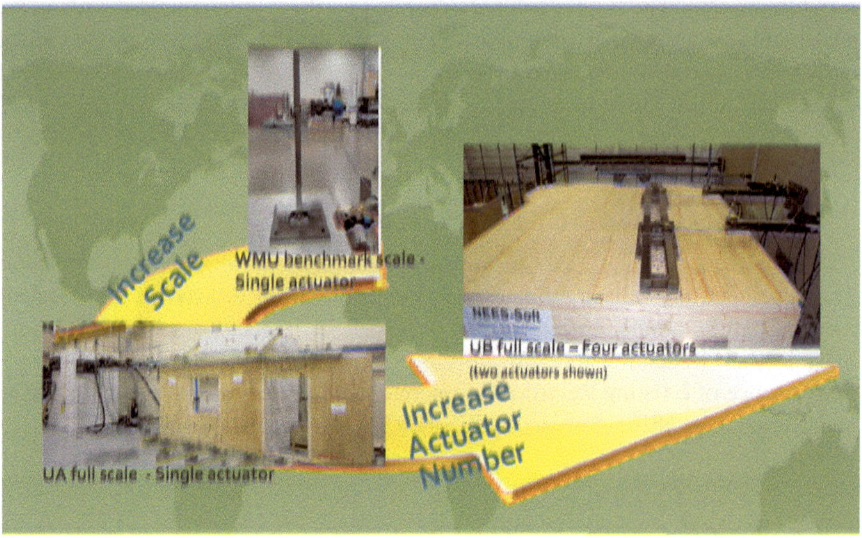

Fig. 9.5 Road map of hybrid simulation method development in the NEES-Soft project

integration. The challenges associated with the UB large scale hybrid simulation are two-fold. One is that a real-time hybrid testing system (Shao 2007) would be used for the slow hybrid simulation purpose. Another challenge is that the NEES@UB facility is fully scheduled with research projects allowing a very limited onsite period for hybrid simulation controller development. To address these two challenges, a roadmap for NEES-Soft project hybrid simulation controller development was proposed as illustrated in Fig. 9.5. Benchmark scale and full scale PSD hybrid simulations were conducted at LESS of WMU and the Structural Engineering Laboratory at The University of Alabama (UA) respectively (see Fig. 9.5) using single actuator test setup. The experiences gained in these two tests were therefore utilized in the development of the full-scale UB hybrid simulation controller that commands four actuators. This section discusses the development work along the roadmap.

9.4.1.1 Benchmark Scale Pseudodynamic Hybrid Simulation at WMU

The NEES-Soft hybrid simulation controller development took place at WMU using the developed hybrid testing system discussed above. The test specimen was a small-scale steel cantilever column designed and fabricated for the controller development purpose. Through the easily replaced nonlinear structural coupons, this specimen provided predictable resistance for the numerical model which is desirable in the development and verification phases of the hybrid simulation controller. The test specimen and a schematic diagram of the hybrid testing system used in this controller development phase are shown in Fig. 9.6.

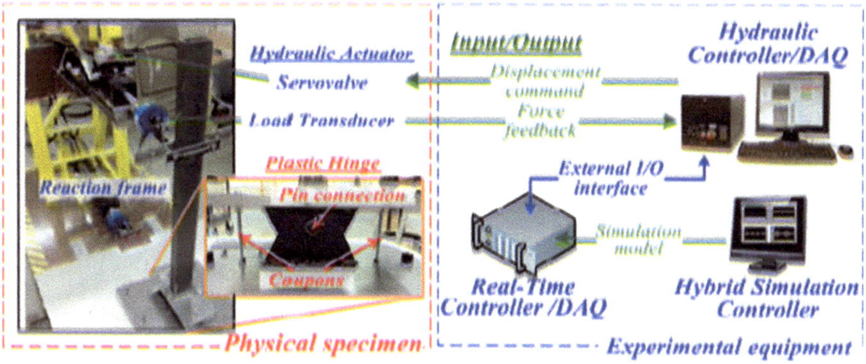

Fig. 9.6 Small scale WMU specimen and the LESS hybrid testing system

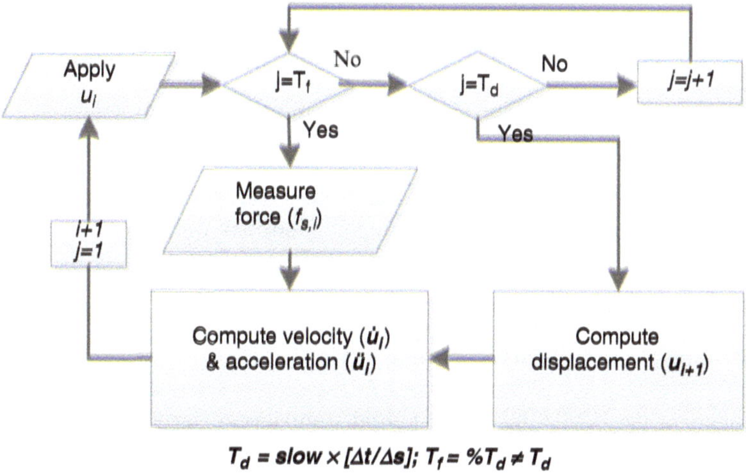

Fig. 9.7 Flowchart of step/hold controller and experimental results

The hybrid simulation model was created in the Matlab/Simulink template shown in Fig. 9.3, which consisted of the numerical model of the test structure, the integration algorithm solving the dynamic response and the hydraulic compensations for inherent actuator delay and its command tracking errors. The slow hybrid simulation controller developed at this phase adopted a step/hold loading pattern (see Fig. 9.7) using double triggered functions. The first trigger was used to slow down the hybrid simulation at any desired loading rate "slow". When the simulation step j reached the predefined T_d value, the controller would trigger the next step integration. The second trigger function triggered the force measurement reading (when j reaches T_f) to ensure the measured force used in the next integration step correspond to the target displacement of the previous step at which the actuator has reached.

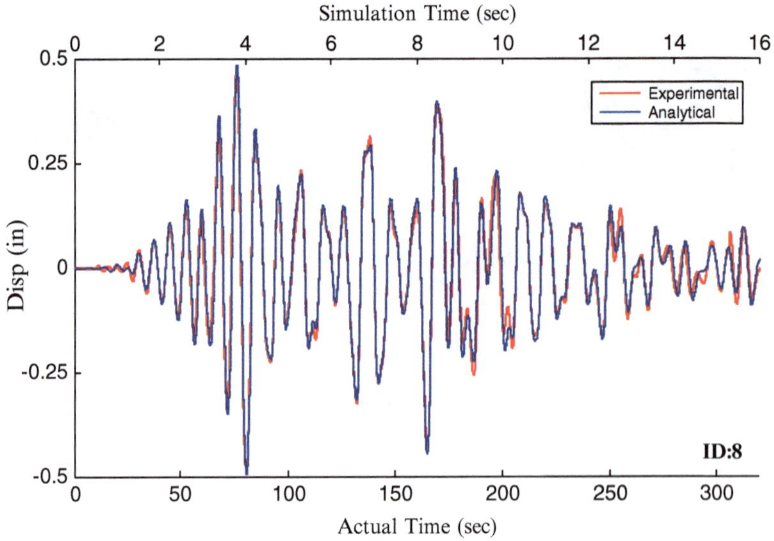

Fig. 9.8 Comparison of hybrid simulation results (experimental) with analytical results

The agreement between the analytical and the experimental responses is shown in Fig. 9.8 demonstrating that the developed controller was capable of providing reliable hybrid simulation results.

9.4.1.2 Full Scale Hybrid Simulation at UA

To implement the WMU developed hybrid simulation controller in the Structural Engineering Laboratory at UA and eventually at the NEES@UB facility, a complete study on the equipment available at UA and UB was conducted focusing on identifying the functional hybrid simulation controller components and hydraulic loading equipment. Since the UB hybrid testing system is very similar to the UA's system, Table 9.2 only lists the system comparison between WMU and UA. As can be seen from Table 9.2, both facilities use Matlab/Simulink to set up the hybrid testing model allowing a fast migration of the WMU controller to the UA laboratory. Other controller components such as the real-time controller, hydraulic controller and input/output interfaces between these two hybrid testing systems are similar.

However the capacities (i.e. force capacity, stroke limit and maximum velocity) of the UA hydraulic loading equipment are much greater compared to the WMU equipment, as the UA equipment is mainly for real-time hybrid simulation. Undesirably, excessive vibration was observed when the WMU step/hold slow hybrid simulation controller was directly implemented at UA. To address the excessive vibration, a ramp/hold command pattern was developed to "smooth" the loading (see Fig. 9.9). Within each integration step, the command was ramped to a predetermined

Table 9.2 Hybrid simulation testing system at WMU and UA

	Controller component	WMU	UA
Hydraulic equipment	Linear hydraulic actuator	Shore Western 910D Force: ±3.2 kips Stroke: ±3 in. Maximum vel.: 9 in./s	MTS 244.31 Force: ±55 kips Stroke: ±20 in. Maximum vel.: 50 in./s
	Load transducer	2.5 kip	55 kip
	Servovalve	2.5 gpm at 1,000 psi	250 gpm at 2,800 psi
Controller components	Hydraulic controller	SW SC6000 w/DAQ and user interface, operated at 1,000 Hz	MTS Series 793 w DAQ and testing software, operated at 4,096 Hz
	Hybrid testing controller simulation software	Matlab/Simulink	
	Hybrid testing controller interface to Real time controller	NI-VeriStand	Matlab/xPC Target
	External I/O interface	SCB-68 Connector Block	SCRAMNet GT150
	Real-time controller	NI 2.53 GHz Dual-Core PXI 8108 controller	3.0 GHz Xeon Dual-Core Real-time Target PC

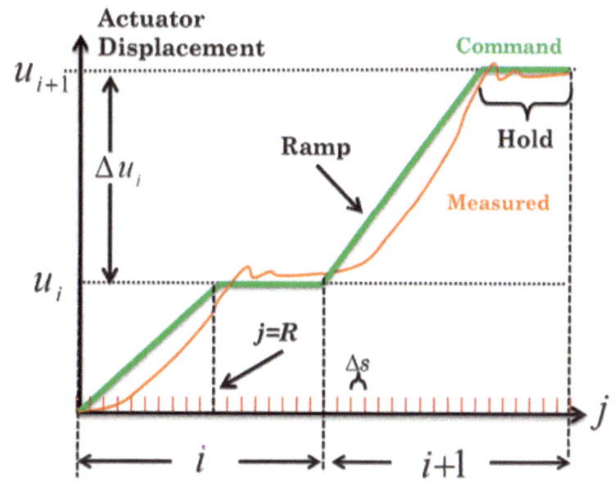

Fig. 9.9 Ramp/hold load pattern for slow PSD hybrid simulation

simulation step, '*R*', followed by a holding phase for the remaining sub-steps. Similarly, the slow PSD can be slowed at any desired rate using the developed ramp/hold load pattern.

9.4.1.3 Full Scale Hybrid Simulation at UB

It was decided that for the hybrid simulation at NEES@Buffalo the focus would be on the damage to the upper stories when the bottom story or bottom two stories are retrofitted (via stiffening/strengthening or damping). The hybrid simulation allowed evaluation of the impact of various retrofit options on the performance

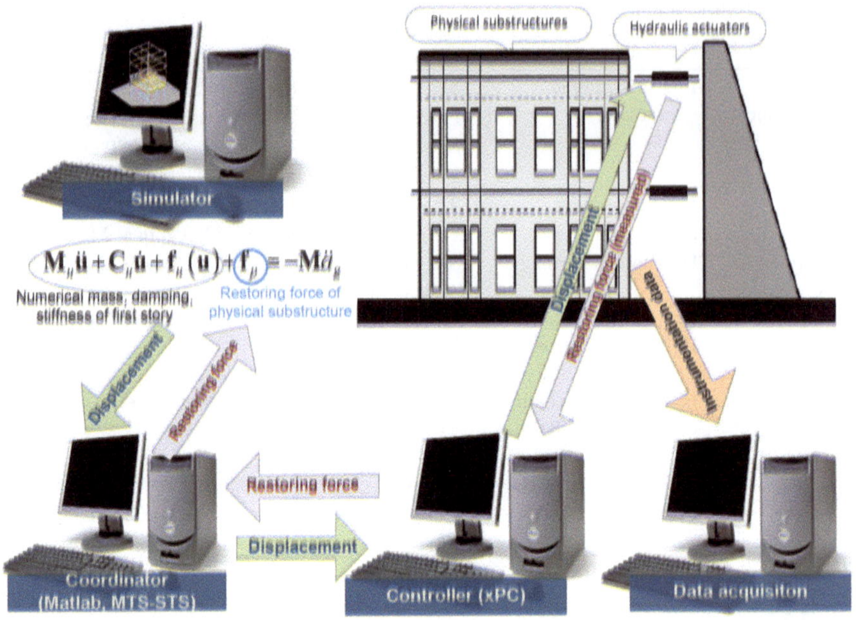

Fig. 9.10 Schematic diagram of UB hybrid simulation

of the upper stories without requiring physical construction of the retrofits. The prototype structure in the UB slow PSD hybrid simulation is a full-scale three–story woodframe building that has a two-bedroom apartment on floors 2 and 3, and parking on floor 1. The existing first soft floor with retrofits is the numerical substructure and the upper two floors are the experimental substructure (see Fig. 9.10) with a fixed base to the lab strong floor. Interactions between these two substructures are simulated and applied using four actuators, two at each story level. The ramp/hold load pattern developed during the UA hybrid simulation was implemented in the UB hybrid simulation controller and was expanded to control four actuators. Hybrid simulation has been successfully carried out for five retrofits so far and the results are being investigated.

9.4.2 NEES-RCCollapse

Reinforced concrete (RC) frame buildings built before 1976 are particularly vulnerable to seismic attack due to the lack of adequate seismic provisions in building codes at that time. Evaluating their survivability following a major earthquake requires a full understanding of their structural collapse mechanism. However, current collapse acceptance criteria for existing structures in ASCE/SEI 41-06 (American Society of Civil Engineers 2007) are defined at the element level rather than the system level, despite the fact that element failure does not necessarily constitute structural collapse. Therefore, hybrid simulation, with the capability of capturing system level structural response, was adopted in the NEES-RCCollapse project to provide insight on the structural system level response near collapse.

In the proposed hybrid simulation experiment, the prototype structure is a representative 10-story RC frame building designed according to the 1970 Uniform Building Code. During a preliminary analysis, the three columns expected to fail first were identified and will be tested as *experimental substructures* at the NEES Multi-Axial Full Scale Sub-Structured Testing & Simulation (MUST-SIM) facility at the University of Illinois at Urbana-Champaign (UIUC). The remaining building components will be simulated as the *numerical substructure* in OpenSees (2002) at Northeastern University (NEU). It is expected that after failure of the experimental columns, loads will be redistributed to surrounding columns through diaphragms and beams, which may or may not ultimately lead to structural collapse. In this manner, the effects of element failure on the entire structural system can be examined, and new analytical tools and collapse acceptance criteria can be developed.

Hybrid simulation is essential for this project due to the highly nonlinear conditions that will occur in the three critical columns near failure, which would be very difficult to capture through pure numerical simulation. In addition, hybrid simulation allows for full scale experimentation in an economical manner. Since the experimental and numerical substructures will be located at different sites, the hybrid simulations in this project will be geographically distributed. Therefore, the challenge is to provide reliable communication between the numerical model in OpenSees and the control and data acquisition system at MUST-SIM. Currently, two options are being investigated for this purpose: UI-SimCor (Kwon et al. 2007) and OpenFresco (2009).

9.4.2.1 Benchmark Scale Distributed Test Between WMU and HIT

An international distributed hybrid simulation was conducted between WMU in the US and Harbin Institute of Technology (HIT) in China with two objectives: (1) to further verify the versatility of the benchmark scale hybrid testing system; and (2) to assess the available options for distributed hybrid simulation, specifically within the context of the NEES-RCCollapse project. The knowledge gained from this test is currently being applied to the preparation of the hybrid simulation in the NEES-RCCollapse tests.

The prototype structure was a simple one-bay, one-story portal frame with linear elements and two lumped masses, each consisting of one horizontal degree of freedom. The left column (Module 1) was represented by a cantilever column specimen at WMU, same as the specimen in the NEES-Soft project (see Fig. 9.6). The beam (Module 2) was simulated numerically in OpenSees at HIT, and the right column (Module 3) was represented by a spring specimen at HIT. Coordination of the three modules during the distributed hybrid simulation was carried out by

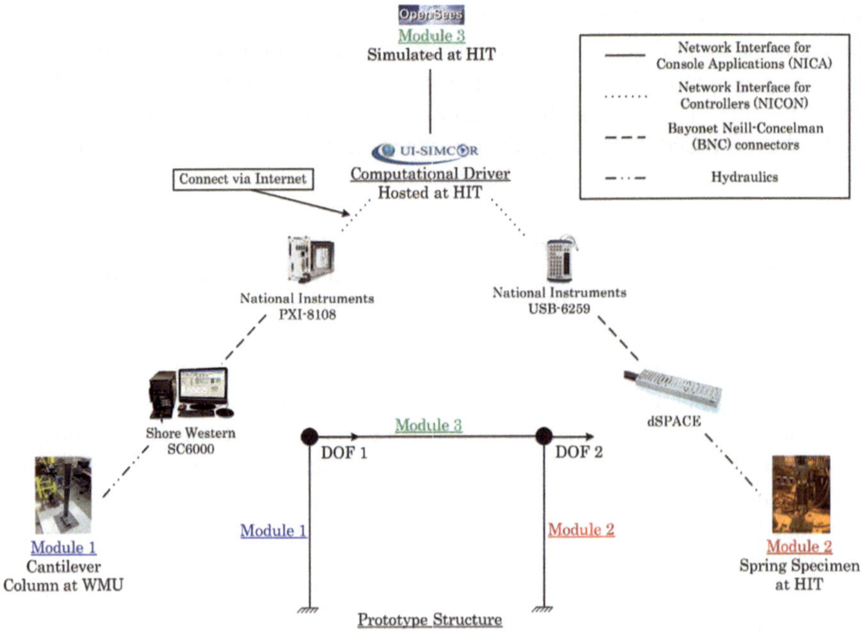

Fig. 9.11 Geographically distributed hybrid simulation between WMU and HIT

UI-SimCor through (1) NICA to the numerical simulation in OpenSees (Kwon et al. 2007) and; (2) NICON to the National Instruments (NI) hardware at both sites that were used to control their respective hydraulic controllers (Kammula et al. 2012) (see Fig. 9.11). The whole simulation lasted 1 h and 7 min and completed 1,500 numerical integration steps. No disruption was observed during the hybrid simulation.

Reliable communications between the two experimental sites using UI-SimCor were achieved in this benchmark scale distributed hybrid simulation, indicating that UI-SimCor is an option for the NEES-RCCollapse distributed tests. However, UI-SimCor, when working with numerical simulation in OpenSees, may only be able to handle small models with a limited number of DOFs due to the fact that static equilibrium analysis and dynamic stepping analysis are being conducted in OpenSees and UI-SimCor separately. Separation of these two analyses during a hybrid simulation of a large numerical model requires large amount of data to be transferred between the two software, which is not only computationally inefficient but also leads to numerical stability issues due to the incompatible way of these two software defining their numerical models. OpenFresco, on the other hand, allows OpenSees to become the computational driver. It can be directly incorporated with OpenSees as it is written in the same language and is essentially an extension of OpenSees. Currently, OpenFresco is being investigated for the distributed testing purpose and will soon be implemented at LESS.

9.5 Conclusions

A benchmark hybrid testing system was developed for the study of seismic effects of structural systems through its ability of conducting various types of hybrid simulation experiments. The components of the system as well as the integration among these components were discussed. In addition, several hybrid simulation examples were presented. This hybrid testing system has been applied to two NEESR projects to develop their respective hybrid simulation controllers including the slow pseudodynamic hybrid simulation controller and the geographically distributed hybrid simulation controller. It is demonstrated through these two cases that the developed hybrid testing system is versatile and is particular suitable for the development of hybrid simulation techniques.

Acknowledgments This work was supported in part by the U.S. National Science Foundation under Grant CMMI-1135005 (NEESR-RCCollapse project), CMMI-1314957 (NEES-Soft project) and the Research Development Award by the office the Vice president for Research of Western Michigan University.

References

American Society of Civil Engineers (2007) Seismic rehabilitation of existing buildings. American Society of Civil Engineers, Reston

Bahmani P, van de Lindt JW (2012) Numerical modeling of soft-story woodframe retrofit techniques for design. In: Proceedings of the 2012 structures congress, Chicago, pp 1755–1766

Griffith C (2013) Implementation of a versatile pseudodynamic hybrid simulation for seismic evaluation of structural systems. Master thesis, Western Michigan University

Kammula V et al (2012) Performance assessment of the Self Centering Energy Dissipative (SCED) bracing system using hybrid simulation. In: Proceedings of the 15th world conference on earthquake engineering, Lisbon

Kwon OS, Nakata N, Park KS, Elnashai AS, Spencer BF (2007) User manual and examples for UI-SIMCOR v 2.6 NEES-SM v2.0. Department of Civil and Environmental Engineering, University of Illinois at Urbana-Champaign, Urbana

OpenFresco (2009) Open-source framework for experimental setup and control. http://openfresco.berkeley.edu/. Pacific Earthquake Engineering Research Center, University of California, Berkeley

OpenSEES (2002) Open system for earthquake engineering simulation. http://opensees.berkeley.edu/. Pacific Earthquake Engineering Research Center, University of California, Berkeley

Pang W, Ziaei E (2012) Nonlinear dynamic analysis of soft-story light-frame wood buildings. In: Proceedings of the 2012 structures congress, Chicago, pp 1767–1777

Shao X (2007) Unified control platform for real time dynamic hybrid simulation. Ph.D dissertation, University of Buffalo, The State University of New York

Shao X, Enyart G (2012) Development of a versatile hybrid testing system for seismic experimentation. Exp Tech. doi:10.1111/j.1747-1567.2012.00837

Shao X, Reinhorn AM (2011) Development of a controller platform for general force-based real time hybrid simulation. J Earthq Eng 16(2):274–295

Sivaselvan MV, Reinhorn AM, Shao X, Weinreber S (2008) Dynamic force control with hydraulic actuators using added compliance and displacement compensation. Earthq Eng Struct Dyn 37(15):1785–1800

Tian J, Symans MD (2012) High-performance seismic retrofit of soft-story wood-framed buildings using energy dissipation systems. In: Proceedings of the 2012 structures congress, Chicago, pp 1790–1801

Part IV
Resilience-Based Design

Chapter 10
Introduction to Resilience-Based Design (RBD)

Gian Paolo Cimellaro, C. Renschler, and M. Bruneau

Abstract The chapter is presenting a holistic framework for defining and measuring disaster resilience for a community at various scales. Seven dimensions characterizing community functionality have been identified and are represented by the acronym PEOPLES: Population and Demographics, Environmental/Ecosystem, Organized Governmental Services, Physical Infrastructure, Lifestyle and Community Competence, Economic Development, and Social-Cultural Capital. The proposed framework provides the basis for development of quantitative and qualitative models that measure continuously the functionality and resilience of communities against extreme events or disasters in any or a combination of the above-mentioned dimensions. Over the longer term, this framework will enable the development of geospatial and temporal decision-support software tools that help planners and other key decision makers and stakeholders to assess and enhance the resilience of their communities.

G.P. Cimellaro, Ph.D., P.E. (✉)
Department of Structural, Geotechnical & Building Engineering (DISEG),
Politecnico di Torino, 10129 Turin, Italy
e-mail: gianpaolo.cimellaro@polito.it

C. Renschler
Department of Geography, University at Buffalo (SUNY), 116 Wilkeson Quad,
Buffalo, NY 14261, USA
e-mail: rensch@buffalo.edu

M. Bruneau
Department of Civil, Structural and Environmental Engineering, University at Buffalo (SUNY),
130 Ketter Hall, Buffalo, NY 14260, USA
e-mail: bruneau@buffalo.edu

G.P. Cimellaro et al. (eds.), *Computational Methods, Seismic Protection, Hybrid Testing and Resilience in Earthquake Engineering*, Geotechnical, Geological and Earthquake Engineering 33, DOI 10.1007/978-3-319-06394-2_10,

© Springer International Publishing Switzerland 2015

10.1 Introduction

Over the past years, the concept of resilience has gained attention recognizing the fact that not all threats or disasters can be averted. In fact, communities around the world are turning their attention to efforts and ways that can enhance their resilience against extreme events of any kind. Resilience is becoming increasingly important for modern societies as States start accepting the fact that they cannot prevent every risk from being realized, but rather they must learn to adapt and manage risks in a way that minimizes impact on human and other systems.

While studies on the disaster resilience of technical systems have been undertaken for quite some time (Chang and Shinozuka 2004), the societal aspects and the inclusion of various and multiple types of extreme events are new developments. In this regard, communities around the world are increasingly debating ways to enhance their resilience.

In earlier work by Bruneau et al. (2003), resilience was defined including technical, organizational, economic, and social aspects and with four main properties of robustness, redundancy, resourcefulness, and rapidity. The quantification and evaluation of disaster resilience was based on non-dimensional analytical functions related to the variations of functionality during a "period of interest," including the losses in the disaster and the recovery path (Cimellaro et al. 2005, 2006, 2009). This evolution over time, including recovery, differentiates the resilience approach from other approaches addressing only loss estimation and its momentary effects.

The objective of this chapter is to describe a holistic framework for defining and measuring disaster resilience for a community at various spatial scales (local, regional etc.), minimizing all the possible consequences and reaching as soon as possible the initial conditions again. Seven dimensions characterizing community functionality have been identified and are represented by the acronym PEOPLES (Renschler et al. 2010a). A new more general design methodology is proposed called "Resilience-Based Design" (RBD) which can be considered as an extension of Performance-Based Design which is only a part of the total "design effort". The goal of RBD is to make individual structures and communities as "Resilient" as possible, developing technologies and actions that allows each structure and/or community to regain its function as promptly as possible. Over the longer term, this framework will enable the development of geospatial and temporal decision-support software tools that help planners and other key decision makers and stakeholders to assess and enhance the resilience of their communities.

10.2 Development of PBEE

In order to understand RBD, this paragraph summarizes in the bullet list below the steps of Performance-Based design in US in the last two decades:

- **1990s**, Performance-based earthquake engineering;
- 1997, First-generation of Performance-based Earthquake Engineering (FEMA 273: NEHRP Guidelines for the Seismic Rehabilitation of Buildings);

- 2000, Second-generation Performance-Based Earthquake Engineering (FEMA 356: Prestandard and Commentary for the Seismic Rehabilitation of Buildings);
- 2006, Next generation Performance-based Earthquake Engineering (FEMA 445: Next generation Performance-Based Seismic Design guidelines);

The main equation of PBD is based on the total probability theorem de-aggregating the problem into several interim probabilistic models (namely seismic hazard, demand, capacity and loss models) and it is defined as follows (Cornell and Krawinkler 2000)

$$\underbrace{\lambda\left(dv < DV\right)}_{Seismic\ \ risk} = \iiint \underbrace{G\left(dv|\,dm\right)}_{Loss\ Analysis}\underbrace{dG\left(dm|\,edp\right)}_{Damage\ Analysis}\underbrace{dG\left(edp|\,im\right)}_{Response\ Analysis}\underbrace{|d\lambda(im)|}_{PSHA} \quad (10.1)$$

where im = intensity measure (e.g. $Sa(T_1)$, epsilon, $S_{dinelastic}$, duration etc.); dm = damage measure (e.g. physical condition and consequences/ramifications); edp = engineering demand parameters (e.g. drift ratio (peak, residual), acceleration, local indices etc.); dv = decision variable (e.g. loss, functionality, downtime, casualties etc.); $\lambda(dv)$ = mean annual frequency of a decision variable (dv); $G(a/b)$ is the probability of exceedance $a > a_0$ given b. Each component of Eq. (10.1) need to be determined statistically.

A very schematic step-by-step procedure of the PEER methodology is the following (Yang et al. 2009):

1. *Define structural and nonstructural component performance groups (PGs)*;
2. *Conduct seismic hazard analysis and ground motion selection*;
3. *Evaluate the response of the building*;
4. *Generate additional correlated Engineering Demand Parameters (EDP) vectors*;
5. *Compute the total repair cost*;
6. *Compute different expressions of the total repair cost data*;

Currently the method has been implemented in ATC-58 and ATC-63, but also in other countries like in China where PBEE has been added to the new version of the code "Seismic Design for Building Structures" (GB50011-2010), to design tall buildings and innovative systems, while majority of buildings could be still designed with traditional RSA.

10.3 Limitations of PBEE

Although this methodology is rapidly spreading, there are parts that the PBEE does not cover such as:

1. The portfolio assessment;
2. Community assessment;

The concept of Performance-Based Design/Engineering can be applied to describe the behaviour of a single building or structure, but the performance of an individual structure is not governed by its own performance, but interacts heavily with the performance of other entities within the same community. A clear example of these interdependencies between the building and the community is for example a hospital, which it will not be able to perform without electricity and water even if the structure has no structural damage. Another example of the limitations of PBD is given by 2009 L'Aquila earthquake (Cimellaro et al. 2010c), during which the small town of Castelnuovo was completely destroyed, except a single housing unit that was standing after the earthquake and suffered minor damage. According to PBD the building is ok, but in RBD the housing unit would be not operational, because it is not able to interact with other entities inside the same community.

10.4 Resilience-Based Design

Today, designers and engineers approach a structure as if it stands alone, without considering the interaction with the community, which instead should be considered as an integrated part of the design process. There is now a new fundamental way of looking at all the problem. The building is not considered alone, but as a group of buildings using the "Portfolio Approach" which will allow regional loss analysis. So it will be moved from housing units to housing blocks.

This concept is borrowed from the financial industry, where Modern Portfolio Theory (MPT) was developed in the 1950s through the early 1970s and was considered an important advance in the mathematical modelling of finance. MPT is defined as a theory of investment which attempts to minimize risk for a given level of expected return (performance), by carefully choosing the proportions of various assets. MPT is a mathematical formulation of the concept of diversification in investing, with the aim of selecting a collection of investment assets that has collectively lower risk than any individual asset.

Analogously the concept of diversification can be applied in the field of disaster resilience, where diversification in retrofit of different buildings in a given region can increase resilience collectively more than any individual retrofit.

MPT models an asset's return as a normally distributed function, defines risk as the standard deviation of return, and models a portfolio as a weighted combination of assets so that the return of a portfolio is the weighted combination of the assets' returns. Similarly the risk in RBD can be defined as the standard deviation of the performances of each housing unit; therefore the performances of a given community can be defined as the weight combination of the performance of each housing unit. It is important to mention that community is a complex system where losses as well as the recovery process are coupled dimensions which involve several parameters which are not only engineering parameters such as drift and accelerations, but also other parameters such as socio-economic gender, age of the population etc. All these parameters are used to define the recovery that becomes part of the design process in RBD and therefore should be planned upfront.

10.4.1 Defining Resilience

The concept of resilience does not have a unique definition, because of its broad utilization in the field of ecology, social science, economy, and engineering with different meanings and implications.

As Klein et al. stated (2003), the root of the term has to be found in the Latin word '*resilio*' that literary means 'to jump back'. The field, in which it was originally used, first, is still contested, however, it has been claimed that the study of resilience evolved from the disciplines of psychology and psychiatry in the 1940s, and it is mainly accredited to Garmezy (1973).

The concept of resilience was originally established in the field of ecology by Holling (1973) who stated that for ecological systems resilience is "a measure of the persistence of systems and of their ability to absorb change and disturbance and still maintain the same relationships between populations or state variables. Stability represents the ability of a system to return to an equilibrium state after a temporary disturbance; the more rapidly it returns to equilibrium and the less it fluctuates, the more stable it would be".

The researches in resilience have forced to study it deeper and in a wider way. An extended literature review has been elaborated about resilience for years (Table 10.1), each contribution has added new nuances. Primarily resilience has been defined in context to the speed of systems to go towards equilibrium (Adger 2000), capability to cope and bounce back (Wildavsky 1991), ability to adapt to new situations (Comfort 1999), be inherently strong and flexible and adaptive (Tierney and Bruneau 2007), ability to withstand external impacts and recover with least outside interferences (Mileti 1999).

After the original definition of resilience in ecological systems, the word expanded its meaning to *engineering*, *social* and *economical* fields.

In *engineering*, resilience is defined as the capability of a system to maintain its functions and structure in the face of internal and external change and to degrade gracefully when it must (Allenby and Fink 2005). The main difference in defining and understanding resilience arises between the engineering approach that resilient recovery occurs by moving towards the previous stable state (Bruneau et al. 2003), and the ecological approach that resilience is developed to move towards a different system state (Handmer and Dovers 1996).

Social resilience, explained by Adger (2000), is the ability of groups or communities to cope with external stresses and disturbances as a result of social, political, and environmental change.

Economic resilience was first defined by Rose and Liao (2005) as the inherent ability and adaptive response that enables firms and regions to avoid maximum potential losses. It has mainly been studied in context to seismic response and recovery (Tierney 1997; Bruneau et al. 2003), community behavior (Chang and Shinozuka 2004) and disaster hazard analysis (Rose 2004a), among others (Rose 2009). From the literature described above it appears that even though there are different opinions in defining resilience, there is some consensus in the measurement

Table 10.1 Literature review about resilience definitions

Author	Definition
Holling (1973)	Ecological systems resilience is a measure of the persistence of systems and of their ability to absorb change and disturbance and still maintain the same relationships between populations or state variables
Wildavsky (1991)	Resilience is the capacity to cope with unanticipated dangers after they have become manifest, learning to bounce back
Horne and Orr (1998)	Resilience is the ability of a system to withstand stresses of 'environmental loading'... [it is] a fundamental quality found in individuals, groups, organizations, and systems as a whole
Haimes et al. (1998)	Resilience is the ability of system to return to its optimal condition in a short period of time. Considering resilience one of four strategies for hardening a system, together with security, redundancy and robustness
Mileti (1999)	Local resiliency with regard to disasters means that a locale is able to withstand an extreme natural event without suffering devastating losses, damage, diminished productivity, or quality of life and without a large amount of assistance from outside the community
Comfort (1999)	Resilience is the capacity to adapt existing resources and skills to new situations and operating conditions
Adger (2000)	Social resilience is the ability of groups or communities to cope with external stresses and disturbances because of social, political, and environmental change
Gunderson et al. (2002)	Engineering resilience [...] is the speed of return to the steady state following a perturbation [...] ecological resilience [...] is measured by the magnitude of disturbance that can be absorbed before the system is restructured....
Fiksel (2003)	Resilience is the essence of sustainability [...] the ability to resist disorder
Bruneau et al. (2003)	Resilience is defined in terms of three stages: the ability of a system to reduce the probability of an adverse event, to absorb the shock if the adverse event occurs, and to quickly re-establish normal operating conditions. So resilience thus encompasses the four characteristics of robustness, redundancy, resourcefulness, and rapidity. Are considered four types of resilience: technical; organizational; economic; and social
Allenby and Fink (2005)	Resiliency is defined as the capability of a system to maintain its functions and structure in the face of internal and external change and to degrade gracefully when it must.
Rose and Liao (2005)	Regional economic resilience is the inherent ability and adaptive response that enables firms and regions to avoid maximum potential losses
Hollnagel (2006)	Resilience is defined as the intrinsic ability of an organization (system) to maintain or regain a dynamically stable state, which allows it to continue operations after a major mishap and/or in the presence of a continuous stress
Manyena (2006)	Evaluating all the possible definitions provided from the 1990s to nowadays, resilience could be viewed as the intrinsic capacity of a system, community or society predisposed to a shock or stress to adapt and survive by changing its non essential attributes and rebuilding itself

(continued)

Table 10.1 (continued)

Author	Definition
Woods (2006)	Evaluating all the possible definitions provided from the 1990s to nowadays, resilience could be viewed as the intrinsic capacity of a system, community or society predisposed to a shock or stress to adapt and survive by changing its non essential attributes and rebuilding itself
Holmgren (2007)	Resilience is the ability of the system to return to a stable condition after a disruption. Distinguishing robustness and resilience, using robustness to imply that the system will remain (nearly) unchanged even in the face of disruption
Tierney and Bruneau (2007)	Resilience is both the inherent strength and ability to be flexible and adaptable after environmental shocks and disruptive events
DHS (2008)	Resilience is the ability of systems, infrastructures, government, business, and citizenry to resist, absorb, recover from, or adapt to an adverse occurrence that may cause harm, destruction, or loss of national significance
Haimes (2009)	Resilience is defined as the ability of the system to withstand a major disruption within acceptable degradation parameters and to recover within an acceptable time and composite costs and risk.
Vugrin et al. (2010)	Given the occurrence of a particular disruptive event (or set of events), the resilience of a system to that event (or events) is the ability to efficiently reduce both the magnitude and duration of the deviation from targeted system performance levels

of system resilience. Generally, resilience is measured in terms of the amount by which a system is able to avoid maximum impact (static resilience (Rose 2004a)/robustness (McDaniels et al. 2008)) and the speed at which the system recovers from a disruption (dynamic resilience (Rose 2004b)/rapidity (Zobel 2010)).

As the research advances, one realizes that resilience must be studied on a global level and not individually. Bruneau et al. (2003) consider four types of resilience: *technical*; *organizational*; *social*; and *economical*, (TOSE). They note that different measures of resilience are needed to address these different dimensions. Technical and economic, are related to the resilience of physical systems, and organizational and social, are more related to the community affected by the physical systems. *Technical resilience* concerns the ability of a system to function. Some measures of technical resilience for electric power systems are the percentage of the demand, the ratio between the supply and the demand, the restoration time, the time to full recovery, etc. *Organizational resilience* concerns the ability of the organization(s) to manage the system. For example, measures of organizational resilience could include how well emergency units function, how quickly spare parts are replaced, how quickly repair crews are able to reach the affected components of a system, etc. *Social resilience* concerns how well society copes with the loss of services because of a blackout. For severe blackouts, social resilience can be the most critical dimension of resilience. Finally, *economic resilience* concerns the ability to reduce direct and indirect economic losses. Rose and Liao (2005) note that

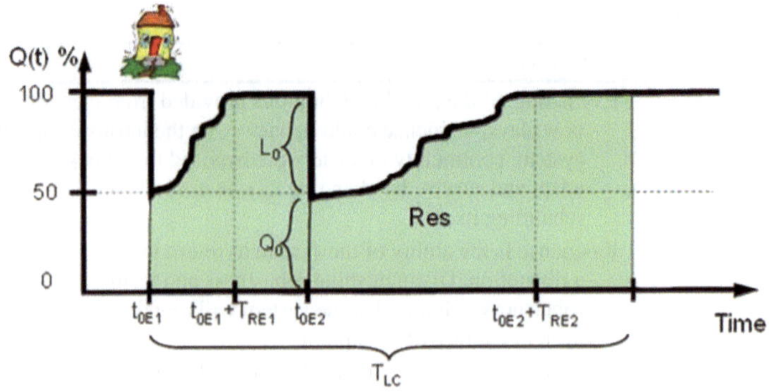

Fig. 10.1 Resilience (Cimellaro et al. 2010a)

direct costs manifest themselves in four ways: lost sales; equipment damage/restart costs; spoilage of variable inputs; and idle labor costs (in addition to the costs of measures to reduce potential losses, such as backup generators and capacity expansion). Indirect costs are multipliers that ripple through the economy, such as impacts on the customers and suppliers of a disrupted firm, decreased consumer spending, decreased investments in the disrupted firm, public-health problems (such as dysfunctional sewage treatment), and economic disorder (looting, etc.)

After the four dimension framework provided by Bruneau et al. (2003) various studies have been carried out, with the goal of practically evaluate the concept of resilience and identify the main units of measurement of it (Miles and Chang 2011). In this paper, is intended to expand the holistic resilience approach with a seven dimension framework known by the acronym PEOPLES: *Population and Demographics*, *Environmental/Ecosystem*, *Organized Governmental Services*, *Physical Infrastructure*, *Lifestyle and Community Competence*, *Economic Development*, *and Social-Cultural Capital*. The seven dimensions are used to characterize the community functionality for defining and measuring disaster resilience at various scales.

10.4.2 Analytical Formulation

The resilience index is illustrated graphically in Fig. 10.1 as the normalized shaded area underneath the functionality function of a system *Q*(*t*). Analytically, *Resilience* is defined as

$$R\left(\vec{r}\right) = \int_{t_{OE}}^{t_{OE}+T_{LC}} Q_{TOT}(t)/T_{LC}\,dt \tag{10.2}$$

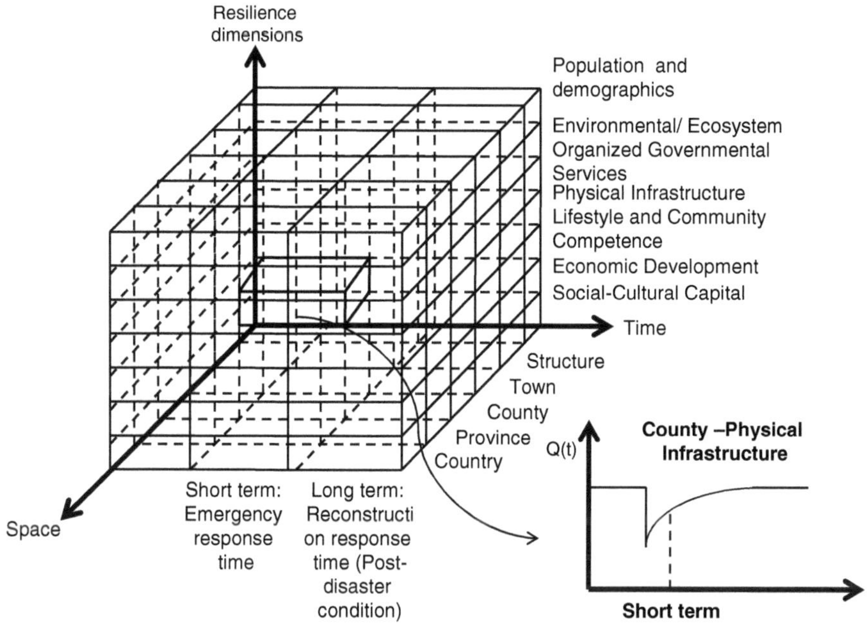

Fig. 10.2 Spatial and temporal dimension of Resilience-Based design (*RBD*) using PEOPLES approach

where $Q_{TOT}(t)$ is the global functionality of the region considered which will be described in the next paragraph; T_{LC} is the control time of the period of interest; $\vec{r}$ is a position vector defining the position P in the selected region where the resilience index is evaluated (Cimellaro et al. 2009, 2010a, b). The community functionality is the combination of all functionalities related to different facilities, lifelines, etc.

10.4.3 Spatial vs. Temporal Scale of Community Resilience

Resilience can be considered as a dynamic quantity that changes over time and across space. It can be applied to engineering, economic, social, and institutional infrastructure, and it can be used for various geographic scales.

The first step to quantify the resilience index (R) is to define the *spatial scale* (e.g. building, structure, community, city, region, etc.) of the problem of interest (Fig. 10.2). It is important to mention that the entire recovery process is affected by the *spatial scale* of the disaster. Huge disasters will take longer recovery process. The *spatial scale* will also be used to define the performance measures for the global functionality of the system. The second step is to define the *temporal scale* (short term emergency response, long term reconstruction phase, midterm reconstruction phase, etc.) of the problem of interest. The selection of the control period T_{LC} will affect the R index, therefore when comparing different scenarios the same control period should be considered.

10.5 The Seven Dimensions of Community Resilience: The PEOPLES Framework

In order to emphasize the primary role of the human system in community sustainability, the acronym "PEOPLES" (Renschler et al. 2010a, b, 2011) has been adopted to describe a framework that is built on and expands previous research at the Multidisciplinary Center of Earthquake Engineering Research (MCEER). This framework linked several previously identified resilience characteristics (technical, organizational, societal, and economic) and resilience attributes (r4: *robustness, redundancy, resourcefulness, and rapidity*) (Bruneau et al. 2003; Bruneau and Reinhorn 2007; Cimellaro et al. 2010b). These are the four attributes along which resilience can be improved. Further details about the description of these attributes can be found in Cimellaro et al. (2010b, c).

PEOPLES incorporates MCEER's definitions of service functionality, and its components (assets, services, demographics) and parameters influencing resilience.

The seven dimensions of the PEOPLES framework are the following:

1. **P**opulation and demographics;
2. **E**nvironment/ecosystem;
3. **O**rganized government services;
4. **P**hysical infrastructure;
5. **L**ifestyle and community competence;
6. **E**conomic development;
7. **S**ocial-cultural capital

In Table 10.2 is shown the complete list of components and sub-components of the "PEOPLES Framework". The dimensions will be explained in the next points but further details about the description of each one can be found in Renschler et al. (2010a, b, 2011).

10.5.1 Population and Demographics

The first dimension *Population and demographics* is used to describe and differentiate communities using for example the *median income* and *age distribution* which might be critical for understanding its economic, health and potential resilience. One measure of the functionality of this dimension (Q_p) can be quantified for example by using the social vulnerability index (SoVI) proposed by Cutter (1996). Social

Table 10.2 Complete list of components and subcomponents of PEOPLES framework

(1) Population and demographics			
(a) Distribution/density	(b) Composition	(c) Socio-economic status	
(i) Urban	(i) Age	(i) Educational attainment	(iv) Home ownership
(ii) Suburban	(ii) Gender	(ii) Income	(v) Housing vacancies
(iii) Rural	(iii) Immigrant status	(iii) Poverty	(vi) Occupation
(iv) Wildland	(iv) Race/ethnicity		
(2) Environmental/ecosystem			
(a) Water quality/quantity	(b) Air quality	(c) Soil quality	(d) Biodiversity
(e) Biomass (vegetation)	(f) Other natural resources		
(3) Organized governmental services			
(a) Executive/administrative		(b) Judicial	(c) Legal/security
(i) Emergency response and	(ii) Health and hygiene		
(4) Physical infrastructure			
(a) Facilities		(b) Lifelines	
(i) Residential		(i) Communications	
(1) Housing units		(1) Internet (2) Phones (3) TV (4) Radio (5) Postal	
(2) Shelters		(ii) Health care	
(ii) Commercial		(1) Acute care (2) Long-term acute care (4) Psychiatric	
(1) Distribution facilities	(3) Manufacturing facilities	(3) Primary care (5) Specialty	
(2) Hotels – accommodations	(4) Office buildings	(iii) Food supply	
(iii) Cultural		(iv) Utilities	
(1) Entertainment venues	(4) Schools	(1) Electrical (2) Fuel/gas/energy (3) Waste	
(2) Museums	(5) Sports/recreation venues	(v) Transportation	
(3) Religious institutions		(1) Aviation (2) Bridges (3) Highways	
		(4) Railways (5) Transit (6) Vehicles (7) Waterways	
(5) Lifestyle and community competence			
(a) Collective action and decision making		(b) Collective efficacy and empowerment	(c) Quality of life
(i) Conflict resolution	(ii) Self-organization		

(continued)

Table 10.2 (continued)

(6) Economic development			
(a) Financial services	(b) Industry – employment – services		(c) Industry – production
(i) Asset base of financial institutions	(i) Agriculture	(x) Number of corporate headquarters	(i) Food supply
(ii) Checking account balances (personal and commercial)	(ii) Construction	(xi) Other business services	(ii) Manufacturing
	(iii) Education and health services	(xii) Professional and business services	
(iii) Consumer price index	(iv) Finance, insurance and real estate	(1) Employment services	
(iv) Insurance		(a) Flexibilities	
(v) Number and average amount of loans	(v) Fortune 1000	(b) Opportunities	
	(vi) Fortune 500	(c) Placement	
(vi) Number of bank and credit union members	(vii) Information, professional business, other	(2) Transport and utilities	
(vii) Number of banks and credit unions	(viii) Leisure and hospitality	(3) Wholesale and retail	
	(ix) Manufacturing		
(viii) Savings account balances (personal and commercial)			
(ix) Stock market			
(7) Social/cultural capital			
(a) Child and elderly services	(b) Commercial centers	(c) Community participation	(d) Cultural and heritage services
(e) Education services	(f) Non-profit organizations	(g) Place attachment	

vulnerability is defined as the inability of people, organizations, and societies to withstand adverse impacts from multiple stressors to which they are exposed. These impacts are due in part to characteristics inherent in social interactions, institutions, and systems of cultural values. Social vulnerability is a pre-existing condition of the community that affects the society's ability to prepare for and recover from a disruptive event.

Resilience focuses on the quality of life of the people at risk and develops opportunities to enhance a better outcome, while *vulnerability* places stress on the production of nature to resist the natural hazard. Manyena (2006) evaluates all the possible definitions provided from the 1990s up until the present, and compares the concept of *resilience* as the opposite of *vulnerability*.

This dimension of vulnerability can be measured using a social index that describes the socioeconomic status, the composition of the population (elderly and children), development density, rural agriculture, race, gender, ethnicity, infrastructure employment, and county debt/revenue. The social index described is based on Cutter's Hazards-of-Place Model of Vulnerability framework that integrates exposure to hazards with the social conditions that make people vulnerable to them (Cutter 1996; Cutter and Mitchell 2000). High SoVI indicates high vulnerability, and conversely, low SoVI indicates low vulnerability. Analytically, functionality of population can be given as follow:

$$Q_P(\mathbf{r}, t) = 1/(f_1 + f_2 + f_3 + f_4 + f_5 + f_6 + f_7 + f_8 + f_9 + f_{10} + f_{11}) \tag{10.3}$$

where $f_1, f_2, \ldots, f_n$ are the 11 independent factors considered. Among the 11 independent factors are *socioeconomic status*, *elderly and children*, *development density*, *rural agriculture*, *race*, *gender*, *ethnicity*, *infrastructure employment*, *and county debt/revenue*. Additionally, qualitative and quantitative measures about population and demographics from the US Census database are an essential component for this dimension of the PEOPLES Resilience Framework. Key indicators include educational attainment, marital status, annual income, age, gender, race/ethnicity distribution, and other data that describe and differentiate the focal population.

10.5.2 Environmental/Ecosystem

The *Environmental/Ecosystem* dimension is typically measured by the amount of disturbance an ecosystem can absorb without drastically altering its functions, processes, and structures, or by the ability of an ecosystem to cope with disturbance.

In the context of the *PEOPLES Resilience Framework*, environmental and ecosystem resources serve as indicators for measuring the ability of the ecological system to return to or near its pre-event state. One such indicator is the Normalized Difference Vegetation Index (NDVI), which is calculated from satellite-derived remote sensing imagery that analyzes the density of green vegetation across a region. NDVI can be used in the framework as a proxy for ecosystem productivity and is calculated using the red (Red) and near infrared (NIR) absorption bands:

$$\text{NDVI} = (\text{NIR}-\text{Red}) / (\text{NIR} + \text{Red}) \tag{10.4}$$

NDVI correlates strongly with above-ground net primary productivity (NPP) (Pettorelli et al. 2005; Olofsson et al. 2007; Prince 1991), which measures biomass accumulation and can be an indicator of ecosystem resilience. Simoniello et al. (2008) characterized the resilience of Italian landscapes using a time series to calculate NDVI trends, and Diaz-Delgado et al. (2002) used NDVI values derived from Landsat imagery to monitor vegetation recovery after fire disturbance.

Building on previous research, the *PEOPLES Resilience Framework* quantifies a portion of ecological resilience through a comparison of stable-state NDVI trends to post-disturbance NDVI trends to determine differences in ecosystem productivity across spatial-temporal scales. NDVI is applicable for quantifying ecosystem structure following disturbances such as fire, flooding, and hurricanes. In other types of disasters such as terrorist attacks or blizzards, vegetation density and ecosystem structure may not be altered. In these instances, ecological resilience quantification through NDVI would be negligible and other indicators would be more relevant. As with the other dimensions, ecological resilience is the integration of all key indicators that include air, water and soil quality, biodiversity, and other natural resources.

10.5.3 Organized Governmental Services

Organized governmental services dimension include traditional legal and security services such as police, emergency, and fire departments and increasingly, the military. In this dimension, are also included the services provided by public health and hygiene departments as well as cultural heritage departments. Each of these organized government services plays a key role in sustaining communities both before and after extreme events. A good example of the necessity of a well-functioning government may be seen in the devastating January 12, 2010 earthquake in Haiti. In the aftermath, the news media reported a lack of government services and orderly control, and a general perception that the government is not in a position to help its people. In contrast, the Darfield earthquake in New Zealand was followed by quick response on the part of local, territorial, and national government services.

Spontaneous helping behavior, convergence, mass volunteering, and emergent groups are sources of resilience, in that they infuse resources and creativity into disaster response activities (Drabek and McEntire 2002). At the level of organizations and networks, organizational responses during crisis are most likely to be effective – and resilient – when they successfully blend discipline and agility (Harrald 2006). Pre-existing plans, training, exercises, mutual aid agreements, and other concepts of operations help ensure disciplined and appropriate responses, but they do so not because they encourage the playing out of pre-determined scripts but rather because they facilitate collective sense-making and inspire action toward shared goals (Weick 1995; Weick et al. 2005). Flexibility, adaptability, and improvisation among responding entities make their own distinctive contributions to resilience. Organizational expansion, extension, and emergence are key bases of resilient disaster responses (Sutton and Tierney 2006).

The concept of collaborative emergency management seeks to engage all critical community sectors in preparing for and responding to disasters, including local elected and appointed officials; subject matter experts; community-based, faith based and other non-governmental organizations, the general public, including both community members that belong to groups such as community emergency response teams and volunteers; the private sector and business networks; and the mass media. Collaborative management, as opposed to top-down direction, is another characteristic of resilient systems. Hierarchies tend to stand in the way of upward information flow, the form of communication that is most essential during disasters. Less hierarchical forms of organization work best in all types of turbulent environments, including disasters, in part because they encourage a free flow of ideas, but also because flatter organizations and decentralized networks are more nimble in responding to those environments.

Key indicators for this dimension include the number of available response units and their capacity. Population and demographic numbers would be used to normalize the number and capacity of these services. In addition to assessing the availability of government services in terms of personnel and equipment, this dimension also includes an evaluation of emergency preparedness planning. For example, surveys may reveal the extent to which organized government services have developed memoranda of understanding (MOUs) and other types of mutual aid agreements, and the extent to which various organized government services participate in emergency and evacuation drills and table-top exercises (Tierney 2009).

10.5.4 Physical Infrastructure

The *physical infrastructure dimension* focuses on a community's built environment. It incorporates both facilities and lifelines while different performance indicators are available in literature.

Within the category of facilities, we include housing, commercial facilities, and cultural facilities. Within the category of lifelines, we include food supply, health care, utilities, transportation (Arcidiacono et al. 2012), and communication networks (Scura et al. 2013; Vallero and Cimellaro 2014). Lifelines are those essential utility and transportation systems that serve communities across all jurisdictions and locales. Lifelines are thus components of the nation's critical infrastructure, which also includes medical (Cimellaro et al. 2010a), financial, and other infrastructure systems that create the fabric of modern society. For clarity, lifeline infrastructures are simply called in short *lifelines* in this report. Lifelines include: (a) energy utilities and companies (electric power and natural gas (Cimellaro et al. 2013) and liquid fuel pipelines); (b) transportation systems (roads and highways, railroads, airports, and seaports); (c) water, storm-water, and sewerage; (d) communication systems; and (e) health care facilities (Cimellaro et al. 2011) (hospitals, emergency facilities, etc.), most distributed in well linked networks.

Next to impacts on people, the physical infrastructure is often the most compelling "story" in the immediate aftermath of a disaster, as organized government services work to restore needed utilities and clear roadways of structural and other debris. After people had been evacuated from New Orleans after Hurricane Katrina in 2005, people focused on the physical infrastructure. Everywhere one looked, one saw destroyed houses, commercial buildings, and cultural and other critical facilities such as churches, schools, and hospitals. Photographs of destruction are used to communicate the devastating effects of the hurricane and subsequent flooding to the world outside New Orleans.

Without water and electricity, critical facilities such as hospitals cannot perform effectively their primary functions. Inaccessible roads make surface transportation impossible, creating an obstacle for supply chain management and efficient movement. When streets and buildings are cordoned off because of damage, businesses may be open, but will not be "in business." Even when businesses relocate for the short-term due to damage to facilities, customers may not find the businesses. Damaged schools shake a community's confidence in itself to overcome disasters and recover.

In terms of housing, key indicators may include proportion of housing stock not rated as substandard or hazardous and vacancy rates for rental housing (Tierney 2009). In terms of communication networks, key indicators may include adequacy (or sufficiency) of procedures for communicating with the public and addressing the public's need for accurate information following disasters, adequacy of linkages between official and unofficial information sources, and adequacy of ties between emergency management entities and mass media serving diverse populations (Tierney 2009).

In the aftermath of a disaster, the restoration and recovery of physical infrastructure remain by-and-large technical issues; however, those are tightly related and often driven by organizations, economics and socio-political events. Resilience must consider these interactive dimensions in order to be relevant to the system; therefore, interdependencies among different lifelines should be taken in account during the analysis (Cimellaro et al. 2014a).

10.5.5 Lifestyle and Community Competence

Lifestyle Community competence dimension deals with community action, critical reflection and problem solving skills, flexibility and creativity, collective efficacy, empowerment, and political partnerships (Norris et al. 2008).

This dimension reflects the reality that community resilience is not simply a passive "bouncing back" to pre-disaster conditions (Brown and Kulig 1996) but rather a concerted and active effort that relies on peoples' ability to creatively imagine a new future and then take the requisite steps to achieve that desired future. It captures both the raw *abilities* of the community (e.g., ability to develop multifaceted solutions to complex problems, ability to engage in meaningful political networks) and the community's *perceptions* of its ability to effect positive change. Communities that collectively believe that they can rebuild, restructure, and revive themselves are more likely to be persistent in the face of environmental, governmental, and other obstacles.

Quality of life surveys often reveal whether members of a given community are committed to that community and willing to engage in the activities necessary to sustain the community, regardless of whether a disaster strikes. Less soft general indicators of community competence may include measures of migration, measures of citizen involvement in politics, and others. Disaster-specific indicators may include the comprehensiveness of community warning plans and procedures, and the extensiveness of citizen and organizational disaster training programs (Tierney 2009).

10.5.6 Economic Development

For many years, disaster loss estimation has focused on structural damage, while all other types of impacts (economic, sociological, psychological, etc.) were termed "indirect" or "secondary" losses. By the mid-1990s, there was a growing appreciation of the role of business interruption losses, which refer to the reduction in the flow of goods and services produced by property (capital stock). This stock/flow distinction is a basic concept in economics which has been applied in earthquake engineering to distinguish between direct and indirect losses.

Direct structural damage relates to the effects of natural phenomena, such as earthquakes, while *indirect structural damage* is exemplified by ancillary fire caused by ruptured pipelines, frayed electrical wires, etc., and exacerbated by loss of water services. Direct structural damage takes place during the extreme event, while the indirect structural damage takes place in fairly short time span after the event.

Direct Business interruption refers to the immediate reduction or cessation of economic production in a damaged factory or in a factory cut off from at least one of its utility lifelines. *Indirect Business interruption* stems from the "ripple," or "multiplier," effects associated with the supply chain or customer chain of the directly affected business. Business interruptions are time-dependent, because it begins when the extreme event starts and continues until the built environment is repaired and reconstructed to some desired or feasible level (not necessarily pre-disaster status) and a healthy business environment is restored.

The economic impacts of the September 11th attacks on the World Trade Center have been estimated at $25 billion in structural damage and $110 billion in business interruption (Grossi 2009).

One important factor which affected the economic response is resilience, or inherent and adaptive actions that mute the losses by using resources more efficiently and recovering more quickly (Wein and Rose 2011). In fact it is impossible to protect the general population against all natural disasters and terrorist attacks. However, individuals, firms, and government agencies can protect themselves from the negative impacts of business and infrastructure disruptions by implementing various types of resilience actions.

Resilience actions can be identified that are low cost (e.g. production rescheduling, conservation etc.). Both mitigations and resilience actions can significantly result in cost-savings to society as a whole. Indeed, both September 11th and hurricane Katrina were characterized by several forms of resilience, such as business relocation and substitution of imported goods, respectively, that reduced the losses to levels lower than might otherwise have occurred.

A community needs to have access to resources to grow, react to changes and reduce losses. The difference between resilient and non-resilient resources is that the former focus on addressing local needs and are often locally based sources of employment, skills, and finances. There are six points to this dimension of resilience:

1. Employment in the community is diversified beyond a single employer or employment sector;
2. Major employers in the community are locally owned;
3. The community has a strategy for increasing independent local ownership;
4. There is openness to alternative ways of earning a living and economic activity.
5. The community looks outside itself to seek and secure resources (skills, expertise, finance) to address areas of identified weakness;
6. The community is aware of its competitive position in the broader economy.

Economic development dimension includes both the static assessment of a community's current economy (economic activity) and the dynamic assessment of a community's ability to continuously sustain economic growth (economic development).

As described in the RICSA Poverty Project (2010), economic *activity* takes into account the supply of labor for the production of economic goods and services, which includes:

> All production and processing of primary products whether for market, for barter or for own consumption, the production of all other goods for the market and, in the case of households which produce such goods and services for the market, the corresponding production for own consumption.

Economic *development* addresses the future and growth. It addresses a community's efforts to increase its:

> productive capacities ..., in terms of technologies (more efficient tools and machines), technical cultures (knowledge of nature, research and capacity to develop improved technologies), and the physical, technical and organizational capacities and skills of those engaged in production.

Resilient communities are characterized by their involvement in a diverse array of products and services that are both produced in and available to the community. Diversity in production and employment is linked to a community's ability to substitute goods and services and shift employment patterns as the situation demands. The *PEOPLES Resilience Framework* incorporates three illustrative subcategories within this dimension: Industry – Production, Industry – Employment Distribution, and Financial Services. Primary indicators of this dimension include the proportion of the population that is employed within the various industries, and the variability that might characterize a community's industrial employment distribution. This dimension is closely interconnected with the Population and Demographics dimension. For example, key indicators of economic development beyond employment and industry distribution include literacy rates, life expectancy, and poverty rates. Disaster-specific indicators related to economic development include extent of evacuation plans and drills for high-occupancy structures, adequacy of plans for inspecting damaged buildings following disasters, and adequacy of plans for post-disaster commercial reconstruction (Tierney 2009).

10.5.7 Social/Cultural Capital

Social/cultural capital dimension incorporates several subcategories, including education services, child and elderly services, cultural and heritage services, and community participation. Measuring social/cultural capital requires acquisition of tallies, such as the number of members belonging to various civil and community organizations. It also requires surveys of community leaders and their perceptions (e.g., quality of life surveys).

For example, social support underlies many of the services associated with social/cultural capital. It includes both the "helping behaviors within family and friendship networks" and the "relationships between individuals and their larger neighborhoods and communities" (Norris et al. 2008, p. 139). People choose to provide social and cultural services that manifest and extend their sense of community, defined as an attitude of bonding with other members of one's group Norris et al. 2008). They may feel an emotional connection to their neighborhood or city, which may or may not relate to the people who inhabit those places (Manzo and Perkins 2006). For example, after Hurricane Katrina, many displaced residents of New Orleans expressed a strong desire to return home, irrespective of the people they knew or the jobs they once had. It seems likely that people with a strong "place attachment" would be more willing to act in order to help their community bounce back after a disaster, assuming that other essential factors such as employment and housing were available. Citizen participation takes into account the "engagement of community members in formal organizations, including religious congregations, school and resident associations, neighborhood watches, and self-help groups" (Norris et al. 2008, p. 139). Participation in community organizations is a means of demonstrating one's care for one's community. Pragmatically, participation in community organizations is a means for meeting and understanding one's fellow citizens. It increases individuals' circle of influence and perception of control.

Measuring social/cultural capital requires acquisition of tallies, such as the number of members belonging to various civil and community organizations. It also requires surveys of community leaders and their perceptions (e.g., quality of life surveys). Disaster-specific indicators include existence of community plans targeting transportation-disadvantaged populations, adequacy of post-disaster sheltering plans, adequacy of plans for incorporating volunteers and others into official response activities, adequacy of donations management plans, and the community's plans to coordinate across diverse community networks (Tierney 2009).

10.6 General Framework at the Community Level: The Layer's Model

The general framework at the community level is described by the equations below, where for each dimension a performance indicator and/or functionality is defined by combining different functionality dimensions:

$$Q_{TOT}(t) = Q_{TOT}\left(Q_P, Q_{Env}, Q_O, Q_{Ph}, Q_L, Q_{Eco}, Q_S\right) \tag{10.5}$$

where Q_{TOT} = global functionality; and Q_x = functionality of each of the seven dimensions defined above. Within each dimension, functionality is defined as a combination of functionalities of their respective subsystems. For example, the functionality of the physical infrastructure Q_{ph} is defined as follows:

$$Q_{Ph}(t) = Q_{Ph}\left(Q_{Hosp}, Q_{Ele}, Q_{Road}, Q_{Water}, \ldots\right) \tag{10.6}$$

where Q_{hosp} = functionality of health care facilities; Q_{Ele} = functionality of the electric network; Q_{Road} = functionality of the road network; Q_{Water} = functionality of the water network; etc. Once the geographic scale is defined, it is possible to plot the global functionality Q_{TOT} over the region of interest in a contour plot at a given instant of time t, so *time-dependent functionality maps* of the region can be obtained. When also the temporal scale is defined through the control time T_{LC}, then the *resilience contour map* of the region of interest can be plotted (e.g. Fig. 10.3).

The *Resilience contour map* is obtained by integrating functionality maps over time using Eq. (10.7), therefore they will be *time independent*, but they will vary in space from point to point in the selected region. Finally, the community resilience index R_{com} is given by the double integral over time and space as follows:

$$R_{com} = \int_{A_C} R\left(\vec{r}\right)/A_C dr = \int_{A_C} \int_{t_{OE}}^{t_{OE}+T_{LC}} Q_{TOT}(t)/\left(A_C T_{LC}\right) dt dr \tag{10.7}$$

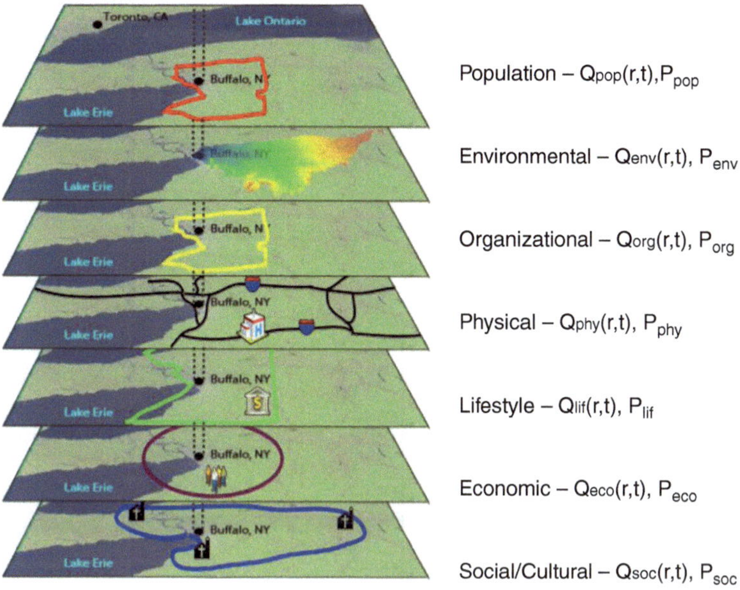

Fig. 10.3 Layer model of PEOPLES framework

where A_c is the total area of the selected region. For each dimension, a contour plot can be determined and combined using a layered approach as shown in Fig. 10.3.

Then a radar graph (Fig. 10.4) can be plotted and the area will determine the final value of the resilience score for the region of interest. This will identify gaps as well as priority actions, which will enter in the decision process.

In summary, a schematic step-by-step procedure of the MCEER methodology described in Fig. 10.5 is the following:

Define extreme event scenarios (e.g. PSHA and ground motion selection);

1. Define the system model;
2. Evaluate the response of the model;
3. Compute performance measures (e.g. losses, recovery time, functionality, resilience);
4. Identify remedial mitigation actions (e.g. advanced technologies) and/or resilience actions (e.g. resourcefulness, redundancy, etc.);

This design approach has analogies with the feedback loop taken from control theory. The same framework can be used for a region as well as a single structure (e.g. hospital). In this case, functionality reduces the functionality of a single hospital Q_{hosp} which can be evaluated for example with the procedure described in Cimellaro et al. (2011), where the waiting time of a patient before receiving assistance is the main parameter of response to measure resilience. The hospital performance is described using a double exponential function, called metamodel which is able to estimate the hospital capacity and the dynamic response in real time incorporating the influence of damage of structural and non-structural components.

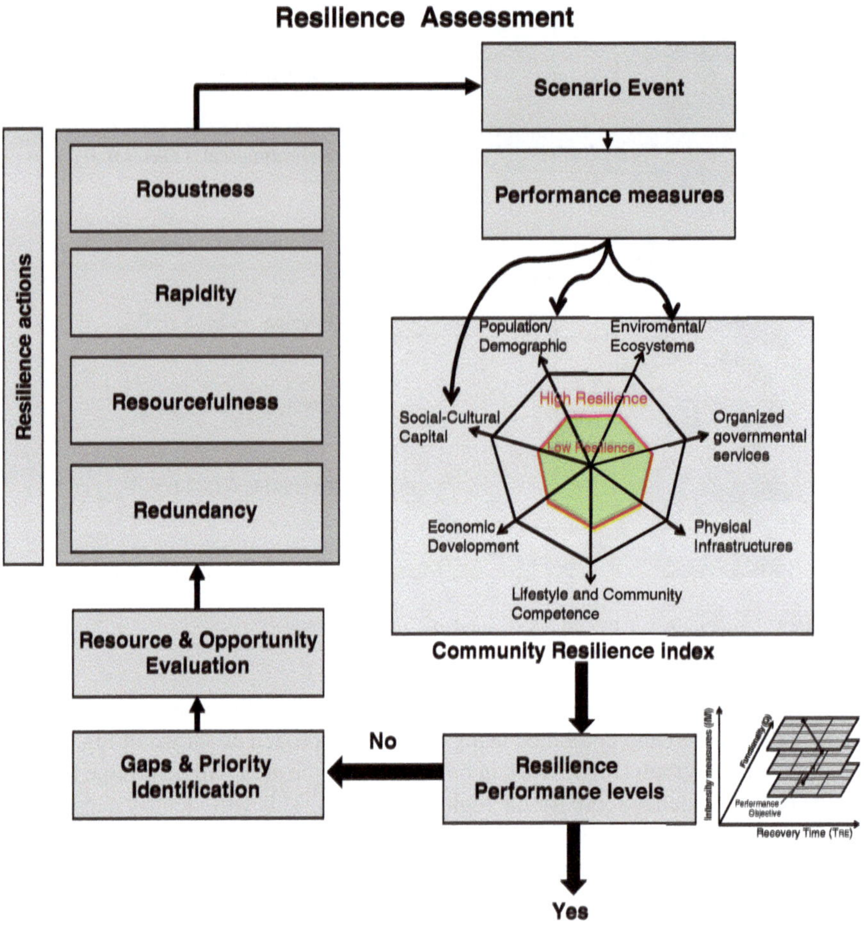

Fig. 10.4 Resilience scoring using PEOPLES methodology

10.7 Recovery Models

In general, the performance measure of a community and a system during transient analysis is a function of time t and other parameters that depend on the type of a community considered. Therefore, at time t after the crisis, functionality is given by

$$Q(t) = f\,(t, x_1, \ldots, x_n) \tag{10.8}$$

where $x_1, \ldots, x_n$ are the parameters involved in describing the recovery model. Several models have been presented in Cimellaro et al. (2010a) to describe the recovery function which can be either ***empirical*** or ***analytical*** depending on the source of data and the type of analysis.

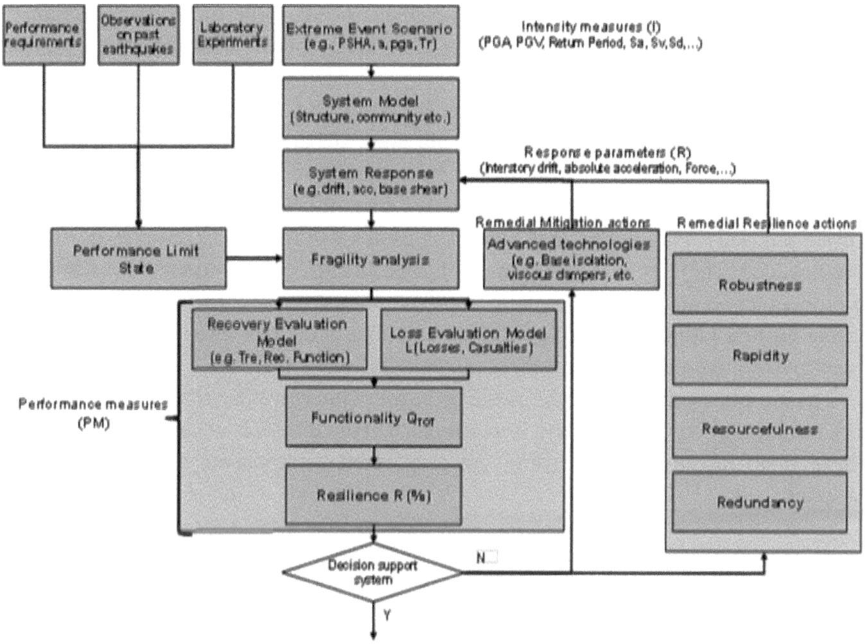

Fig. 10.5 MCEER methodology for Resilience-based design (*RBD*) based on control (feedback loop) approach applied to the Physical infrastructure dimension

Empirical recovery functions are based on test or field data interpretation and engineering judgment. They can be built using the maximum likelihood method based on data reported from past extreme events as well as Monte Carlo simulations of specified community models. Since the complexity of the problem changes case by case, no specific model is presented in this part.

Analytical recovery functions are developed from community response data obtained through analysis of the system using numerical simulations. For example, for the case of earthquake events, they can be obtained from nonlinear time history analysis, response spectral analysis, etc.

Since the recovery process is characterized by uncertainties, the parameters considered in the model are modeled as random variables in order to quantify the uncertainties in the system. These uncertainties can be divided in aleatoric and epistemic uncertainties.

Several models can be fitted to the observed data, and subsequently, model selection can be carried out using as goodness of fit measure, such as the r^2 value. The essential requirement of the analytical recovery models is the simplicity, therefore the model should be selected so that it is easy to fit to real or numerical observation data and the number of parameters involved should be as low as possible. Below are reported five different recovery models which are grouped according to the two control periods (*short term* vs. *long term*). *Long term recovery models* are used when the reconstruction phase needs to be modeled, while *short term recovery models* are used when the emergency phase after the extreme event needs to be focused upon.

Several *long term recovery models* are proposed in Cimellaro et al. (2010a, b, c). They can be grouped according to the number of parameters (one, two, or three parameters). Complex recovery models with more parameters can be proposed, but simpler mathematical models have benefits over more complex ones. They have fewer unknown parameters, and thus it is easier to fit to data (fewer experiments are needed). There is also less chance of "overfitting". With more free parameters, a model can be made to fit any data; however, at best the exercise is little more than curve fitting (with little meaningful understanding gained), while at worst the model may give an overconfidence in its predictive ability.

The simplest recovery model is the *uniform cumulative distribution* (cdf) recovery function (also known as the *linear model*). This model is usually adopted when there is no information regarding the preparedness, resources available, societal response, etc.

$$Q(t) = Q_0 + F\left(t/t_0, t_0 + T_{RE}\right) \cdot \left[Q_R - \left(Q_0 - L_0\right)\right] \tag{10.9}$$

where Q_0 is the initial functionality after the drop; L_0 is the initial total loss of functionality after the drop; Q_R is the residual functionality after the recovery process ends; and $F(t/t_0, t_0 + T_{RE})$ is the uniform cumulative distribution function which is given by

$$F\left(t/t_0, t_0 + T_{RE}\right) = \frac{(t - t_0)}{T_{RE}} I\left(t_0, t_0 + T_{RE}\right) \tag{10.10}$$

where $I(t_0, t_0 + T_{RE})$ is the interval step function. The model is characterized by only one parameter (Fig. 10.6a) which defines the slope of the curve and it represents *rapidity* (Cimellaro et al. 2010a, b, c). The model can also be generalized by dividing the recovery process in several time intervals using a *multilinear model* that is given by

$$Q(t) = Q_i + \sum_i H\left(t - t_i\right) \frac{(t - t_i)}{(t_{i+1} - t_i)} \left(Q_{i+1} - Q_i\right) \tag{10.11}$$

where Q_i is the residual functionality at the step i and Q_{i+1} is the residual functionality at the step $i + 1$; $H(\cdot)$ is the Heaviside step function.

Alternatively, *lognormal cumulative distribution* (cdf) recovery function, can be adopted, having three parameters (L_0, θ, β), and it is given by

$$Q(t) = Q_0 + F\left(t/\theta, \beta\right) \cdot \left[Q_R - \left(Q_0 - L_0\right)\right] \tag{10.12}$$

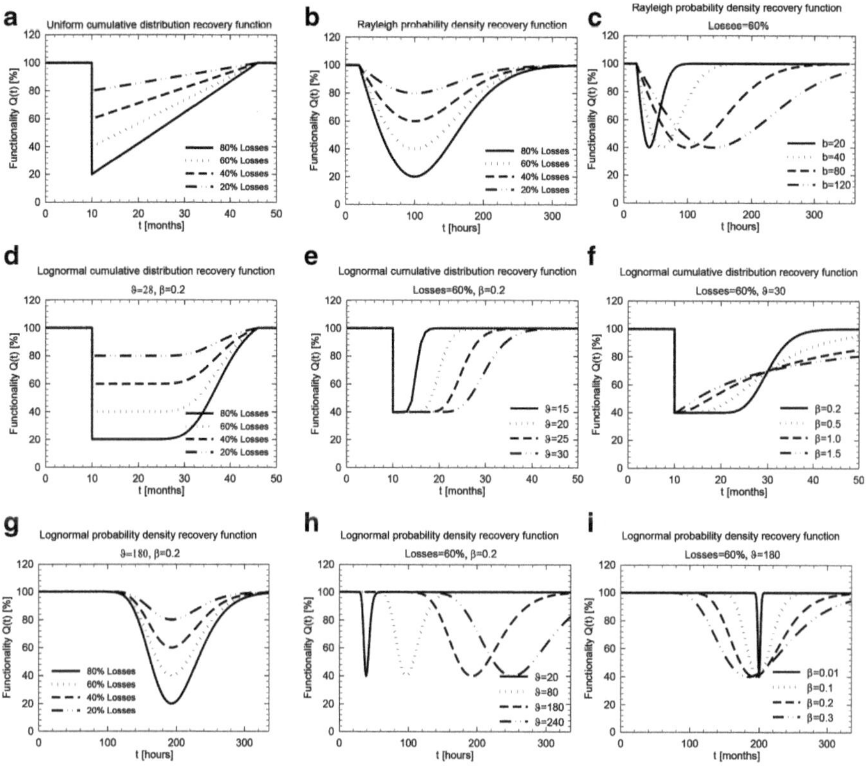

Fig. 10.6 Long term and short term recovery function

where

$$F\left(t/\theta,\beta\right)=\frac{1}{\beta\sqrt{2\pi}}\int_{-\infty}^{t}\frac{e^{-\frac{(\log(x)-\theta)^2}{2\beta^2}}}{x}dx \tag{10.13}$$

This model combines both the exponential recovery model proposed by Kafali and Grigoriu (2005) and the trigonometric recovery model proposed by Chang and Shinozuka (2004). The parameter L_O in Eq. (10.12) can be used to define the initial total loss of functionality after the drop (Fig. 10.6d). The parameter θ can be used to define the time frame (Fig. 10.6e) when the societal response and recovery are driven by lack or limited organization and/or resources. The parameter β defines the rapidity of the recovery process (Fig. 10.6f).

The second group of recovery models is called *short term recovery models* and instead of using *cdf* shape models such as in the long term recovery models, they use the probability density functions (*pdf*) shape models. The simplest recovery model after the linear model proposed in Eq. (10.9) is the Rayleigh probability density function recovery model, and it is defined as

$$Q(t) = 1 - L_0 \frac{f\,(t|\,b)}{\max\,(|\,f\,(t|\,b)|)} \tag{10.14}$$

where

$$f\,(t|\,b) = \frac{t}{b^2} e^{\left(\frac{-t^2}{2b^2}\right)} \tag{10.15}$$

The model is calibrated using two parameters: L_0 is related to the robustness dimension (Fig. 10.6b), while b is related to the rapidity and the delay in the recovery process (Fig. 10.6c). Another model is the *lognormal probability density recovery function* which is given by

$$Q(t) = 1 - L_0 \frac{f\,(t|\,\theta,\beta)}{\max\,(|\,f\,(t|\,\theta,\beta)|)} \tag{10.16}$$

where

$$f\,(t|\,\theta,\beta) = \frac{1}{x\beta\sqrt{2\pi}} e^{-\frac{(\log(x)-\theta)^2}{2\beta^2}} \tag{10.17}$$

The sensitivity of the three parameters on the recovery process is shown in Fig. 10.6g–i.

In the short term emergency response, more complex analytical recovery models are available such as the *metamodel* for describing the organizational performance of a hospital facility (Cimellaro et al. 2010a, b, c). The model is based on a *double exponential function* and its parameters are calibrated based on simulated data obtained by a discrete event simulation model. The *metamodel* is capable of estimating the hospital capacity and dynamic response in real-time incorporating the influence of structural and non-structural damaged components on the entire organizational model. It is important to mention that the constants in all the models presented can be continuously updated as soon as more data are available using a Bayesian approach.

10.8 Uncertainties in RBD

Either a deterministic or probabilistic approach can be used within the PEOPLES framework methodology with preference to the latter approach when a particular level of confidence of achieving performance objective is of interest. Five random variables are involved in the probabilistic description of the resilience index when uncertainties are included and variables are dependent. The joint probability density function of resilience, intensity, response, performance, and recovery is given by the following expressions

$$f_{R,T_{RE},Q,X,I}(r,t_{RE},q,x,i)=\begin{cases}f_{R,T_{RE},Q,X,I}\left(r\middle|t_{RE},q,x,i\right)\cdot f_{R,T_{RE},Q,X,I}(t_{RE}|\,q,x,i)\cdot\\ \cdot f_{Q,X,I}\,(q|\,x,i)\cdot f_{X,I}(x|\,i)\cdot f_{I}(i);\end{cases}\tag{10.18}$$

where I = intensity measures; X = response measures; Q = performance measures; T_{RE} = recovery time measures; R = resilience index; m_r = mean resilience index. The marginal PDF of the resilience index is given by

$$f_R(r)=\int_{t_{RE}}\int_{q}\int_{x}\int_{i} f_{R,T_{RE},Q,X,I}\,(r,t_{RE},q,x,i)\,dt_{RE}\cdot dq\cdot dx\cdot di\tag{10.19}$$

Therefore, the expected value of the resilience index is given by

$$m_r=E\{R\}=\int_{-\infty}^{\infty} r\cdot f_{R,T_{RE},Q,X,I}\,(r,t_{RE},q,x,i)\cdot dr\tag{10.20}$$

10.9 Communicating Risk in RBD

Engineers need to know which measures of risk are most meaningful or relevant to decision makers, and then be able to communicate those risks, and the costs and benefits of mitigation, in concise, credible, meaningful terms. Keller and his colleagues (2006) have shown that when problems are formulated in terms of frequencies rather than probabilities, the perceived threat of the risk is increased.

The probabilistic approach described in previous paragraph is more comprehensive and general, but the information provided to the public (e.g. decision makers, politicians, etc.) should be deterministic (scenario basis or event), because it is more simple and easy to understand. In communicating risk effectively, the public has difficulty thinking in probabilistic terms (Patt and Sohrag 2003). In fact according to Kahneman and Tversky (2000), small probabilities (which are frequently associated with natural hazard events) are often underestimated. According to Samant (personal communication, 2011), “By eliminating probability, which is a confusing concept for a lot of people, the [risk] becomes way more impactful for the average person”. Many authors believe the scenario approach may also impact the emotions associated with an event.

10.10 Definitions of Resilience Performance Levels

The objective of Performance Based Seismic Engineering (PBSE) is to design, construct and maintain facilities with better damage control. A comprehensive document has been prepared by the SEAOC Vision 2000 Committee (1995) that includes interim recommendations. The performance design objectives couple expected or desired performance levels with levels of seismic hazard. Performance-Based design levels focus on the performances a building can hold during the shaking and are associated to engineering demand parameters such as deformations. More recently SPUR (SPUR 2009) which is the San Francisco planning and Urban Research Association introduced other definitions of performance levels for physical infrastructures based on recovery target states which take in account the safety as well as the recovery time. Five performance measures for buildings have been identified:

1. Safe and Operational;
2. Safe and usable during repair;
3. Safe and usable after repair;
4. Safe but not repairable;
5. Unsafe;

In Japan, there was a trial to find the desired performance goals in downtime and bridge performance from public interview. This information will be useful guideline to define the boundaries of the different resilience performance levels (Fig. 10.7).

The new proposed Resilience Performance Levels (RPL) focus on building performance after the earthquake stops, recognizing the importance of the temporal dimension (Recovery time T_{RE}) in the assessment of the RPLs of structures and communities in general. A 2-D performance domain consisting of Performance Levels PL(i, j), defined by the combination of Functionality (index j) and recovery

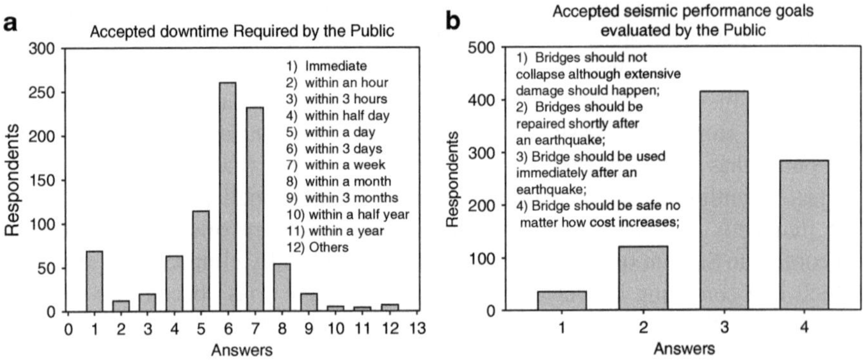

Fig. 10.7 (**a**) Accepted downtime required by the public; (**b**) Accepted seismic performance goals evaluated by the public (Yokohama National University, 2011)

a

	Recovery Time		
Functionality Performance Levels	Short term (Emergency)	Midterm	Long term (Reconstruction)
Fully Operational (Q1)	Basic Objective	Unacceptable	Unacceptable
Operational (Q2)	Essential Objective	Basic Objective	Unacceptable
Partially Operational (Q3)	Critical Objective	Essential Objective	Basic Objective
Near not Operational (Q4)	Not feasible	Critical Objective	Basic Objective

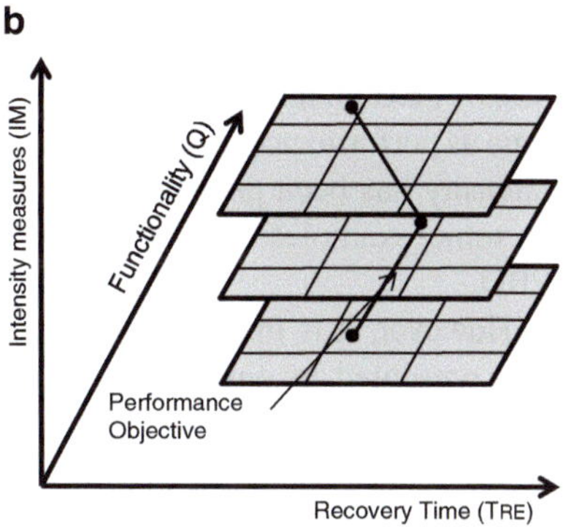

Fig. 10.8 Three-dimensional Resilience Performance objectives matrix for structures, communities, systems etc.

time (index i) is proposed. By accounting for the effect of the temporal dimension, a 3-dimensional performance matrix (Fig. 10.8) can be visualized as a set of predefined joined performance domains ("masks") for different seismic intensity level, IM and different RPLs.

10.11 Conclusions

After all recent extreme events worldwide, the international community became conscious that Resilience is the key to describe earthquake engineering performance.

A new more general design methodology is developed in this chapter called "Resilience-Based Design" (RBD) which can be considered as an extension of PBD which is only a part of the total "design effort". The goal of RBD is to make individual structures and communities as "Resilient" as possible, developing technologies and actions that allows each structure and/or community to regain its function as promptly as possible.

The definition of community resilience combines information from technical and organizational fields, from seismology and earthquake engineering to social science and economy. So it is clear that many assumptions and interpretations are made during the study of seismic resilience, but the final goal is to integrate the information from these fields in a unique function that reach results that are unbiased by uninformed intuitions or preconceived notions of how large or how small the risk is. The goal of this chapter is to provide a quantitative definition of resilience in a rational way through the use of an analytical function that may fit both technical and organizational issues. The fundamental concepts of community resilience are analyzed and a common frame of reference is established. However, even if the performance evaluation of an individual structure is the engineers' goal, the level of performance required should not be determined individually, but on a community basis. A Resilience Performance level matrix combining functionality and recovery time as performance levels and performance objectives at increasing level of seismic intensity has been presented. The resilience-based approach is illustrated in detail in some examples, showing its efficiency and better description with respect to PBD.

However, it is important to mention that the assumptions that are made for the case presented are only representative to illustrate the definitions; for other problems users calculating resilience should focus on the assumptions that most influence the problem at hand.

10.12 Future Research Trends

In order to coordinate future recovery processes and international collaborations, common formats and tools should be applied for rapid resilience assessments, for example the reports from the recent earthquakes should follow the same common format. New open source community software frameworks should be developed through education and training, in order to develop a common research vision to raise awareness of stakeholders worldwide and local.

Acknowledgments The authors would like to thank Dr. Chris Poland for his suggestions and discussions which improved the quality of the chapter. The research leading to these results has also received funding from the European Community's Seventh Framework Programme – Marie Curie International Reintegration Actions – FP7/2007–2013 under the Grant Agreement n° PIRG06-GA-2009-256316 of the project ICRED – Integrated European Disaster Community Resilience.

References

Adger WN (2000) Social and ecological resilience: are they related? Prog Hum Geogr 24(3):347–364

Allenby B, Fink J (2005) Toward inherently secure and resilient societies. Science 309(5737):1034–1036

Arcidiacono V, Cimellaro GP et al (2012) Road network resilience assignment methodology. In: 6th international conference on bridge maintenance, safety and management (IABMAS2012). F. B. D. Frangopol. CRC Press Taylor & Francis Group, Stresa, Lake Maggiore, 8–12 July 2012

Brown D, Kulig J (1996) The concept of resiliency: theoretical lessons from community research. Health Can Soc 4:29–52

Bruneau M, Chang SE et al (2003) A framework to quantitatively assess and enhance the seismic resilience of communities. Earthq Spectra 19(4):733–752

Bruneau M, Reinhorn AM (2007) Exploring the concept of seismic resilience for acute care facilities. Earthq Spectra 23(1):41–62

Chang S, Shinozuka M (2004) Measuring improvements in the disaster resilience of communities. Earthq Spectra 20(3):739–755
Cimellaro GP, Reinhorn AM, Bruneau M (2005) Resilience of a health care facility. In: Proceedings of annual meeting of The Asian Pacific Network of Centers for Earthquake Engineering Research (ANCER), Seogwipo KAL Hotel Jeju, Korea
Cimellaro GP, Reinhorn AM, Bruneau M (2006) Quantification of seismic resilience. In: Proceedings of the 8th national conference of earthquake engineering, vol paper 1094, no., April 2006, San Francisco, pp 18–22
Cimellaro GP, Fumo C, Reinhorn AM, Bruneau M (2009) Quantification of seismic resilience of health care facilities. MCEER technical report-MCEER-09-0009. Multidisciplinary Center for Earthquake Engineering Research, Buffalo
Cimellaro GP, Reinhorn AM, Bruneau M (2010a) Seismic resilience of a hospital system. Struct Infrastruct Eng 6(1–2):127–144
Cimellaro GP, Reinhorn AM, Bruneau M (2010b) Framework for analytical quantification of disaster resilience. Eng Struct 32(11):3639–3649
Cimellaro GP, Christovasilis IP, Reinhorn AM, De Stefano A, Kirova T (2010c) L'Aquila earthquake of April 6th, 2009 in Italy: rebuilding a resilient city to multiple hazard. MCEER technical report – MCEER-10-0010, State University of New York at Buffalo (SUNY), Buffalo
Cimellaro GP, Reinhorn AM et al (2011) Performance-based metamodel for health care facilities. Earthq Eng Struct Dyn 40:1197–1217
Cimellaro GP, Villa O, De Stefano A (2013) Serviceability of natural gas distribution networks after earthquakes. J Earthq Tsunami 7(2):22
Cimellaro GP, Solari D, Bruneau M (2014a) Physical infrastructure interdependency and regional resilience index after the 2011 Tohoku earthquake in Japan. Earthq Eng Struct Dyn. doi:10.1002/eqe.2422
Cimellaro GP, Villa O, Bruneau M (2014b) Resilience-based design of natural gas distribution networks. J Infrastruct Syst ASCE. doi:10.1061/(ASCE)IS.1943-555X.0000204
Comfort LK (1999) Shared risk: complex systems in seismic response. Pergamon, New York
Cornell A, Krawinkler H (2000) Progress and challenges in seismic performance assessment. Peer News 3(No. 2) 1–4
Cutter S (1996) Vulnerability to environmental hazards. Prog Hum Geogr 20(4):529–539
Cutter SL, Mitchell JT (2000) Revealing the vulnerability of people and places: a case study of Georgetown County, South Carolina. Ann Am Geogr 90(4):713–737
Department of Homeland Security (DHS) (2008) DHS risk lexicon, risk steering committee. U.S. Department of Homeland Security, Washington, DC
Díaz-Delgado R, Lloret F et al (2002) Satellite evidence of decreasing resilience in Mediterranean plant communities after recurrent wildfires. Ecology 83(8):2293–2303
Drabek TE, McEntire DA (2002) Emergent phenomena and multiorganizational coordination in disasters: lessons from the research literature. Int J Mass Emergencies Disasters 20:197–224
FEMA (1997) FEMA 273 NEHRP guidelines for seismic rehabilitation of buildings. Federal Emergency Management Agency, Washington, DC
FEMA (2000) FEMA 356 prestandard and commentary for the seismic rehabilitation of buildings. Federal Emergency management Agency, Washington, DC
FEMA (2006) FEMA 445: next generation performance-based seismic design guidelines. Federal Emergency Management Agency, Washington, DC
Fiksel J (2003) Designing resilient, sustainable systems. Environ Sci Technol 37(23):5330–5339
Garmezy N (1973) Competence and adaptation in adult schizophrenic patients and children at risk. In: Dean SR (ed) Schizophrenia: the first ten dean award lectures. MSS Information Corp., New York, pp 163–204
Grossi P (2009) Property damage from the World Trade Center attack. Peace Economics, Peace Science, and Public Policy 15(3):1–18
Gunderson LH, Holling CS et al (2002) Resilience of large-scale resource systems. Island Press, Washington, DC
Haimes YY (2009) On the definition of resilience in systems. Risk Anal 29(4):498–501

Haimes YY, Matalas NC et al (1998) Reducing vulnerability of water supply systems to attack. J Infrastruct Syst 4(4):164–177
Handmer J, Dovers S (1996) A typology of resilience: rethinking institutions for sustainable development. Org Environ 9:482–511
Harrald J (2006) Agility and discipline: critical success factors for disaster response. Ann Am Acad Polit Soc Sci 604:256–272
Holling CS (1973) Resilience and stability of ecological systems. Annu Rev Ecol Syst 4:1–23
Hollnagel E (2006) Resilience – the challenge of the unstable. Ashgate, Hampshire
Holmgren A (2007) A framework for vulnerability assessment of electric power systems. Springer, Berlin/Heidelberg
Horne JF, Orr JE (1998) Assessing behaviors that create resilient organizations. Employ Relat Today 24(4):29–39
Kafali C, Grigoriu M (2005) Rehabilitation decision analysis. In: ICOSSAR'05: proceedings of the ninth international conference on structural safety and reliability, Buffalo
Kahneman D, Tversky A (2000) Editors, choices, values, and frames. Cambridge University Press/Russell Sage Foundation, New York
Keller C, Siegrist M, Gutscher H (2006) The role of the affect and availability heuristics in risk communication. Risk Anal 26(3):631–639
Klein RJT, Nicholls RJ et al (2003) Resilience to natural hazards: how useful is this concept? Glob Environ Change B Environ Hazards 5(1–2):35–45
Manyena SB (2006) The concept of resilience revisited. Disasters 30:434
Manzo L, Perkins D (2006) Finding common ground: the importance of place attachment to community participation and planning. J Plan Lit 20:335–350
McDaniels T, Chang S et al (2008) Fostering resilience to extreme events within infrastructure systems: characterizing decision contexts for mitigation and adaptation. Global Environ Chang Hum Policy Dimens 18(2):310–318
Miles SB, Chang SE (2011) ResilUS: a community based disaster resilience model. J Cartogr GIS (CAGIS) 38(1):36–51
Mileti DS (1999) Disasters by design: a reassessment of natural hazards in the United States. A Joseph Henry Press, Washington, DC
Norris FH, Stevens SP et al (2008) Community resilience as a metaphor, theory, set of capacities, and strategy for disaster readiness. Am J Community Psychol 41:127–150
Olofsson P, Eklundh L et al (2007) Estimating net primary production for Scandinavian forests using data from Terra/MODIS. Space Res 39:125–130
Patt AG, Schrag DP (2003) Using specific language to describe risk and probability. Clim Chang 61(1–2):17–30
Pettorelli N, Vik J et al (2005) Using the satellite-derived NDVI to assess ecological responses to environmental change. TRENDS Ecol Evol 20(9):503–510
Prince S (1991) A model of regional primary production for use with coarse resolution satellite data. Int J Remote Sens 12:1313–1330
Renschler C, Frazier A, Arendt L, Cimellaro GP, Reinhorn AM, Bruneau M (2010a) Developing the "PEOPLES" resilience framework for defining and measuring disaster resilience at the community scale. In: Proceedings of the 9th US national and 10th Canadian conference on earthquake engineering (9USN/10CCEE), Toronto, 25–29 July 2010
Renschler C, Frazier A, Arendt L, Cimellaro GP, Reinhorn AM, Bruneau M (2010b) Framework for defining and measuring resilience at the community scale: the PEOPLES resilience framework. MCEER technical report – MCEER-10-006. University at Buffalo (SUNY), The State University of New York, Buffalo, p 91
Renschler C, Reinhorn AM, Arendt L, Cimellaro GP (2011) The PEOPLES resilience framework: a conceptual approach to quantify community resilience. In: Proceedings of the 3rd international conference on Computational Methods in Structural Dynamics and Earthquake Engineering (COMPDYN 2011), Corfu', 26–28 May 2011
RICSA (2010) RISCA poverty project. http://web.uct.ac.za/depts/ricsa/projects/publicli/poverty

Rose A (2004a) Defining and measuring economic resilience to disasters. Disaster Prev Manag 13:307–314
Rose A (2004b) Economic principles, issues, and research priorities in hazard loss estimation. Springer, New York
Rose A (2009) Economic resilience to disasters. In: Center for Risk and Economic Analysis of Terrorism Events. University of Southern California
Rose A, Liao SY (2005) Modeling regional economic resilience to disasters: a computable general equilibrium analysis of water service disruptions. J Reg Sci 45(1):75–112
Scura G, Arcidiacono V et al (2013) Integrated damage assessment communication system using smartphone network during 2012 Emilia earthquake. In: Vienna congress on recent advances in earthquake engineering and structural dynamics & 13. D-A-CH Tagung (VEESD2013), Vienna, 28–30 Aug 2013
SEAOC (1995) Vision 2000 – a framework for performance based earthquake engineering. Structural Engineers Association of California, Vision 2000 Committee, Sacramento
Simoniello T, Lanfredi M, Liberti M, Coppola R, Macchiato M (2008) Estimation of vegetation cover resilience from satellite time series. Hydrol Earth Syst Sci 12:1053–1064
SPUR Hazard Mitigation Task Force (2009) The dilemma of existing buildings: private property, public risk. San Francisco Planning + Urban Research Association, San Francisco. www.spur.org. Accessed 1 Feb 2009
Sutton J, Tierney K (2006) Disaster preparedness: concepts, guidance, and research. Fritz Institute, San Francisco
Tierney K (1997) Business impacts of the Northridge earthquake. J Conting Crisis Manag 5:87–97
Tierney K (2009) Disaster response: research findings and their implications for resilience measures. CARRI research report 6. Community & Regional Resilience Institute, Boulder
Tierney K, Bruneau M (2007) Conceptualizing and measuring resilience. TR News 250:14–17
Vallero A, Cimellaro GP (2014) EDAM: earthquake damage assessment communication system using smartphones. In: Second European conference on earthquake engineering and seismology (2ECEES), Istanbul, 24–29 Aug 2014
Vugrin ED, Warren DE et al (2010) A framework for assessing the resilience of infrastructure and economic systems. Springer, Berlin
Weick KE (1995) Sensemaking in organizations. Sage, Thousand Oaks
Weick KE, Sutcliffe K et al (2005) Organizing and the process of sensemaking. Organ Sci 16: 409–421
Wein A, Rose A (2011) Economic resilience lessons from the ShakeOut earthquake scenario. Earthq Spectra 27:559–573
Wildavsky A (1991) Searching for safety. Social Philosophy and Policy Center, Transaction Publishers, New Brunswick
Woods DD (2006) Essential characteristics of resilience. Ashgate, Aldershot
Yang TY, Moehle J, Stojadinovic B, Kiureghian AD (2009) Performance evaluation of structural systems: theory and implementation. J Struct Eng ASCE 135(10):1146–1154
Zobel C (2010) Comparative visualization of predicted disaster resilience. In: 7th international ISCRAM conference, Seattle

Chapter 11
Modelling Economic Dimension of Community Resilience

Gian Paolo Cimellaro and D. Martinelli

Abstract Earthquakes and extreme events in general cause direct and indirect economic effects on every major economic sector of a given community. These effects have grown in the last years due to the increasing interdependency of the infrastructures and make the community more vulnerable to natural and human-induced disruptive events. Therefore, there is need for metrics and models which are able to describe economic resilience, defined as the ability of a community affected by a disaster to resist at the shock and bounce back to the economy in normal operating conditions. Several attempts have been made in the past to achieve a better measurement and representation of the economic resilience and to find suitable metrics to help decision planning. The most popular methodologies are based on Computable General Equilibrium models (CGE) and Inoperability Input-Output models (IIM). In this study, we analyze these methods, showing advantages and limitations. Finally, a new method is proposed to evaluate economic resilience which is based on equilibrium growth models and compared with other approaches on a specific case study: the San Francisco Bay Area.

G.P. Cimellaro, Ph.D., P.E. (✉)
Department of Structural, Geotechnical & Building Engineering (DISEG), Politecnico di Torino, 10129 Turin, Italy
e-mail: gianpaolo.cimellaro@polito.it

D. Martinelli
Department of Civil and Environmental Engineering, University of California at Berkeley, Berkeley, CA 94720-1710, USA
e-mail: martinelli.davide5@gmail.com

G.P. Cimellaro et al. (eds.), *Computational Methods, Seismic Protection, Hybrid Testing and Resilience in Earthquake Engineering*, Geotechnical, Geological and Earthquake Engineering 33, DOI 10.1007/978-3-319-06394-2_11,
© Springer International Publishing Switzerland 2015

11.1 Introduction

Resilience, according to the current literature, is defined as the ability of systems to rebound after severe disturbances, disasters, or other forms of extreme events. As suggested by the work of Renschler et al. (2010) seven dimensions of resilience summarized with the acronym PEOPLES have been identified. In his framework, the performance indices are integrated over space and time. Among these dimensions, the economic one is certainly one of the most controversial. In fact the economic aspect has been often not taken in account in the recent studies which have focused mainly in the actual applications and quantification of the other dimensions (Cimellaro et al. 2013, 2014a; Miles et al. 2003; Davidson and Cagnan 2004). However, the possibility to measure the economic resilience of a community after a disaster is increasingly being seen as a crucial step towards disaster risk reduction.

Recent studies focusing on the economic resilience measurement and decision planning after natural/manmade disasters mainly use two approaches: *Inoperability Input-Output models* (*IIM*) (Haimes and Jiang 2001) and *Computable General Equilibrium models* (*CGE*) (Rose and Liao 2005). These models are able to describe the behavioral response to input shortages and changing market conditions by computing the overall changes in economic variables across sectors, and compare the changes with the economy in normal operating conditions. Both models share many common features of the classical *Leontief Input-Output Models* but they differ in some properties and characteristics.

The *IIM* models have been formulated by Haimes and Jiang (2001) to analyze the behavior of interconnected systems and was then expanded by Santos and Haimes (2004) to model the demand reduction due to the terrorism threat of interconnected infrastructures. Later Lian and Haimes (2006) have focused on the risk of terrorism through the dynamic IIM. More recently, Pant et al. (2011) have focused on the interdependent impacts at multimodal transportation container terminals, and offer an overview on the metrics suited to decision support (Pant and Barker 2012). They also developed a specific approach (Pant et al. 2013) for the evaluation of quantitative resilience metrics accounting for interdependencies among multiple infrastructures.

They belong to the second group of models the work by Konovalchuk (2006) which developed a CGE model to analyze the economic effects of the Chernobyl nuclear disaster. Boisvert et al. (2012) depicted the macroeconomic costs of the foot-and-mouth disease in the United States. Rose and Liao (2005) described a CGE methodology able to consider one of the few advantages of I-O over CGE that is the clear distinction between direct and indirect impacts. However, as stated in the work of Benjamin H. Mitra-Kahn (2008) the CGE models can be a very useful policy tool, only for understanding static fixed output models, because they are not built for dynamic analyses. Resilience is a static but also a dynamic process, as affirmed, for example, by Rutter (2012) in the Social field, by Stokols et al. (2013) in the Ecological field and by Rose (2009) in the Economic field. Based on the considerations above, a new methodology for measuring economic resilience based on the *Structural Dynamic Growth model* described by Li (2010) is presented and adapted to evaluate the economic resilience index, using the procedure described by Cimellaro et al. (2010a, b) where the restoration curves are the activity/output curves provided by the model.

The chapter describes and compares the three approaches to evaluate economic resilience showing the advantages and limitations within the resilience framework.

11.2 The Common Origin of IIM and CGE models

Leontief (1966) developed in 1966 the Input-Output model, a quantitative economic model that was capable of describing the interdependencies between systems of a national economy or different regional economies. The model is defined as

$$\boldsymbol{x} = \boldsymbol{A}\boldsymbol{x} + \boldsymbol{d} \leftrightarrow \left\{x_i = \sum_j a_{ij} x_j + d_i\right\} \forall i \tag{11.1}$$

where x_i represents the total production output of the i-th industry, a_{ij} of the ratio of the input of the industry i to the industry j with respect to the total supply requirements of industry j, d_i is the final demand for the ith industry. However, extended approaches have been formulated, starting from I-O Model, in order to address the estimation and planning of economic resilience. The most popular are IIM and CGE. While the first one is more suitable for decision planning, the other is advantageous in the evaluation of economic resilience.

11.3 The Inoperability Input-Output Model (IIM)

The assumptions on which the IIM is based are the same of the classical I-O model. Therefore, it is an equilibrium, time-invariant, deterministic and linear representation subjected to all limitations of classical *Leontief's formulation*. The IIM formulation derives from the metrics of *Inoperability q* given in Eq. (11.2), a vector where each component represents the ratio of production loss with respect to the usual production level of the industry. It applies to represent resilience metrics, and demand perturbation d* given in Eq. (11.3), a vector expressed in terms of normalized degraded final demand. Using the same symbols used in PEOPLES, the metrics can defined as

$$\boldsymbol{q} = \left[\boldsymbol{diag}\left(\hat{\boldsymbol{Q}}\right)\right]^{-1}\left(\hat{\boldsymbol{Q}} - \tilde{\boldsymbol{Q}}\right) \tag{11.2}$$

$$\boldsymbol{d}^* = \left[\boldsymbol{diag}\left(\hat{\boldsymbol{Q}}\right)\right]^{-1}\left(\hat{\boldsymbol{d}} - \tilde{\boldsymbol{d}}\right) \tag{11.3}$$

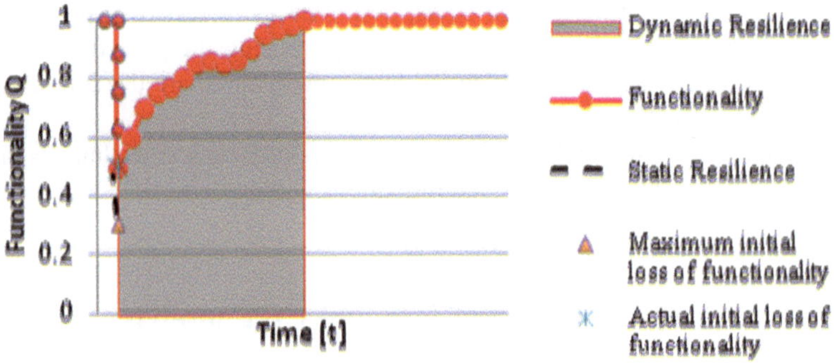

Fig. 11.1 Comparison between static and dynamic resilience

where $\hat{Q}$ and $\hat{d}$ are the equilibrium functionality and demand levels respectively while $\tilde{Q}$ and $\tilde{d}$ are the respective disrupted equilibrium levels. Combining q and d^* Santos and Haimes obtained the IIM formulation in Eq. (11.4), that maintains a form similar to the Leontief I-O model, and that shows how *Inoperability* is driven by perturbations in demand.

$$q = A^*q + d^* \rightarrow q = \left[I - A^*\right]^{-1} d^* \tag{11.4}$$

where A^* represents the normalized interdependency matrix that indicates the degree of coupling of the industry sectors. However, since resilience needs a dynamic formulation, a dynamic extension of the IIM (DIIM) developed by Lian and Haimes (2006) is considered. It is a first-order differential equation that incorporates a rate constant into the static IIM structure, and whose analytical equation is given in Eq. (11.5).

$$q(t) = e^{-K(I-A^*)t} q(0) + \int_0^t e^{-K(I-A^*)(t-z)} K d^*(z)\, dz \tag{11.5}$$

where K is the rate term, a matrix with elements that represent the speed at which sectors attain particular responses to disruptions in outputs or change in demands. Pant et al. (2013) made distinction between metrics which can describe static and dynamic economic resilience. However, in their definition the static resilience value corresponds to the avoided initial loss of functionality as shown in Fig. 11.1. The static resilience index is not able to consider the recovery phase in its formulation therefore the focus has shifted toward the dynamic definition of economic resilience.

The dynamic dimension of economic resilience can be evaluated using DIIM models. A decision space can be generated by varying the values of three resilience metrics that represent the matrix K and reflect investment options. These metrics are given in Eqs. (11.6), (11.7) and (11.8) and are respectively: (i) *the time averaged level of operability* M_l which represents the overall level of functionality maintained by a system, (ii) *the maximum loss of sector functionality* q_i^m, and

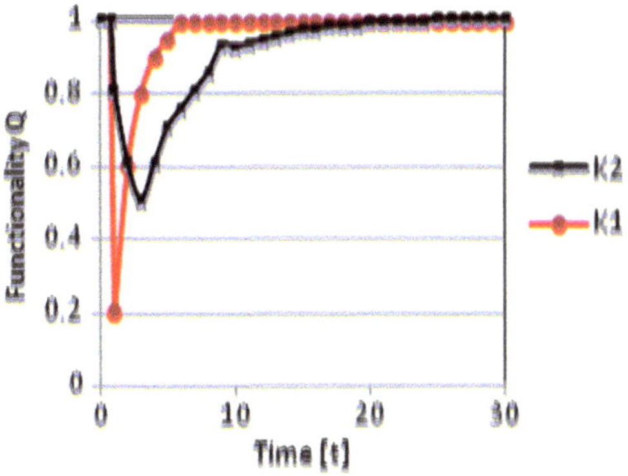

A*	
0.3	0.45
0.15	0.1

q(0)
0
0.5

K1	
1	0
0	1

K2	
0.4	0
0	0.4

Fig. 11.2 Trade-off between time to recovery and maximum inoperability (Adapted from Pant et al. 2013)

(iii) *the recovery time* τ_i which represents the time that the system implies to return to pre-disruption levels of functionality.

$$M_i = 1 - \frac{1}{T}\int_0^T q_i(t)dt \leftrightarrow \boldsymbol{M} = 1 - \frac{1}{T}\int_0^T \boldsymbol{q}(t)dt \tag{11.6}$$

$$q_i^m = \max_{t>0}\left[q_i(t)\right] \leftrightarrow \boldsymbol{q}^m = \max_{t>0}\left[\boldsymbol{q}(t)\right] \tag{11.7}$$

$$\tau_i = \left\{t : t > 0, \left|q_i(t) - q_i^e(t)\right| \leq \varepsilon \ll 1\right\} \tag{11.8}$$

The division of the matrix $\boldsymbol{K}$ into the above metrics allow us considering the multidimensional aspect of dynamic economic resilience, and in particular the trade-off that exists between the recovery time and the maximum loss of functionality, as shown in Fig. 11.2 where both lines represent the behave of the sector having no initial perturbation. The metrics are put in relation each other in Eq. (11.9) under specific assumptions (Pant and Barker 2012). Considering that $\boldsymbol{\tau}_i$ is function of α_i a parameter which is a measure of interdependency (Oliva et al. 2010) and introducing another constant $\tau_i\alpha_i = L_i$, the final dynamic decision space among F_i, q_i^m and τ_i is obtained in Eq. (11.10) after some mathematical manipulations.

$$\boldsymbol{M} = \mathbf{1} - \frac{1}{\mathrm{T}}\left[\mathbf{1} - \mathrm{e}^{-\mathbf{K}[\mathbf{I}-\mathbf{A}^*]\tau}\right]\left[\mathbf{K}\left(\mathbf{I} - \mathbf{A}^*\right)\right]^{-1}\mathbf{q}^{\mathrm{m}} \tag{11.9}$$

$$M_i = 1 - \frac{1}{L_i T}\left[1 - \mathrm{e}^{-L_i}\right]\tau_i {q_i}^m \tag{11.10}$$

The equations generate a decision space through contour curves that can be used to estimate the system performance. Below is described how the method can be used as decision support tool. Considering that L_i is a measure of the amount of recovery

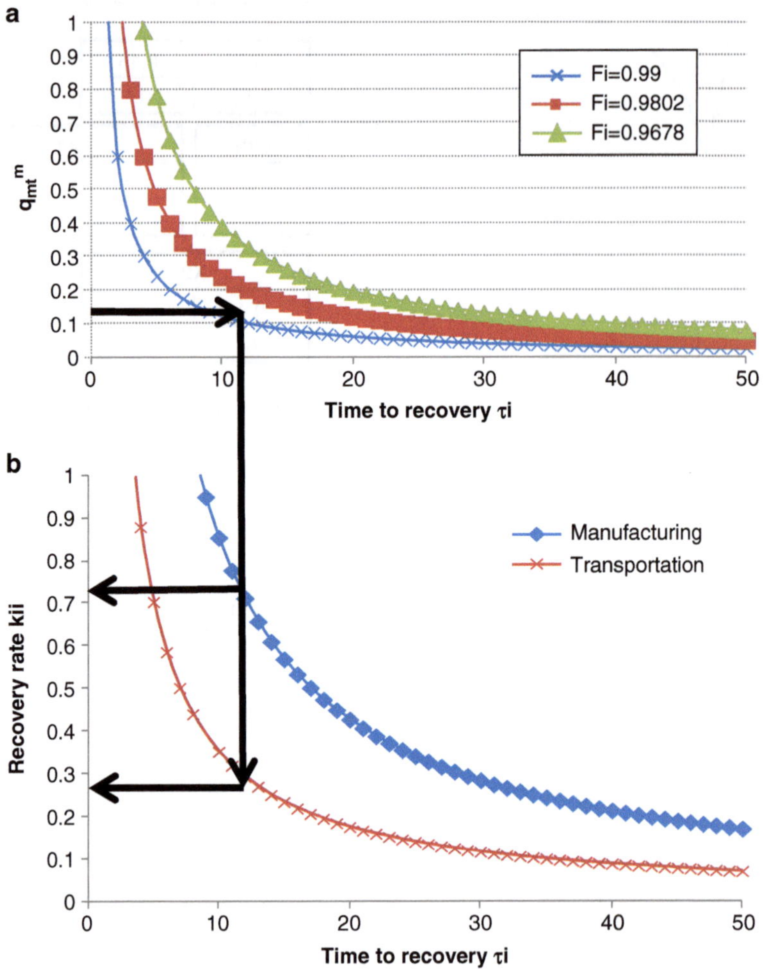

Fig. 11.3 (**a**) Contour curves of decision space for dynamic economic resilience; (**b**) Relationship between time to recovery and recovery rate

and denoting with $r_i \in [0, 1]$ the fraction of recovery from the observed maximum impact, L_i can be reformulated in Eq. (11.11).

$$L_i = \ln\left(\frac{q^m{}_i}{(1-r_i)\, q^m{}_i}\right) \tag{11.11}$$

Figure 11.3 identifies the decision space of an economy where there are only two sectors. The contour curves are equal for both sectors obtained by assuming $r_i = 0.95$ in order to address the time when the system recovers the 95 % of the experienced loss.

Starting from the pair (M_i, q_i^m) that indicates the desired level of overall operability during recovery, the recovery time is evaluated from Fig. 11.3a. Combining α_i and $\tau_i\alpha_i = L_i$, we can graphically represent in Fig. 11.3b the relationship between the recovery rate and the recovery in Eq. (11.12). In Fig. 11.3b the evaluated recovery time are entered and the recovery rate are estimated, so that the sectors that need to more investments to maintain a similar level of functionality compared to the others, are identified.

$$k_{ii} = \frac{L_i}{\tau_i \left(1 - \sum_{j=1}^{n} \alpha^*{}_{ij}\right)} \tag{11.12}$$

11.4 Computable General Equilibrium Models

Computable General Equilibrium models (CGE) are usually based on a system of non-linear equations and are more suitable to represent international and interregional setting with respect to I-O models. However, they are empirical models and users often they are forced to impose strong assumptions regarding the structure of the production and the household behavior for example. The CGE set of equations adopted for the analysis of economic resilience must satisfy specific conditions such as market clearance of commodity and factor, zero-profit and income balance. In particular, the algebraic framework can be derived distinguishing between consumers (households) and industry sectors (producers). The problem of households is to maximize their utilities subject to the constraints of their incomes, while the problem of industries is to maximize the profit subject to the constraints of the production technology. These optimization problems are solved by using the Lagrangian equations for the household's utility and for the producer's profit, that are shown respectively in Eqs. (11.13) and (11.14) for a Cobb-Douglas economy.

$$\mathscr{L}^C = p_U U - \sum_{i=1}^{N} p_i c_i + \lambda^C \left(U - A_C \coprod_{i=1}^{N} c_i^{\alpha_i}\right) \tag{11.13}$$

$$\mathscr{L}_j^P = p_j y_j - \sum_{i=1}^{N} p_i x_{ij} - \sum_{f=1}^{F} w_f v_{fj} + \lambda_j^P \left(y_j - A_j \prod_{i=1}^{N} x_{ij}^{\beta_{ij}} \prod_{f=1}^{F} v_{fi}^{\gamma_{fi}}\right) \tag{11.14}$$

where: N and F are the types of commodities and primary factors respectively; U is an utility good generated by consumption; p_U, p_i, p_j and w_f are respectively the prices of the utility good, the intermediate goods, the outputs, and the primary factors, A_C and A_j are scaling parameters; x_{ij} and v_{fj} are respectively the amount of intermediate good i and the amount of primary factor f for the jth firm; c_i and y_j are the consumption of the commodity i and the output of the jth firm; α_i are the shares of each good in expenditure on consumption while β_{ij} and γ_{fi} are the shares of each input in the cost of production. From Eqs. (11.13) and (11.14),

Wing (2004) demonstrated how we can infer (i) the representative consumer's demand function for the consumption of the *i*th commodity, (ii) the producer *j*'s demands for intermediate inputs of commodities, and (iii) the producer *j*'s demands for primary factor inputs. These demands are connected each other by the economic assumptions of market clearance, zero-profit and income balance listed above. These assumptions yield two excess demand vectors that define the divergence Δ^C between supply and demand in the market for each commodity and the divergence Δ^F between supply and demand in the market for each primary factor, one excess profit vector Δ^π and one excess income vector Δ^m. The absolute values of both sets of differences are minimized to zero to reach the general equilibrium. However, the CGE models need to be calibrated considering real data representing an economy to estimate the economic resilience dimension. These data are usually taken from the SAM matrices (Shoven 1992) that are related to the mentioned I-O matrices. The most acknowledged methodology to compute regional economic resiliency to earthquakes or manmade disasters with CGE models is the one introduced by Rose and Liao (2005), an approach which is able to consider the clear distinction between direct and indirect impacts, which is one of the few advantages of I-O over CGE models. The approach uses a multilayered *Constant Elasticity of Substitution* (CES) *production function* for each sector. The initial values of the elasticity of substitution that represent how easy is to substitute one input for the other are based on information provided from the literature. Each sectoral production function is extracted and the elasticities of substitution are recalibrated with a numerical solution to match the losses found by the empirical estimates. Thereafter, the re-calibrated sectoral production functions are re-inserted in the CGE model and the supply input is reduced to a level consistent with empirical estimates of the computed total regional losses. Finally, the indirect losses are defined subtracting the direct losses from the total losses. The outcome is a measure of resilience through two indices called DRER and TRER (Direct/Total Regional Economic Resilience), given in Eqs. (11.15) and (11.16).

$$DRER = \frac{\%\Delta DQ^m - \%\Delta DQ}{\%\Delta DQ^m} \tag{11.15}$$

$$TRER = \frac{\%\Delta TQ^m - \%\Delta TQ}{\%\Delta TQ^m} \tag{11.16}$$

Where $\%\ \Delta DQ^m$ and $\%\ \Delta TQ^m$ are the maximum percent change in direct and total output, while $\%\ \Delta DQ$ and $\%\ \Delta TQ$ are the estimated percent change in direct and total output.

11.5 The Structural Dynamic Growth Model

Li (2010) developed the structural growth model from the classical growth framework. Even if it was conceived as a *growth model*, it can also be used to compute the *general equilibrium*. The model represents the production processes in the economy through two matrices: the input and the output coefficient matrices. For example, for the economy described in Eq. (11.17) the two matrices are given in Eqs. (11.18) and (11.19).

$$\begin{cases} 280 \text{ quarters wheat} + 12 \text{ tons iron} \rightarrow 575 \text{ quarters wheat} \\ 120 \text{ quarters wheat} + 8 \text{ tons iron} \rightarrow 20 \text{ tons iron} \end{cases} \tag{11.17}$$

$$A = \begin{bmatrix} 56/115 & 6 \\ 12/575 & 2/5 \end{bmatrix} \tag{11.18}$$

$$B = I \tag{11.19}$$

where the ith column in matrix A represents the standard input bundle of agent i. In the classical economic growth framework, the equilibrium price vectors and the equilibrium output vectors are the left and right P-F eigenvectors of matrix $\mathbf{A}$. The Structural Dynamic Growth model tries to integrate the market mechanism into the classical growth model by embedding an exchange process in it which is represented by an exchange vector, to reach equilibrium.

11.5.1 Exchange Process

The exchange process considers the economy as a discrete-time dynamic system and supposes the economic activities such as the price adjustment, exchange and production occur in turn in each period. With reference to the previous economic system, $\mathbf{S}$ in Eq. (11.20) represents the $(n \times m)$ supply matrix, and $\mathbf{s}$ in Eq. (11.21) represents the supply vector in the initial period.

$$S = \begin{bmatrix} 575 & 0 \\ 0 & 20 \end{bmatrix} \quad s = \begin{bmatrix} 575 \\ 20 \end{bmatrix} \tag{11.20}$$

Let call $\mathbf{z}$ the *purchase vector* or *exchange vector* (of standard input bundles), which represents the purchase amounts of m agents, while $\mathbf{Az}$ is called the *sales vector of goods*. It's possible to derive Eq. (11.21) where $\hat{\mathbf{s}}$ represents the diagonal matrix with the $\mathbf{s}$ vector in the main diagonal and $\mathbf{u}$ the n-dimensional sales rate vector indicating the sales rates of n goods.

$$\mathbf{u} \equiv \hat{\mathbf{s}}^{-1}\mathbf{Az} \tag{11.21}$$

Under the given price vector $\mathbf{p}$, the purchase and sales values of m agents are $\boldsymbol{p}'A\hat{\mathbf{z}}$ and $\boldsymbol{p}'\hat{\mathbf{u}}\boldsymbol{S}$ respectively. Equation (11.22) is obtained by assuming that the value of purchases is equal to the value of sales of each agent as in the CGE income balance, which is given by

$$\boldsymbol{p}'A\hat{\mathbf{z}} = \boldsymbol{p}'\hat{\mathbf{u}}\boldsymbol{S} \equiv \boldsymbol{p}'\widehat{\hat{\mathbf{s}}^{-1}Az}\boldsymbol{S} \tag{11.22}$$

When Eq. (11.23) holds and **S'A** can not be decomposed, it's there exists a unique normalized *exchange vector*, and a unique *maximal exchange vector* which can be found by the following steps:

Step 1. Compute the matrix $\mathbf{p}'\widehat{\hat{s}^{-1}\boldsymbol{A}z\boldsymbol{S}}$;
Step 2. Find the normalized right P-F eigenvector of **z**, denoted by **x**;
Step 3. Find the minimal component of $\widehat{\boldsymbol{A}\mathbf{x}^{-1}}\boldsymbol{s}$, denoted by ξ;
Step 4. Compute the exchange vector $z = \xi \boldsymbol{s}$.

So, it is assumed that the state of the economic system at a period t is represented by the variables $\boldsymbol{p}(\mathrm{t})$ = price vector; $\boldsymbol{S}(\mathrm{t})$ = supply matrix; $\boldsymbol{u}(\mathrm{t})$ = sales rate vector; $\mathbf{z}(\mathrm{t})$ = exchange vector and production intensity vector; $\mathbf{Y}(\mathrm{t})$ = Output matrix. The market mechanism is described by Eq. (11.23) considering that at period $t + 1$ the economy runs following Eq. (11.23) until the system reaches the equilibrium.

$$
\begin{aligned}
&- \quad \boldsymbol{p}\,(t+1) = P\,(\boldsymbol{p}(t), \boldsymbol{u}(t)) \\
&- \quad \boldsymbol{S}\,(t+1) = \boldsymbol{B}\widehat{z(t)} + Q\left(\widehat{e - \boldsymbol{u}(t)}\boldsymbol{S}\,(t)\right) \qquad (11.23) \\
&- \quad (\boldsymbol{u}\,(t+1), z\,(t+1)) = Z\,(\boldsymbol{A}, \boldsymbol{p}\,(t+1), \boldsymbol{S}\,(t+1))
\end{aligned}
$$

where P represents price adjustment process, Q is the inventory depreciation function and stands for the depreciation process of inventories and Z is the exchange function depicted above.

11.5.2 Application of Structural Growth Model Models to Evaluate Economic Resilience

Starting from the I-O matrices describing the economy in normal operating condition, a shock is applied to simulate a natural disaster such as an earthquake, modifying the *exchange vector*, which is the driver of the equilibrium process.

After the application of the shock, from the restoration curves of the system obtained in Fig. 11.4, the values of the economic resilience of the system can be quantified using the procedure explained by Cimellaro et al. (2010a) that describes resilience as "*the normalized shaded area underneath the function describing the functionality of a system*". The model incorporates also a series of parameters that control the converging speed in case of availability of after-disruption data. Table 11.1 represents the comparison between the results obtained using the static and dynamic resilience analysis with IIM with the results obtained using SGM for the same representative economy under two different scenarios: great demand disruption in manufacturing (A) and great demand disruption in transportation (B).

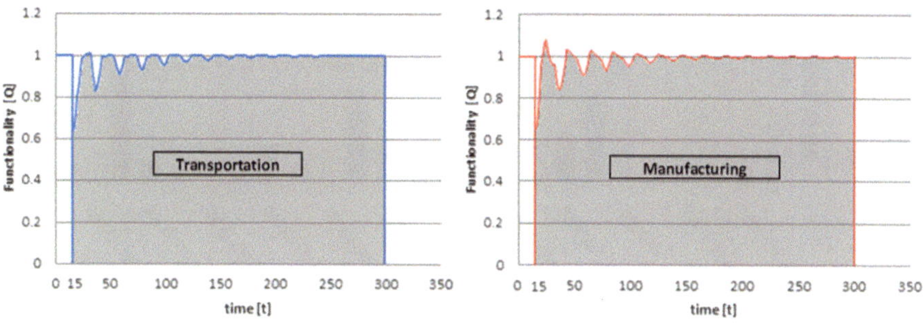

Fig. 11.4 Restoration curves of the different economic sectors

Table 11.1 Outcomes of the IIM and SGM resilience approaches under two scenarios

	SGM		IIM	
	A	B	A	B
Resilience transportation	0.96	0.87	0.99	0.87
Resilience manufacturing	0.97	0.90	0.94	0.92

The outcomes of the two methods are similar, and underline the necessity to protect the transportation sector with respect to the manufacturing one, because of the lower resiliency value of the scenarios B. However, the SGM is able to identify a non-dimensional measure that captures the dynamic dimension of resilience. On the other hand, the IIM can be used to address the comparison between different strategies as described above in its dynamic decision space, but it doesn't give as output a non-dimensional value for the economic resilience.

11.6 Case Study: The San Francisco Bay Area

The two approaches described above have been also applied to the San Francisco Bay Area to identify the critical systems of the area after a disruptive event such an earthquake. Both approaches need as a starting point the Input-Output matrix of the economy.

These matrices aren't public available directly, but are usually published at a national level by the BEA (Bureau of Economic Analysis) under the form of two matrices: the *Make matrix* and the *Use matrix* that can be combined with specific assumptions by the analysts. Considering that San Francisco must achieve certain performance targets to have the city fully operational in case of an earthquake, the analysis starts from the projected Make and Use matrices elaborated by the Bureau of Labor Statistics (2012) for the year 2020. From these matrices, following the procedure described by Chamberlain (2011) the final Input-Output matrix at the

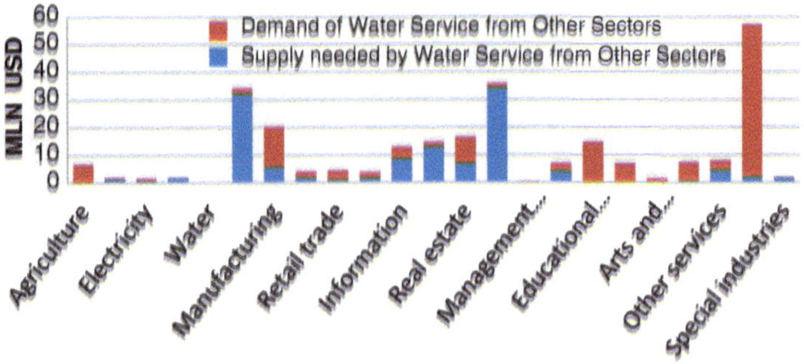

Fig. 11.5 Interdependencies between Water Service and other economic sectors in the Bay Area

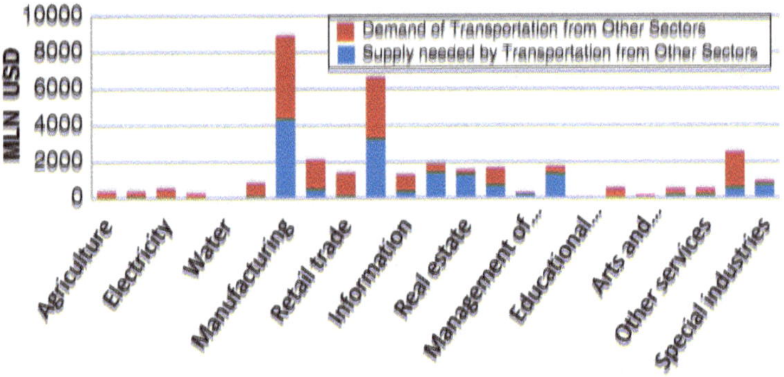

Fig. 11.6 Interdependencies between Transportation and other economic sectors in the Bay Area

national level has been derived. To scale the national matrix at the San Francisco Bay Area level, the RIMS II multipliers published by the BEA every year should be used. However, since the multipliers are not available for 2020 projection neither by the BEA nor by the BLS, the matrix is scaled from the national to the regional level using the ratio between the USA GDP and the San Francisco Bay Area, while keeping separated the utilities like electricity or water. Lifelines play a fundamental role in the aftermath of disruptive events, so attention is given to the Water and the Transportation Service which are also economically interdependent among other sectors in the region. Figures 11.5 and 11.6 show the interdependency among different economic sectors.

Using these data and the previous IIM set of equations, two strategies are analyzed to highlight the difference effects over static economic resilience of similar investment allocation in the economy. The outcomes of the strategies are shown in Table 11.2 assuming a planning function in Eq. (11.24) given by

$$f(\beta_i) = e^{-\beta_i}, \qquad (11.24)$$

Table 11.2 Outcome of two investment allocation strategies

Strategy I		Strategy II	
βw	50.00	βw	2.00
f(βw)	0.00	f(βw)	0.14
βt	2.00	βt	50.00
f(βt)	0.14	f(βt)	0.00
βother	1.00	βother	1.00
f(βother)	0.37	f(βother)	0.37
$\hat{\mathbf{R}}\mathbf{s}$	65.50	$\hat{\mathbf{R}}\mathbf{s}$	65.70

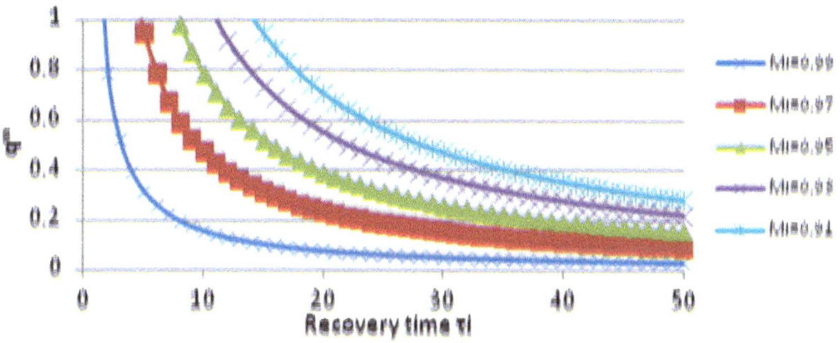

Fig. 11.7 Contour curves of decision space for dynamic economic resilience

$\hat{\mathbf{R}}\mathbf{s}$ is the average static resilience of all the economic sectors which represents a rough estimation of resilience; however, the outcomes can be computed for each sector. The results show that with the same amount of dollars of investment (β_i) using strategy II an increment of 0.2 % of average resilience is obtained with respect to strategy I. Even if this seems to be a small improvement, remember that we are taking in account, for sake of simplicity, the average static resiliency that smoothes the differences among sectors.

On the other side, the Eq. (11.9) generates a dynamic decision space through contour curves that can be used to estimate the system performance for different recovery strategies. Assuming the analysis addresses the time when the system recovers 95 % losses and T = 50 days, is possible to draw the two diagrams below. Figure 11.7 represents the contour curves for the dynamic resilience space, while Fig. 11.8 the relationship between recovery time and recovery rate. The recovery time is evaluated from Fig. 11.4 starting from the pair (M_i,(q˜m)_i) that indicates the desired level of overall operability during recovery. Then the recovery rate can be evaluated entering in Fig. 11.8 with the estimated recovery time, so that the sectors that need more investments to maintain a similar level of functionality can be identified. Table 11.3 shows the outcome of the two strategies. While strategy B requires lower recovery rates achieving lower resiliency, strategy A achieves a faster recovery and a higher resiliency but at the cost of a higher recovery rate for the Agriculture sector, that might be not feasible. The estimation of the economic

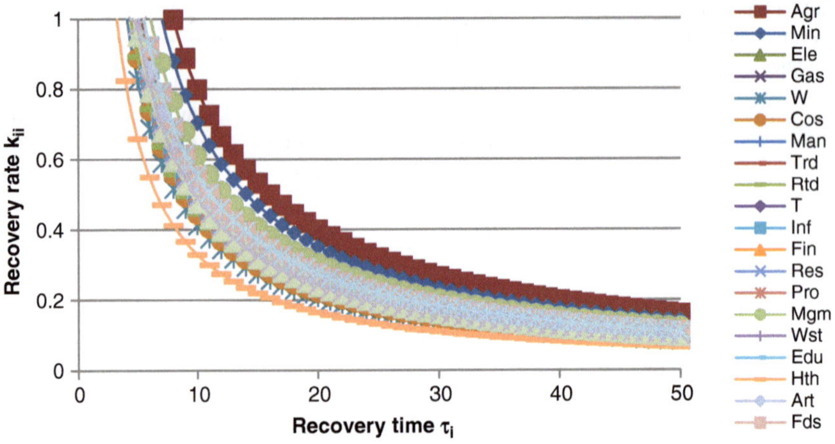

Fig. 11.8 Recovery rate vs. recovery time for different economic sectors

Table 11.3 Comparison between two different strategies

Strategy						τ	k_{agr}	k_W
A	**Mi**	0.97	q^m	0.5	**A**	**10**	**0.8**	**0.45**
B	**Mi**	0.95	q^m	0.5	**B**	**16**	**0.5**	**0.28**

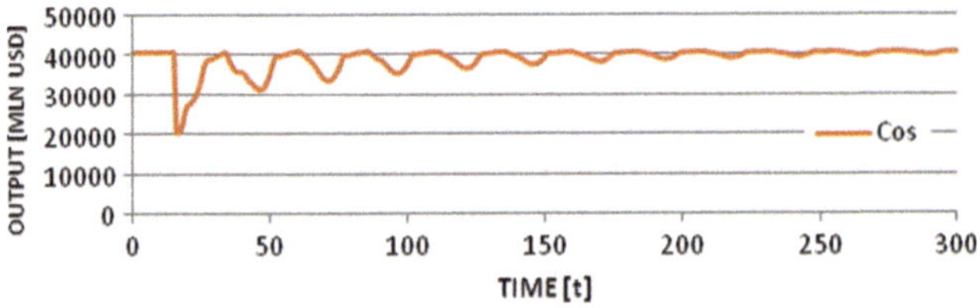

Fig. 11.9 Restoration curve of Construction sector after Water Service disruption

resilience through SGM starts again from the I-O matrices representative of the economy in normal operating condition that is translated in terms of input and output coefficient matrices A and B. Earthquake is simulated modifying the exchange vector, which is the driver of the equilibrium process. Assuming that a shock causes a 50 % disruption in the water service, Fig. 11.9 shows the restoration curve for the Construction sector from which the value of the economic resilience of the system can be quantified using the procedure described by Cimellaro et al. (2010a, b, 2014a, b). Similar curves are obtained for all the other sectors. In the second case, a 50 % Transportation disruption is also considered (not shown).

Computing the resilience for each sector and then considering the average value of all the resiliencies founded in both cases, the effects of the two simulated shocks are compared in Table 11.4.

A 50 % disruption in the Transportation Service would lead to a lower value of average resilience, obtaining the same finding of the IIM approach. So globally, the

Table 11.4 Comparison between two different disruptions

	Water Service disruption	Transportation
$\hat{R}$	96.37	89.25

economy is more dependent on the Transportation than the Water service, and that is why a smart investment allocation would favor the former sector with respect to the latter.

11.7 Future Work

While the economic resilience at the community level can be used as tool by Institutions which make decisions at the regional level, there is also need for a tool to evaluate economic resilience also at the local level when a single sector or a company needs to be analyzed to show if it is economically resilient to disruptive events. To achieve this goal there is need to shift at the individual business level to evaluate the life-cycle costs of individual business and the return of investments for preparedness measures when building structures with higher performance with respect to the minimum required performance levels. This goal can be achieved using the framework proposed by Terzic et al. (2014), where different structural systems are compared to identify the different return of investment in case of a natural disaster strikes the system. The framework they developed represents a combination of business, engineering, and seismology modeling and is divided in four steps. The first is the *Seismic Hazard Analysis* which leads to the selection of the intensity of the representative ground motion. Then, *Response Analysis* is performed through simulations and numerical modeling using OpenSees to identify the Engineering Demand Parameters for the different structures. The third step is the *Damage Analysis* which correlates EDPs with the damage state of all the components of the structures using the fragility curves which are provided in PACT. The final step is the *Loss Analysis*, which correlates damage to losses (e.g. repair cost, downtime etc.). Then the business downtime is evaluated using the downtime model developed by the authors and through a business model, the business interruption costs are evaluated and summed with the repair costs to obtain the total losses, from which the return on investments of different structural systems considering also the different initial costs are evaluated. The framework uses a business model that converts the business downtime to business interruption cost assuming the company owner is just leasing the space to businesses. However, considering the downtime business costs as a rental loss is a simplified way to approach the estimation of costs and can underestimate the return on investments. In fact interdependencies exist among different sectors and direct damage due to disruptive events will always be followed by indirect damage, which are taken in account in the framework, but in a simplified way. Nowadays, different business models have been developed to quantify the losses due to catastrophic events such earthquakes. Further research will be developed in extending the models used at the community level in a smaller scale at the local level using for example the an I-O matrix for the individual sectors or companies. This will allow estimating the interdependencies between business and services, but at the same time among different branch offices of the same company for example.

11.8 Concluding Remarks

The chapter summarizes and compares different methodologies to evaluate economic resilience at the community level which can be applied to any type of disruptive event. The goal is to help decision makers to understand the economic resilience problem of interdependent networks from a community level to a business level. Two different approaches for the evaluation of the economic resilience have been described: the Inoperability Input-Output Model, and the CGE model. A new promising model, called the Structural Growth Model has been proposed to be used to evaluate the economic resilience index. Critical comparison among the models is presented and applied to a specific case study to highlight differences, advantages and limitations of all approaches. The chosen case study is the San Francisco Bay Area. The I-O matrix has been obtained by scaling the matrix from the national to the regional level using the ratio between the USA GDP and the San Francisco Bay Area, while keeping disaggregated the utilities like electricity, water etc. Finally, the chapter gives some insights on how to implement the approach to evaluate the life-cycle costs of structural systems in case of natural disasters, and the return of investments of preparedness measures like building structures with higher performance with respect to the minimum required performance levels given by IBC.

Acknowledgments The research leading to these results has also received funding from the European Community's Seventh Framework Programme – Marie Curie International Reintegration Actions-FP7/2007–2013 under the Grant Agreement n° PIRG06-GA-2009-256316 of the project ICRED – Integrated European Disaster Community Resilience, and by the Marie Curie International Outgoing Fellowship (IOF) Actions-FP7/2007–2013 under the Grant Agreement n°PIOF-GA-2012-329871 of the project IRUSAT – Improving Resilience of Urban Societies through Advanced Technologies.

References

Boisvert RN, Kay D, Turvey CG (2012) Macroeconomic costs to large scale disruptions of food production: the case of foot- and-mouth disease in the United States. Econ Model 29(5):1921–1930

Chamberlain Economics, L. L. C. (2011) Methodology paper: developing square U.S. input-output tables from B.E.A. make and use tables, pp 1–4

Cimellaro GP, Reinhorn AM, Bruneau M (2010a) Framework for analytical quantification of disaster resilience. Eng Struct 32(11):3639–3649
Cimellaro GP, Reinhorn Andrei M, Bruneau M (2010b) Seismic resilience of a hospital system. Struct Infrastruct Eng 6(1–2):127–144
Cimellaro GP, Villa O, De Stefano A (2013) Serviceability of natural gas distribution networks after earthquakes. J Earthq Tsunami 7(2):22
Cimellaro GP, Tinebra A, Renschler C, Fragiadakis M (2014a) Resilience-based design of an urban water distribution system. J Struct Eng ASCE (in review)
Cimellaro GP, Villa O, Bruneau M (2014b) Resilience-based design of natural gas distribution networks. J Infrastruct Syst ASCE. doi:10.1061/(ASCE)IS.1943-555X.0000204
Davidson R, Çagnan Z (2004) Restoration modeling of lifeline systems. Research progress and accomplishments 2003–2004. Multidisciplinary Center for Earthquake Engineering Research, Buffalo
Haimes Y, Jiang P (2001) Leontief-based model of risk in complex interconnected infrastructures. J Infrastruct Syst 7(1):1–12
Konovalchuk V (2006) A computable general equilibrium analysis of the economic effects of the Chernobyl nuclear disaster. The Graduate School College of Agricultural Sciences. The Pennsylvania State University
Leontief W (1966) Input-output economics. Oxford University Press, New York
Li W (2010) A structural growth model and its applications to Sraffa's system. In: 18th international input-output conference, Sydney
Lian C, Haimes YY (2006) Managing the risk of terrorism to interdependent infrastructure systems through the dynamic inoperability input–output model. Syst Eng 9:241–258
Miles SB, Chang SE (2003) Urban disaster recovery: a framework and simulation mode. Technical report MCEER-03-0005. Multidisciplinary Center for Earthquake Engineering Research, University at Buffalo, The State University of New York, Buffalo
Mitra-Kahn BH (2008) Debunking the myths of computable general equilibrium models. SCEPA Work Pap 1:1–93
Oliva G, Panzieri S, Setola R (2010) Agent-based input-output interdependency model. Int J Crit Infrastruct Prot 3(2):76–82
Pant R, Barker K (2012) Building dynamic resilience estimation metrics for interdependent infrastructures. In: Proceedings of the European safety and reliability conference, Helsinki, June 2012
Pant R, Barker K, Grant FH, Landers TL (2011) Interdependent impacts of inoperability at multimodal transportation container terminals. Transp Res Part E Logist Transp 47(5): 722–737
Pant R et al (2013) Static and dynamic metrics of economic resilience for interdependent infrastructure and industry sectors. Reliab Eng Syst Saf
Renschler C, Frazier A, Arendt L, Cimellaro GP, Reinhorn AM, Bruneau M (2010) Framework for defining and measuring resilience at the community scale: the PEOPLES resilience framework. MCEER technical report MCEER-10-006. University at Buffalo, The State University of New York, Buffalo
Rose A (2009) Economic resilience to disasters. Community and Regional Resilience Institute report no 8, Oak Ridge
Rose A, Liao S (2005) Modeling resilience to disasters: computable general equilibrium analysis of a water service disruption. J Reg Sci 45(1):75–112
Rutter M (2012) Resilience as a dynamic concept. Dev Psychopathol 24(2):335–344
Santos JR, Haimes YY (2004) Modeling the demand reduction input-output (I-O) inoperability due to terrorism of interconnected infrastructures. Risk Anal 24:1437–1451
Shoven JB (1992) Applying general equilibrium. Cambridge University Press, Cambridge
Stokols D, Lejano R, Hipp J (2013) Enhancing the resilience of human-environment systems: a social ecological perspective. Ecol Soc 18(1):7

Terzic V, Mahin SA, Comerio MC (2014) Comparative life-cycle cost and performance analysis of structural systems. In: Proceedings of the 10th national conference in earthquake engineering, earthquake engineering research institute, Anchorage

Wing IS (2004) Computable general equilibrium models and their use in economy-wide policy analysis. Technical note, Joint Program on the Science and Policy of Global Change. MIT

Chapter 12
Seismic Performance of Health Care Facilities Using Discrete Event Simulation Models

Gian Paolo Cimellaro and Marta Piqué

Abstract The chapter presents a hospital testbed which aims to help the earthquake engineering community moving another step toward the realization and implementation of resilience-based design strategies. An organizational model describing the response of the Hospital Emergency Department (ED) has been implemented using a discrete events simulation model (DES). The waiting time is the main parameter of response and it is used to evaluate the disaster resilience index of healthcare facilities. It has been considered for patients arriving in an Emergency Department after an earthquake, when no emergency plan is activated, to see the response of the hospital and observe its capacity of performing in emergency conditions. The hospital analyzed is the Mauriziano "Umberto I" Hospital, located in downtown Turin, in Italy. The DES model is an important tool in the hospital decision process either for the engineering profession or for the policy makers.

12.1 Introduction

Healthcare facilities have been recognized as strategic buildings in hazardous events and they play a key role in the disaster rescues. During a dramatic event, the Emergency Department (ED) has a critical role in the health care facilities, as they must provide immediate assistance request 24 h. Even if they are properly organized, a change in the external environment, such as a natural disasters, a train accident etc.,

G.P. Cimellaro, Ph.D., P.E. (✉)
Department of Structural, Geotechnical & Building Engineering (DISEG),
Politecnico di Torino, 10129 Turin, Italy
e-mail: gianpaolo.cimellaro@polito.it

M. Piqué
Faculty ETSEIB, Universitat Politècnica de Catalunya, Barcelona, Spain
e-mail: martapique24@gmail.com

G.P. Cimellaro et al. (eds.), *Computational Methods, Seismic Protection, Hybrid Testing and Resilience in Earthquake Engineering*, Geotechnical, Geological and Earthquake Engineering 33, DOI 10.1007/978-3-319-06394-2_12,
© Springer International Publishing Switzerland 2015

can modify patients' arrival rates and lead to a change in their performance that could end up with a worse service level for all the patients. This kind of "patients' satisfaction" can be measured with different parameters; among these parameters, the most representative one is the waiting time, which gives also an idea of how busy a hospital is. In this chapter, the waiting time (Cimellaro et al. 2011) of the patients arriving in an Emergency Department during a dramatic event such an earthquake has been considered. This parameter can be used to measure the response and the capacity of the hospital during emergency. The hospital analyzed is the Mauriziano "Umberto I" Hospital, located in downtown Turin, in Italy.

12.2 State of Art

Several studies for the health care systems have been performed in the recent years, in particular related to the patients' waiting time. In fact, the study of multiple patient routings and flow processes is important to reduce the waiting time at various stages in the hospital. This was generally made using simulation models which analyze various scenarios with different performance indicators. Models are used in problem solving because their use is generally cheaper, faster and less disruptive than manipulating the real world system (Boxerman 1996). In particular, discrete event simulation (DES) has been widely used in modeling healthcare systems since the 1970s. Martin et al. (2003) studied the elder patients' admission into a Geriatric Department and different scenarios to reduce the waiting time. Van der Meer et al. (2005) analyzed a case study with the goal of reducing elective patients' waiting times for the Department of Orthopedics. Although only for a single specialty, they found DES models to be a good communication tool between hospital administration and modelers. However, attempts to model whole hospitals are rare. This is because of the difficulty to represent the complexity of hospital activity within a simulation model that must be a simplification (Pidd 2003). Appropriate simplification can be a complex process, and it may be easier to segment off a discrete part of the hospital, such as a single ward. In this context, Emergency Departments are one of the most popular areas for DES modeling because they are relatively self-contained with easily observable processes. McGuire (1994) discussed the use of DES models to compare different alternatives for selecting a method to reduce the hospitalization of patients. The same goal was followed later by Samaha et al. (2003). Takakuwa and Shiozaki (2004) constructed a simulation model of the ED to examine patients' flow and waiting times. The reduction of patients' waiting time was made also by analyzing different "what-if" scenarios (Komashie and Mousavi 2005). Davies (2007) developed a computer simulation in which the triage process was eliminated and a simplified service was obtained by eliminating queues between patients that generally causes an increase of the waiting time.

12.3 Overview of the Methodology for Evaluating Resilience of Health Care Facility

The term resilience is used as the capacity of engineering and socio-economic systems to rebound after a severe disaster such as an earthquake (Cimellaro et al. 2010a, b). As it is known, a complete resilience framework must consider both the technical and the organizational aspects, that are the two parts of the whole framework and that are dependent each other. In this chapter, attention is given mainly to the organizational dimension of resilience for an ED, evaluated by means of a response variable. This important variable is patients' waiting time before receiving assistance, as it is the most significant parameter to identify the ED response. Based on this assumption, we can observe that resilience has an inverse trend with respect to the waiting time, so if resilience increases, the waiting time decreases.

12.4 Description of the Hospital

The hospital analyzed in this chapter is the Mauriziano "Umberto I" Hospital located in Turin, in the north of Italy. The hospital is almost 3 km far from the city centre and it contains 17 units which represent different wards. It was built in 1881 but several buildings were added during the twentieth-century and nowadays it covers an overall surface of 52,827 m^2. In this study, only the Emergency Department which is located in the building 17 was modeled. As in the other Italian emergency departments, all patients who enter are divided in four different color codes, according to the type of injury, using a procedure called "triage". *Red codes* (emergency) are patients with at least one of the vital functions compromised and in immediate danger of life. *Yellow codes* (urgency) have a partial impairment but there is not an immediate danger of life. *Green codes* (minor urgency) have lesions that do not affect vital functions while *White codes* (no urgency) are all patients who don't really need to be in the emergency department.

As consequence of this classification, the ER itself is divided in three main areas. An emergency area is located immediately in front, where the ambulance stops and contains two rooms in which red codes are stabilized. Parallel to this area, there is the yellow and green codes' area composed of five clinical labs. White codes are seen in their own clinical lab, during its working hours; otherwise, they have to wait until yellow and green patients' clinical labs are available. Inside the emergency department, there are also recovery rooms in which patients can stay before being discharged or recovered in another part of the hospital.

12.5 Data Analysis

The first step for developing the simulation model is to collect all necessary data involved in the modeling. Some of these data are statistical information provided by the hospital; other information was collected by interviews with hospital personnel. The data needed were the ED flow charts, the procedures inside the different rooms, as well as the personnel (known as resources) needed for each action and the patients' arrival rate to the ED.

As mentioned before, there are four different codes arriving to the ED and, for this reason, four different arrival rates of patients (Yi 2005) were calculated for the simulation, by using, the hospital's register statistics. Figure 12.1 shows the arrival rate for two types of code, known as the probability of having a fixed number of patients arriving during the day, divided into injury codes and into weekdays and weekend. The distribution of the patients' arrivals in a day is defined by the arrival cycle shown in Fig. 12.2. It represents the percentage of daily patients, defined by the arrival rate, that arrive in a certain instant of the day. It was also calculated by using hospital registers.

The patients' flow chart was both determined by interviews with personnel and statistical data provided by them. The different paths that a patient could follow after entering in each room are determined. Then, by analyzing the hospital statistics, the percentage of patients following each path is determined as well as the percentage of patients that changed their injury code because of an improvement or deterioration of their injury. Due to lack of space, this information is not shown in this chapter. The times and the resources involved in each process as well as the time table of the personnel are also used. The total resources used in the analyzed hospital are seven doctors, nine nurses, four assistants and two health workers.

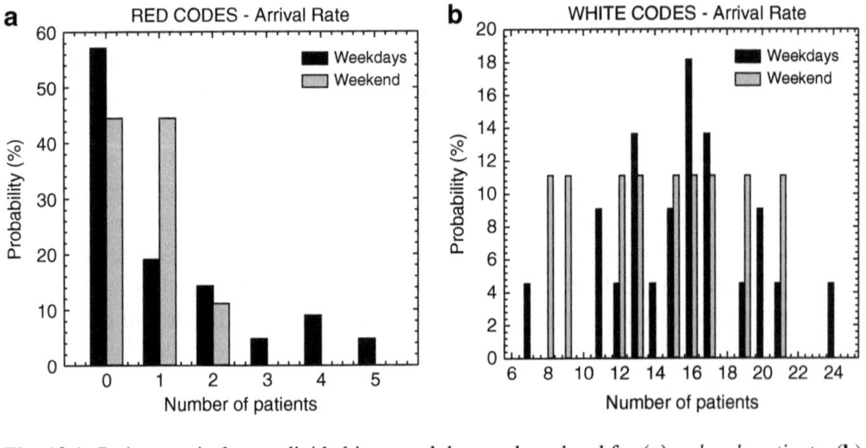

Fig. 12.1 Patients arrival rates divided into weekdays and weekend for (**a**) *red code* patients, (**b**) *white code* patients

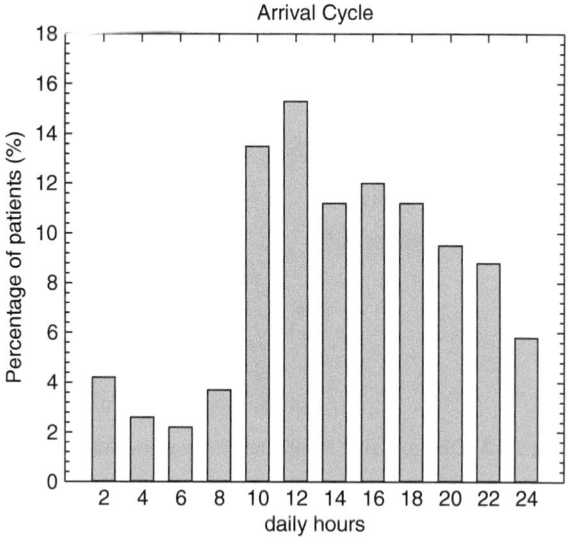

Fig. 12.2 Daily arrival cycle

12.6 Discrete Event Simulation Models

For the simulation of the hospital, a discrete event simulation (DES) model is used. The model is implemented using a commercial software called Promodel by defining the parameters needed as the patients' arrivals, the paths through the ED, the different rooms where patients could be attended, the processing times and the procedures that take place in each location, as well as all the resources such nurses or doctors. All these parameters are simulated in a DES model which is represented in Fig. 12.3, in which the different emergency rooms, paths, counters for the number of patients, patients and hospital personnel are shown.

For the study of the Emergency Department, some assumptions of the DES model are done to simplify the study. The model is run using the patient arrival rate in normal operating conditions and in emergency operating conditions after a seismic event. In the last case, it is assumed that the hospital structural and non-structural parts remain undamaged, meaning that the difference between the normal operating conditions and the emergency operating conditions is just in the patients' arrival rate. Moreover, it is assumed that, even after a dramatic event strikes, the system behavior remains the same with respect to the normal operating conditions. This is an unrealistic assumption because, in dramatic conditions, many characteristics of the system change, but these modifications have not been considered in this work. The current research wants to focus on how the system could react to an extreme event while remaining in normal operating condition mode without changing the initial properties of the simulated model.

As explained above, the arrival rates are divided into codes, while in the reality the injury code is determined once the patient has the first treatment at "triage".

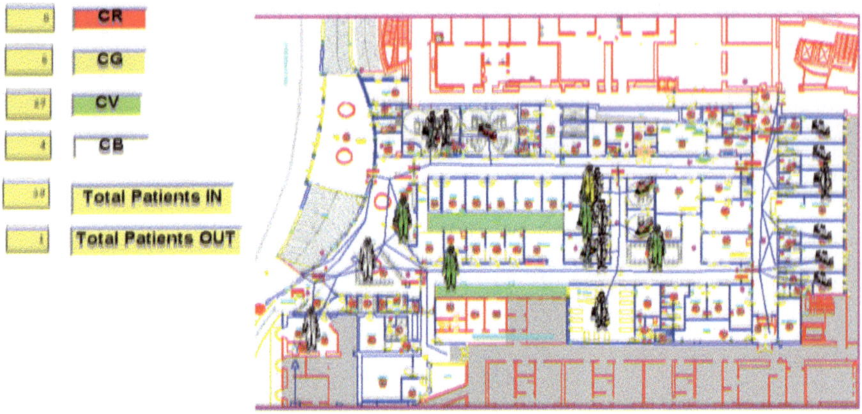

Fig. 12.3 DES model of Mauriziano Emergency Department

This assumption was done to distinguish the type of patients into codes from the first moment they arrive in the hospital. In the above paragraphs, it was explained that patients could change their injury code during their stay at the ED. It was considered that all patients that changed their code, they did it in the same point of their treatment.

12.7 Simulation of Hospital Operations

All the actions that patients could do inside the ED and how much time they spend in each location should be taken into account during the simulations, to build a model which represents properly the ED. Furthermore, it is essential to have a clear idea of how all the resources are used during the patients' path. For example, if a patient goes from a location to another, the resource utilized during this change of position has to be specified. In addition, when a patient enters a room, the time he spends with a doctor as well as with a nurse has to be indicated in the model. Each color code follows different paths and makes different actions in the ED. The process for each location describes every resource utilized and what happens inside the model. The red codes use two main rooms: the shock room and the intensive reanimation room. When red patients arrive at the ED, they can't wait because of their injury, so they are transferred immediately in one of these rooms. Therefore, a process describes which actions are performed inside the first location and another does the same for the second location. Similarly, all the processes related to the other color codes inside all the rooms are described. They include processes for triage, clinical labs and recovery rooms. Each location has a certain capacity so when it is full, patients have to wait in the waiting rooms before entering.

12.8 Results in Normal Operating Conditions

Particular attention is given during the simulations to the evaluation of the waiting time which is the time patients spend in the waiting rooms before being processed by the resources. The model can be calibrated by comparing the real data given by the hospital staff with the numerical results obtained by the numerical simulations. The results of the comparison are shown in Fig. 12.4, where the experimental values are daily averaged over the three time range considered. As can be seen in the bar charts, in Fig. 12.4, the numerical output fit the real data, showing the goodness of the model realized. The trend of waiting time for the 7 days simulation period is obtained for each code. In each day, a downward trend follows an upward trend because of the daily arrival cycle. In addition to patients' waiting times, the program

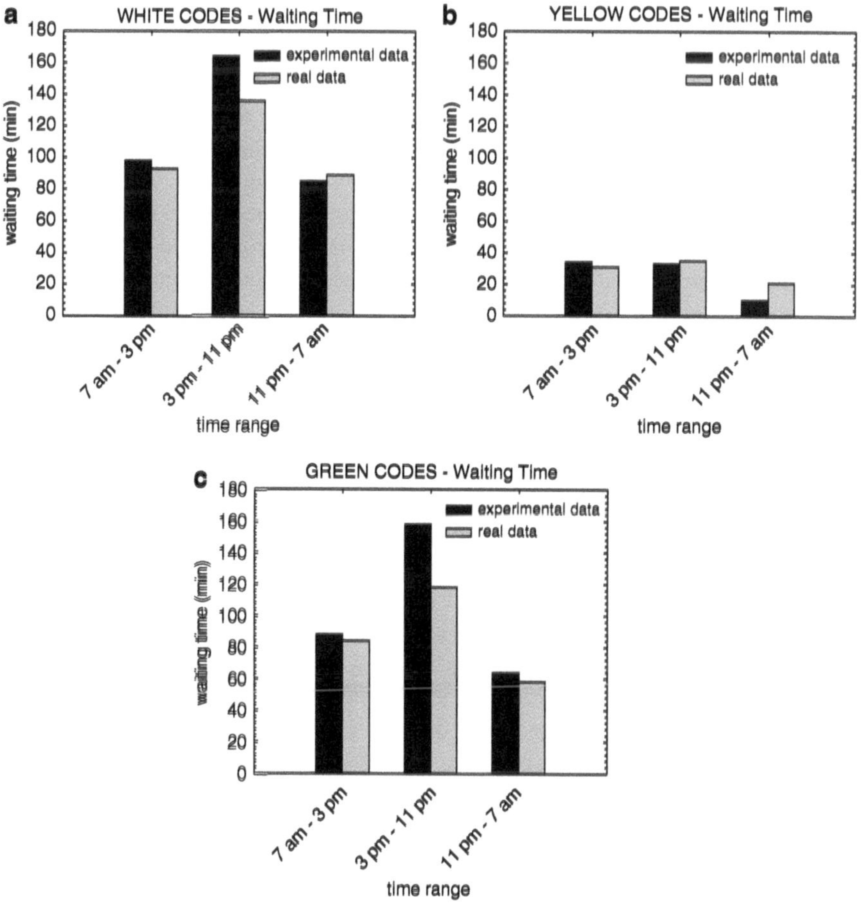

Fig. 12.4 Comparison of real vs. simulated data

provides as output other parameters that can be evaluated to verify the goodness of the model. Examples of these outputs are locations and resources utilization, entities and resources states, rooms' contents and entries trends for all the codes.

12.9 Seismic Input

The data used as input in the model is the patient arrival rate due to a catastrophic event such an earthquake. The arrival rate is determined by scaling the patient inflow of Northridge hospital calculated during the earthquake which occurred in 1994. It is known, from the collected data, that this event had severe impact on people and structures even if it was of modest magnitude. In fact, the epicenter was located in a high populated area. The pattern of the Northridge arrival rate is given in the paper of Cimellaro (Cimellaro et al. 2011). First, the PGA at the site of Mauriziano hospital is determined using the Italian seismic standard (NTC-08 2008), assuming a nominal life for building of strategic importance of 100 years and a 2,500 years return period, which corresponds to the Collapse Limit State (SLC). The final PGA value of the horizontal component is 0.123 g. Initially the arrival rate of Northridge earthquake was scaled to the PGA value measured in Turin, obtaining a ratio of 0.217. However, this scaling procedure has some limitations. In fact, the scaling procedure based on the PGA value doesn't take into account the real level of damage related to the area in which the earthquake happens. The population density and the urbanization level, which represent important indices for the evaluation of the effect of an earthquake and for the calculation of the number of patients arriving in a hospital, are not taken in account. Following the considerations above, the Modified Mercalli Intensity (MMI) scale which takes in account all these features above has been used as second scaling procedure. The MMI scale ranges from I (no felt) to XII (total destruction) and it is able to quantify the effects of an earthquake on the Earth's surface, and all aspects connected to humans and man-made structures. The MMI scale was determined using the relationships available in literature which associate the MMI values to the horizontal average and maximum PGA at the site. The average value obtained by the different relationships has been considered, so MMI is 9 for Northridge and 7 for Turin, so the scale factor used in this case is 0.778. Figure 12.5 shows the plot of the arrival rate for the 3 days period after the earthquake strikes related to Northridge, scaled with respect to the ratio between the PGA and the MMI values to obtain the arrival rate in Turin.

Even if the patient/treatment data is usually, a low priority activity during emergency, Northridge earthquake is a well-documented event, so the seismic arrival rate shown in Fig. 12.5 was then divided in the different color codes following the same distribution proposed by Pengfei (Yi 2005). In particular, knowing 1 week's injury data after the earthquake, patients have been grouped in six categories and then, analyzing the type of injury, each category has been directly compared with one color code. In Table 12.1 are shown the results of this aggregation of data and the changes in the percentages when the system moves from normal to emergency operating conditions.

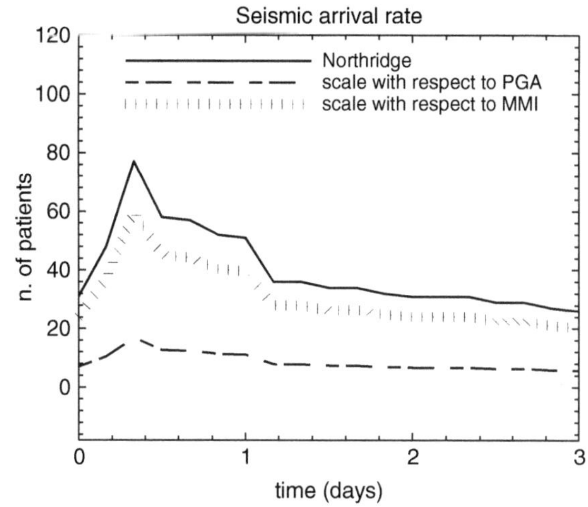

Fig. 12.5 Graphic representation of thc arrival rates for Northridge earthquake and Mauriziano arrival rate calculated using the PGA and the MMI scale factor

Table 12.1 Percentage of patients arriving to the Emergency Department in both normal operating conditions and emergency conditions

	Normal scenario	Emergency scenario	Number of patients
Red codes	0.56 %	3.7 %	21
Yellow codes	16.78 %	40.1 %	271
Green codes	71.19 %	48.49 %	224
White codes	11.47 %	7.81 %	43

As shown in Table 12.1, the emergency operating condition is characterized by a dramatic increase in the percentages related to the major codes, while minor codes reach lower percentages with respect to the normal operating condition.

12.10 Numerical Results During the Emergency

The numerical simulations of the DES hospital model are performed by dividing the seismic input in three main parts. The system starts to work in normal operating conditions for a 7 days period; these days are needed to achieve a steady state condition which is not influenced by the initial conditions. Then, they are followed by 3 days in emergency operating conditions determined using the scaled arrival rate evaluated in previous paragraph. The simulation ends with 3 days of normal operating conditions in which the system goes back to its condition before the earthquake. Globally, a 13 days period is run in each simulation. Different output variables have been collected to have a clear picture of how the overall model by means of a net variation of the patients' inflow. Monte Carlo simulations were run by performing the 13 days simulation multiple times. In this way, the random differences between multiple runs are taken into account. The average waiting time

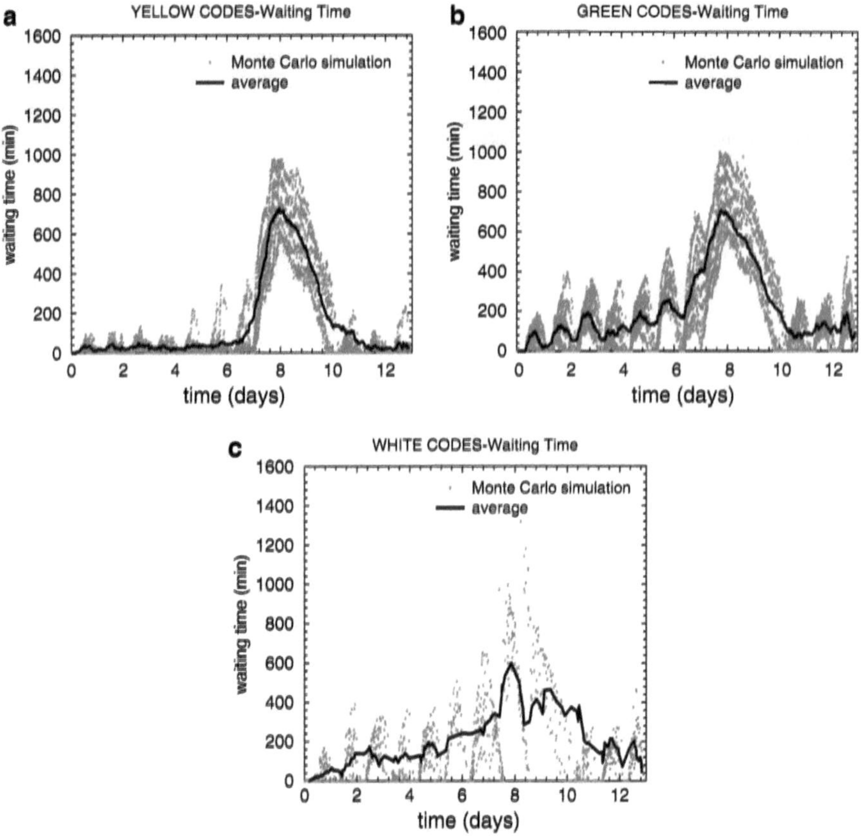

Fig. 12.6 Waiting time for (**a**) *yellow codes*, (**b**) *green codes*, (**c**) *white codes* under seismic input

is plotted in Fig. 12.6, for yellow, green and white codes respectively. As can be observed, the waiting time increases when the earthquake strikes and goes back to the initial value after the 3 days of emergency period.

12.11 Metamodel

The simulation model that has been built is a simplification of real-world system in which, with few parameters, it is possible to have an idea of how a complex system such as the Emergency Department is. However, due to the necessity to model all the hospitals in the disaster area and run the respective DES models multiple times to have statistically meaningful results, it is impossible to use these simulations for evaluating real-time hospital capacity. For these reasons, there is need to create a simpler model, called *metamodel*, which is a set of equations that do not require many parameters such as in the case of discrete events simulation models (DES), therefore it becomes a good candidate for modeling operations for any general hospital in disaster conditions. The purpose of the metamodel is describing the system behavior by means of few response variables. As aforementioned, the response variable chosen in this work is the waiting time, because it directly impacts the survivability defined as the maximum amount of time allowed before the patient is treated to avoid fatality. After a dramatic event strikes, the waiting time value changes with respect to normal operating conditions. The transient behavior of waiting time is the result of a change in patient arrival rate and can be represented for instance, with a double exponential function expressed by means of several parameters (Cimellaro et al. 2011; Yi 2005). The evaluation of all the parameters requires calibration. The parameters of the metamodel must be correlated to the hospital characteristics, such as the number of beds, the number of doctors, the efficiency of the emergency rooms etc. The current research is trying to define a metamodel for all the various color codes, improving previous existing models (Yi 2005) which are based on the entire ED. The reason of this choice derives from the different priority assigned to each code that leads to a totally different output. To obtain four different equations, different parameters for each code have to be selected. Then, it is possible to use the curves for each code, and applying those to different hospitals by changing the parameters of the model. In this way, an organizational model for different hospitals could be obtained, without passing through the complex world of DES models.

12.12 Future Work

The next step will be including in the organizational model the damage of all the content inside a given hospital. The PACT software will be used because it contains more than 700 fragility curves of structural and nonstructural vulnerable elements in the buildings. A new hospital business model will be developed which will consider not only the structural and nonstructural content, but also the organizational aspects in the evaluation of the initial investment. Interdependencies (Cimellaro et al. 2014) among structural and organizational dimension will also be taken in account in probabilistic terms. Information about the hospital is needed for better seismic life cycle assessment. Prices of equipment and fragility functions of content and structure would allow the inclusion of the medical equipment in the assessment together with the organizational dimension with the goal to find an optimal structural/organizational system which makes healthcare facilities more resilient. The goal is to show a reduction of annualized repair costs if resilience-based design is used. The final goal is trying to answer to the following questions. Where are we willing to invest money and research resources to reduce the risk and where should we invest first? What is, finally, the risk the society can afford?

12.13 Concluding Remarks

Critical buildings such as hospitals have a critical role in our society and this is the reason why they must remain operational during catastrophic events. In this context, modeling the organizational part of the hospital is becoming more and more important. Several studies are taking into account this aspect and particular attention is given in identifying a fundamental parameter that can describe the organizational performance of the hospital. The hospital analyzed in this work is the emergency department of an Italian hospital and the patients' waiting time is the parameter that more represents its behavior. An organizational model was built using the commercial software "Promodel" which contains all the processes that are generally needed inside the ED in normal operating conditions. Then the system is subjected to a seismic input which is characterized by a higher patient arrival rate at the ED. The trend of the waiting time is found and a clear picture of how the hospital response changes is finally obtained. However, building a discrete event simulation model is time consuming; therefore, other simplified options should be considered which can easily represent the transient behavior of the waiting time from normal to emergency operating conditions. This simplified model is called *metamodel*. It consists of an equation which is able to describe the waiting time increasing during a catastrophic event for each color code.

Acknowledgments The authors would like to thank Dr. Chris Poland for his suggestions and discussions which improved the quality of the chapter. The research leading to these results has also received funding from the European Community's Seventh Framework Programme – Marie Curie International Reintegration Actions – FP7/2007–2013 under the Grant Agreement n° PIRG06-GA-2009-256316 of the project ICRED – Integrated European Disaster Community Resilience and by the Marie Curie International Outgoing Fellowship (IOF) Actions-FP7/2007–2013 under the Grant Agreement n°PIOF-GA-2012-329871 of the project IRUSAT— Improving Resilience of Urban Societies through Advanced Technologies.

References

Boxerman SB (1996) Simulation modeling: a powerful tool for process improvement. Best Pract Benchmarking Healthc 1(3):109–117

Cimellaro GP, Reinhorn AM, Bruneau M (2010a) Framework for analytical quantification of disaster resilience. Eng Struct 32(11):3639–3649

Cimellaro GP, Reinhorn AM, Bruneau M (2010b) Seismic resilience of a hospital system. Struct Infrastruct Eng 6(1–2):127–144

Cimellaro GP, Reinhorn AM, Bruneau M (2011) Performance-based metamodel for health care facilities. Earthq Eng Struct Dyn 40:1197–1217

Cimellaro GP, Solari D, Bruneau M (2014) Physical infrastructure interdependency and regional resilience index after the 2011 Tohoku earthquake in Japan. Earthq Eng Struct Dyn. doi:10.1002/eqe.2422

Davies R (2007) "See and treat" or "see" and "treat" in an emergency department. In: Proceedings of the 2007 winter simulation conference, Washington, DC, pp 1519–1522

Komashie A. Mousavi A (2005). Modeling emergency departments using discrete event simulation techniques. In: Proceedings of the winter simulation conference, Orlando, pp 2681–2685
Martin E, Grønhaug R, Haugene K (2003) Proposals to reduce over-crowding, lengthy stays and improve patient care: study of the geriatric department in Norway's largest hospital. In: Proceedings of the 2003 winter simulation conference, New Orleans, pp 1876–1881
McGuire F (1994) Using simulation to reduce length of stay in emergency departments. In: Proceedings of the 1994 winter simulation conference, Orlando, pp 861–867
NTC-08 (2008) Nuove Norme Tecniche per le Costruzioni (NTC08) (in Italian). Consiglio Superiore dei Lavori Pubblici, Ministero delle Infrastrutture, Gazzetta Ufficiale della Repubblica Italiana, 4 febbraio 2008, n. 29
Pidd M (2003) Tools for thinking: modelling in management science, 2nd edn. Wiley, Chichester
Samaha S, Wendy AS, Darrell S (2003) The use of simulation to reduce the length of stay in an emergency department. In: Proceedings of the 2003 winter simulation conference, New Orleans, pp 1876–1881
Takakuwa S, Shiozaki H (2004) Functional analysis for operating emergency department of a general hospital. In: Proceedings of the 36th conference on winter simulation, IEEE, Washington, DC, 5–8 Dec 2004
Van der Meer RB, Rymaszewski LA, Findlay H et al (2005) Using OR to support the development of an integrated musculoskeletal service. J Oper Res Soc 56:162–172
Yi P (2005) Real-time generic hospital capacity estimation under emergency situations. Department of Industrial Engineering. State University of New York, Buffalo

Part V
Seismic Protection

Chapter 13
On the Seismic Behaviour of Viscously Coupled Shear Walls

O. Lavan

Abstract In the last couple of decades the use of energy dissipation devices for earthquake mitigation has gained much attention. Most research in that field, however, focused on frame buildings. This chapter is based on (Lavan, Earthq Eng Struct Dyn 41(12):1673–1692) and discusses the possibilities of using energy dissipation devices, or more specifically viscous dampers, in wall structures to result in viscously coupled shear walls (VCSWs). In turn, insight to their behaviour, as well as simplified approximate tools for their analysis and initial design are presented. While very simple to use, those rely on a strong theoretical background.

13.1 Introduction

The use of energy dissipation devices, and in specific viscous dampers (VDs), for seismic mitigation is becoming widespread (Soong and Dargush 1997; Constantinou et al. 1998; Christopoulos and Filiatrault 2006; Takewaki 2009). Those were shown to effectively reduce both displacement related responses and force related responses due to the out-of-phase effect in various types of structural systems (Constantinou and Symans 1992; Lavan and Dargush 2009; Lavan 2012). In parallel to their development (Constantinou and Symans 1992), and to their growing application in practice (Soong and Dargush 1997; Constantinou et al. 1998; Christopoulos and Filiatrault 2006), methodologies for the analysis and design using such VDs have been proposed and studied as well (Agrawal and Yang 1999; Attard 2007; Aydin et al. 2007; Dargush and Sant 2005; Fujita et al. 2010; Gluck et al. 1996; Kanno 2013; Lavan and Avishur 2013; Lavan and Levy 2005, 2006; Lavan et al. 2008; Levy and Lavan 2006; Lopez and Soong 2002; Sonmez

O. Lavan (✉)
Technion – Israel Institute of Technology, Technion City, Haifa 32000, Israel
e-mail: lavan@tx.technion.ac.il

G.P. Cimellaro et al. (eds.), *Computational Methods, Seismic Protection, Hybrid Testing and Resilience in Earthquake Engineering*, Geotechnical, Geological and Earthquake Engineering 33, DOI 10.1007/978-3-319-06394-2_13,
© Springer International Publishing Switzerland 2015

et al. 2013; Takewaki 1997, 2000; Shukla and Datta 1999; Whittle et al. 2012; Zhang and Soong 1992). Most of the applications of VDs and the design methodologies proposed focused on frame buildings.

The extension of the use of VDs for the seismic mitigation in wall buildings is very limited. While in frame structures the efficient locations for potentially allocating such VDs are obvious, in wall structures some creativity is required. Examples for such creativity could be found in the conceptual designs presented by (Trombetti and Silvestri 2004, 2006; Silvestri and Trombetti 2007; Smith and Willford 2007; Madsen et al. 2003; Marko et al. 2004; Rahimian 2002; Taylor 2002). Trombetti and Silvestri (2004, 2006) and Silvestri and Trombetti (2007) proposed allocating viscous dampers between frames and walls at every floor in wall-frame structures. Since the deflected shapes of walls and frames are different, the dampers experience large relative velocities between their ends, hence, efficiently dissipating energy. Another efficient configuration, "the damped outrigger concept", locates dampers between outriggers attached to the core, and the perimeter structure (e.g. Smith and Willford 2007). The large relative velocities expected in those locations lead to efficient energy dissipation. Another approach allocated dampers within openings (cut-out sections) in shear walls (Madsen et al. 2003; Marko et al. 2004). This approach was found to be efficient, in particular when the dampers were allocated at cut-out sections of the shear wall at the lower stories. The use of viscous dampers as coupling elements between two "bending" elements was originally proposed by Rahimian (2002) in the form of "Coupled Truss Systems with Damping". In that steel structure, vertical trusses served as the bending elements.

The latter conceptual design seems promising, and has even been implemented in practice in the Torre Mayor building in Mexico City (Taylor 2002). Nonetheless, some insight on the behaviour of such systems, their controlling parameters and their effect on the responses of interest, as well as tools for initial design, are required. The importance of such insight and such tools should never be underestimated. While advanced tools for analysis are readily available these days and could be easily used in practice, insight and simplified "back of the envelope" analysis and initial design tools can never be replaced. Simplified analysis tools are important for the following reasons:

- They enable a simple estimation of responses of interest to validate results from more complex analyses.
- They enable the identification of the main structural elements.
- They provide means for engineers to know what results to search for in the pile of results attained by complex analyses, and where to look for potential issues in given designs.

While analysis starts with a given structure and seeks for its response. Design answers the reverse question. Given the desired response, what are the structural properties that would achieve it? A reverse procedure to answer this question could rarely be formulated in the context of earthquake engineering. Furthermore, suppose such a procedure existed, insight and initial design tools would further enable to:

- Choose efficient structural systems for consideration and compare the different alternatives.
- Check instantly if a design alternative is feasible.
- Estimate the sizes of structural elements required and estimate the cost.
- Understand why a given design may not perform well and identify and correct the conceptual mistakes that prevent them from performing efficiently.

This chapter discusses a work that answers exactly these questions (Lavan 2012). The main findings of this work will be presented herein, and flowcharts for their use for practical approximate analysis and initial design will be given.

Section 13.2 presents the VCSWs system and the modelling assumptions. This is followed in Sect. 13.3 by the study results where the controlling parameter on the design of such systems is identified ant a parametric study is presented. Insight to the effect of the controlling parameter on various responses is given in Sect. 13.4. Sections 13.5 and 13.6 present simplified analysis and design methods, respectively. These are based on the findings of Sects. 13.3 and 13.4. Finally Sect. 13.7 presents the conclusions.

13.2 Description of the VCSWs System and Modelling Assumptions

The VCSWs investigated by Lavan (2012) consists of two walls connected by viscous dampers that can only transfer shear forces due to shear velocity (Fig. 13.1a, b). As the floors assumed relatively rigid in their own plane, the relative horizontal displacement between the walls is restrained. Note that the same equations to represent the VCSWs also represent another system. This consists of two parallel systems: A shear wall and a viscously braced frame that has zero stiffness to shear deformation but does resist shear velocity (for more details see Lavan 2012). We will focus herein only on the VCSWs system.

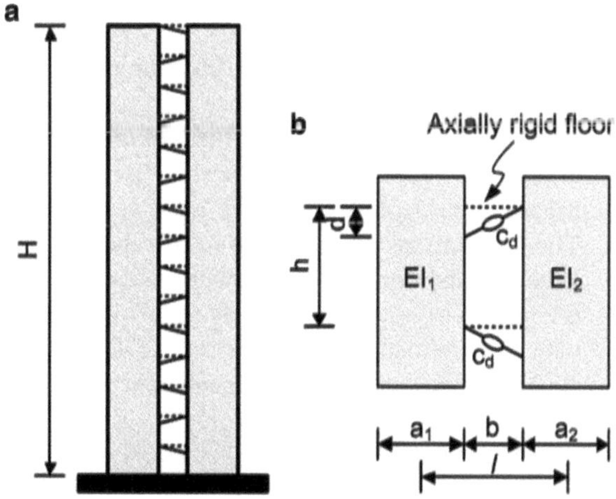

Fig. 13.1 Viscously coupled shear walls: (**a**) layout, and (**b**) parameters (Lavan 2012)

In order to attain approximate and simplified solutions the following assumptions were made:

- **A plane model is considered.** This, of course, limits the discussion to structures where the torsional response is limited.
- **All model parameters are assumed constant along the height.** While mass and stiffness are indeed close to uniform in most structures, the discussion is limited to constant damping along the height. Nonetheless, the conclusions drawn regarding feasibility of using viscous dampers in wall buildings, and the nature of the controlling parameter, are valid for other distributions of damping.
- **The behavior of the system can be approximated by linear analysis**. As use is made of viscous dampers, whether in the design of new structures or for the purpose of retrofitting, the design objective would usually limit plasticity in the structure. Hence, the damped structure is expected to behave linearly or close to it, and linear analysis is expected to lead to reasonable approximation of behavior. Moreover, as the considered structures are "regular", the "equal displacements approximation" is expected to lead to reasonable results in that range.
- **Axial deformations are neglected**. In the viscously coupled shear wall system it is assumed that the axial deformation of the walls is small. In the wall-viscous frame system axial deformations of the frame on which dampers are installed are neglected. This leads to an approximate and simple analysis method with a single controlling parameter. This assumption is valid in cases where the damping is not too large and the deformations in the "damping system" (to be defined subsequently) concentrate in the dampers rather than mostly in axial deformations. For the range of damping typically used in buildings, this assumption seems reasonable.
- **The viscous dampers are linear and are assumed to be installed on relatively rigid diagonals**. Viscous dampers can be designed, in general, as nonlinear. Linear viscous dampers, however, have advantages over nonlinear dampers since the forces they produce are out of phase with the forces due to deformations. Hence, they are adopted in this study. Those are usually mounted on relatively axially stiff braces to fully utilize their efficiency.
- **The walls deform in bending only, i.e. shear deformations are neglected**. This assumption is often used for the analysis of elements of large span to depth ratio and has been extensively used for the analysis of conventional coupled shear walls.
- **The out-of-plane bending stiffness of the slabs is neglected**. In general, the out of plane bending stiffness of the slabs is indeed small. Note that their effect may become considerable in very tall buildings.
- **The foundations are assumed to be relatively rigid**.
- **When added dampers are considered, the inherent damping is neglected**. It is common to assume a relatively small inherent damping in structures retrofitted with viscous dampers. As the added damping would usually result in large damping ratios, the effect of the inherent damping on the responses is negligible.

13.3 Solution Strategy and Study Results

This section summarises the derivations and steps taken in Lavan (2012) and presents the main equations and results. For more details the reader is referred to Lavan (2012).

13.3.1 An Equivalent Continuum Model and Its Governing Equations

In many problems in engineering, where the structure is natively modeled using a discrete model, a continuum equivalent model may well replace the discrete one (see e.g. Reinhorn et al. 1977). In the context of the problem at hand, this is the case when the number of stories is relatively large. For the continuum models, analytical or semi-analytical solutions could be attained. Such a continuum model was also used by Prof. Reinhorn in some of his works (see e.g. Wang et al. 1992; Gluck et al. 1979; Reinhorn et al. 1977; Kusumastuti et al. 2005; Rutenberg et al. 1978; Cimellaro et al. 2006).

The equation of motion of the continuum model of the system assuming constant values of the parameters throughout the height of the structure is given by (Lavan 2012):

$$m\frac{\partial^2 u(z,t)}{\partial t^2} - c\frac{\partial^3 u(z,t)}{\partial z^2 \partial t} + EI\frac{\partial^4 u(z,t)}{\partial z^4} = -m\frac{\partial^2 u_g(t)}{\partial t^2} \tag{13.1}$$

where u = horizontal relative displacement as a function of time, t, and height, z; m = mass per unit length; EI = flexural rigidity of the walls; c = added shear damping as given below; and u_g = ground displacement. The sum of the first expression on the left hand side and the expression on the right hand side of Eq. (13.1) represents the external forces required to maintain the absolute accelerations of the system at time t. The second expression on the left hand side represents the external forces required to maintain the velocities of the system at time t. The system resisting those forces is the "damping system". The third expression on the left hand side of Eq. (13.1) represents the external forces required to maintain the displacements of the system at time t. The system resisting those forces is the "stiffness system". The parameters for viscously coupled shear walls take the values $EI = EI_1 + EI_2$ and $c = c_d d^2 l^2 / h(h^2 + b^2)$ (see parameters' description in Fig. 13.1b).

Equation (13.1) is accompanied by the boundary conditions that require zero displacement and angle at the base relative to the ground, zero wall moment and zero total shear at the top ($z = H$ = wall height).

Once a solution of the governing equations is attained, various responses of interest could be derived. In the case of VCSWs those are the horizontal displacements, *D*; inter-story drifts, *ID*, from which the damper's stroke could be computed as $stroke = ID \cdot (d \cdot l) / \left(\sqrt{d^2 + b^2}\right)$; total accelerations, *A*; total horizontal shear, *TS*; horizontal shear in the stiffness system (the one to resist displacements), *WS*; horizontal shear in the damping system (the one to resist velocities), *DS*, from which forces in the dampers could be computed as $F_D = DS \cdot \left(h \cdot \sqrt{d^2 + b^2}\right) / (d \cdot l)$; total bending moment, *TM*; bending moment in the stiffness system, *WM*; and bending moment in the damping system, *DM*, from which axial forces in the walls could be computed as $N = DM/l$.

13.3.2 Dimensional Analysis and Identification of the Controlling Parameter

To enable a rigorous parametric study, dimensional analysis could be used. In these cases the number of controlling parameters could be reduced, and the results could be generalized to represent many cases that were not analyzed with their dimensional properties. For more details on that the reader is referred to Makris and Black (2004a, b) and references therein. In the context of the problem at hand the governing equations are brought to a non-dimensional form. This enabled reducing the number of controlling parameters to one while identifying this parameter, as follows:

Equation (13.1) and the corresponding boundary conditions were brought to a non-dimensional form. This was done using the following non-dimensional parameters: $\delta = u/H; \eta = z/H; \tau = t\sqrt{EI/mH^4}; 2\xi = \sqrt{c^2/mEI}; \delta_g = u_g/H$. It was further found that the response is controlled by ξ only. It was further noted that, in contrast to coupled shear walls, the controlling parameter in the viscously coupled shear walls, $\xi = 1/2\sqrt{c^2/mEI}$, does not depend on the height of the structure. This is a very important observation since it implies that, in contrast to the conventional coupled shear walls system, the VCSWs system can be efficient also for low-rise buildings.

13.3.3 Eigenvalue Solution and Complex Modal Spectral Analysis

In Lavan (2012) the eigenvalue problem of the non-dimensional governing equations was solved. The non-dimensional natural frequencies and damping ratios for the first three modes and various values of ξ are given in Table 13.1.

The non-dimensional properties could be transformed to their dimensional counterparts using:

Table 13.1 Non-dimensional natural frequencies and damping ratios for various values of ξ (Lavan 2012)

	$\xi=0.0$		$\xi=0.1$		$\xi=0.2$		$\xi=0.3$		$\xi=0.4$	
Mode	$\overline{\omega}_i$	$\overline{\xi}_i$	$\overline{\omega}_i$	$\overline{\xi}_i$	$\overline{\omega}_i$	$\overline{\xi}_i$	$\overline{\omega}_i$	$\overline{\xi}_i$	$\overline{\omega}_i$	$\overline{\xi}_i$
1	3.516	0.000	3.525	0.132	3.553	0.266	3.606	0.403	3.697	0.546
2	22.035	0.000	22.067	0.148	22.181	0.300	22.454	0.462	23.197	0.649
3	61.697	0.000	61.704	0.126	61.730	0.255	61.808	0.391	62.098	0.551

$$\omega_i = \sqrt{\frac{EI}{mH^4}}\overline{\omega}_i; \quad \xi_i = \overline{\xi}_i \tag{13.2}$$

It was further shown (Lavan 2012) that the peak contribution of the mode i and its conjugate to the response of interest can be estimated by means of a response spectrum analysis (similar to the one proposed by Song et al. (2008) for discrete systems) as

$$x_{0,\max,i}(\eta) = R_{0,i} \cdot S_{a,i} \tag{13.3}$$

where $x_{0,\max,i}(\eta)$ is the peak contribution of the mode i and its conjugate to the response of interest, $R_{0,i}$ the modal contribution of the mode i and its conjugate to the response of interest and $S_{a,i}$ is the spectral pseudo acceleration. The modal contributions, $R_{0,i}$, were derived for the various responses of interest in the non-dimensional domain. Figure 13.2 presents graphs of these coefficients as function of height. Each graph presents the coefficient using 5 values of ξ: 0, 0.1, 0.2, 0.3, 0.4. The first row in Fig. 13.2 presents the contribution of the first pair of conjugate modes while the second and third rows present the contributions of the second and third pairs of modes, respectively. Those can then be brought to the dimensional domain as follows: $R_{D,i} = mH^4/EI \cdot R_{d,i}$; $R_{ID,i} = mH^3/EI \cdot R_{id,i}$; $R_{A,i} = R_{a,i}$; $R_{TS,i} = mH \cdot R_{ts,i}$; $R_{WS,i} = mH \cdot R_{ws,i}$; $R_{DS,i} = mH \cdot R_{ds,i}$; $R_{TM,i} = mH^2 \cdot R_{tm,i}$; $R_{WM,i} = mH^2 \cdot R_{wm,i}$; $R_{DM,i} = mH^2 \cdot R_{dm,i}$. As the periods of the various modes are widely spread (see Table 13.1), thy can be combined using the SRSS rule leading to a good approximation of the peak response.

13.4 Insight to the Behaviour of VCSWs

The tables and graphs given in the previous section present a powerful tool for an approximate analysis of VCSWs systems. Nonetheless, it is hard to gain insight to the effect of damping by using them, as presented. This is because the R values extracted from those graphs are to be multiplied by the pseudo acceleration to attain the responses. Nonetheless, for each ξ the values extracted from the graphs are associated with different response spectra. In order to bypass this issue and enable

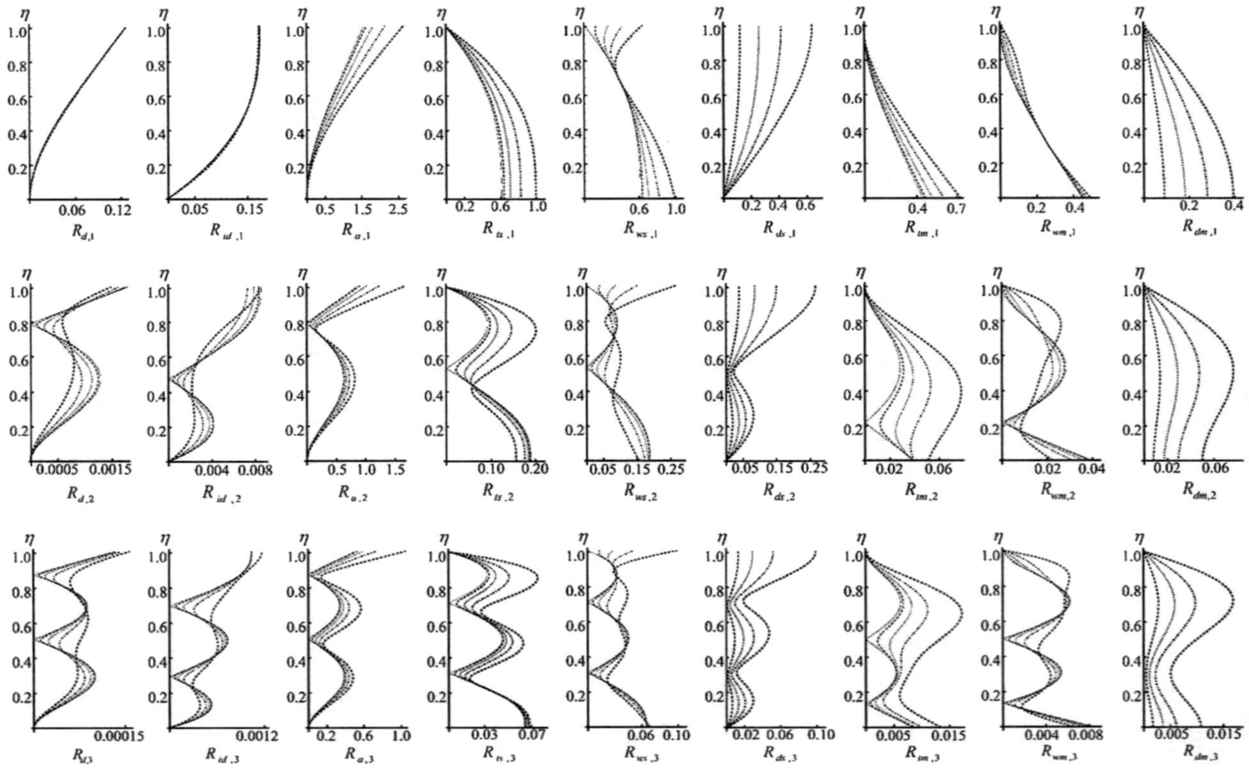

Fig. 13.2 *R* coefficients for various responses (columns) for the first three modes (rows) using various damping values ($\xi = 0$ *continuous*, $\xi = 0.1$ *dashed*, $\xi = 0.2$ *dotted*, $\xi = 0.3$ *dashed-dotted*, $\xi = 0.4$ *dashed thick*) (Lavan 2012)

a clear comparison between the responses of systems with different ξ values, the responses R where multiplied by the value of

$$R_{\xi,i} = \frac{S_a(\xi_i)}{S_a(0.05)} = \sqrt{\frac{0.07}{0.02 + \xi_i}} \tag{13.4}$$

where $R_{\xi,i}$ is the conversion factor from pseudo-acceleration computed for 5 % damping, $S_a(0.05)$, to pseudo-acceleration for a damping ratio ξ_i, $S_a(\xi_i)$. This expression is adopted from CEN (1998). To obtain the contribution of each pair of conjugate modes to the desired response, the plots in Fig. 13.3 are to be multiplied by the pseudo acceleration at the relevant period, with 5 % damping, irrespective of ξ_i. The plots in Fig. 13.3 for the various values of ξ can thus be directly compared to gain some insight on the effect of ξ. The values presented for $\xi = 0$ represent the structure with no added damping, but with 5 % Caughey damping, i.e. 5 % damping in every mode. Note that here the analysis of the viscously damped structures ignore the inherent damping in each mode. This approximation is often made when damping is added to the structure and its deformations considerably reduce.

As can be seen, a ξ value of 0.2 leads to a reduction of circa 40–50 % in all responses except those in the damping system. Values of $\xi > 0.2$ may lead to a further slight reduction in those responses. This, however, will be accompanied by a large increase in the forces in the damping system. In turn, this would lead to a large increase in the peak forces in the dampers and axial forces in the columns/walls. Hence it is suggested using values of $\xi \leq 0.2$.

13.5 Simplified Analysis Approach

Following is the outline for the simplified analysis approach:

1. Estimate the building's mass per unit height, m, and the cracked bending stiffness of the walls, EI;
2. Estimate the damping parameter as $c = \frac{c_d d^2}{n \cdot h \left(d^2 + b^2\right)} \cdot l^2$ (see Fig. 13.1 for more details on the notation);
3. Calculate $\xi = \frac{1}{2}\sqrt{c^2/mEI}$ and evaluate the natural frequencies and damping rations $\omega_i = \sqrt{\frac{EI}{mH^4}}\overline{\omega}_i$ and $\xi_i = \overline{\xi}_i$ where $\overline{\omega}_i$ and $\overline{\xi}_i$ are taken from Table 13.1;
4. Calculate values of pseudo-acceleration corresponding to ω_i and ξ_i of the relevant modes;
5. Calculate the contributions of each relevant mode to the responses of interest using Fig. 13.2 and the corresponding transformations to the dimensional plane (end of Sect. 13.3.3). Combine the contributions of all modes using the SRSS rule.

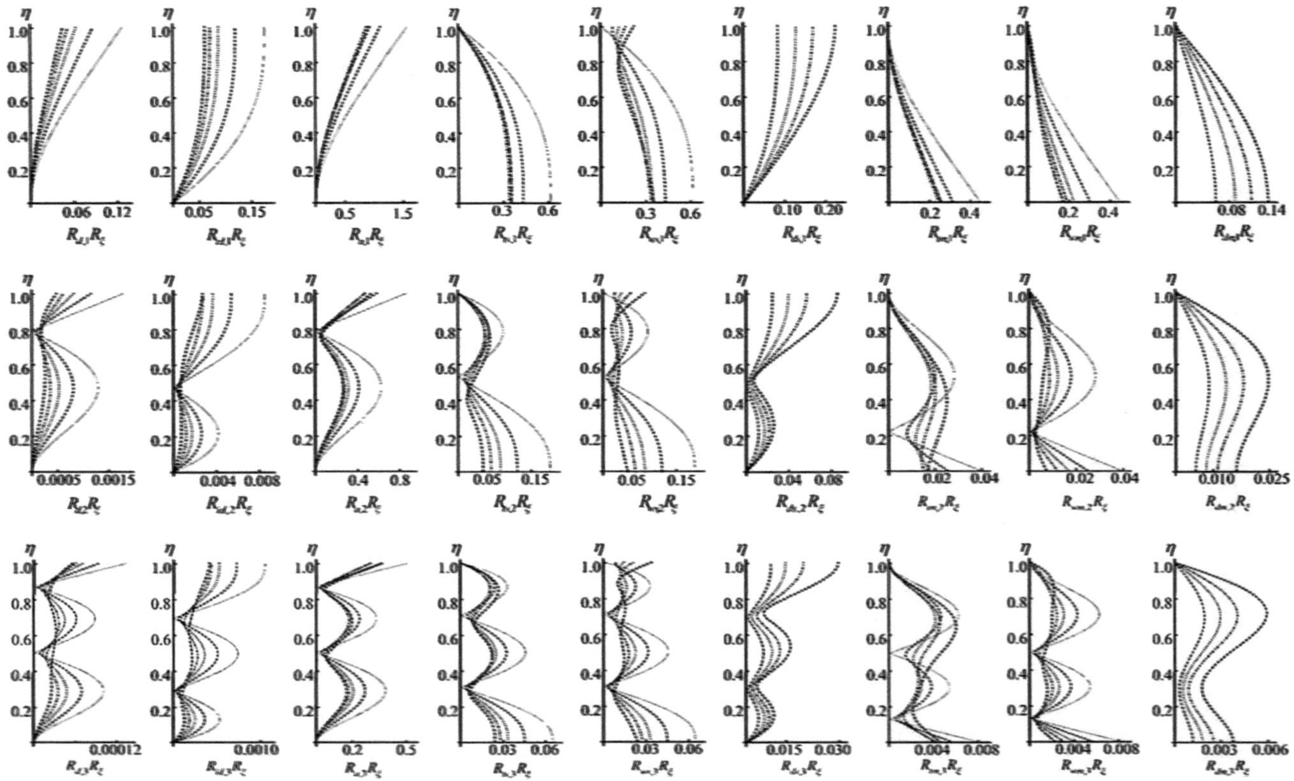

Fig. 13.3 R coefficients multiplied by R_{ξ} for various responses (columns) for the first three modes (rows) using various damping values ($\xi = 0$ *continuous*, $\xi = 0.1$ *dashed*, $\xi = 0.2$ *dotted*, $\xi = 0.3$ *dashed-dotted*, $\xi = 0.4$ *dashed thick*) (Lavan 2012)

Note that the analysis is approximate and in any case one should verify the design according to code requirements and analyses.

13.6 Initial Design Approach

Following is the outline for the initial design approach:

1. Perform an approximate modal analysis to the bare building as indicated in Sect. 13.5 with zero added damping;
2. Determine the required response reduction in relevant responses of interest;
3. Using Fig. 13.3, estimate the required ξ value;
4. Perform an approximate modal analysis to the damped structure as indicated in Sect. 13.5;
5. Evaluate the damping coefficient for the single damper so as to attain the required ξ value from step 3 (see equations in step 2 of Sect. 13.5). Estimate the peak force expected in the damper and its expected stroke (see transformation formulae from *DS* and *ID*, respectively, at the end of Sect. 13.3.1).

13.7 Conclusions

This chapter summarizes the findings of Lavan (2012) regarding the efficiency and the way VCSWs work: It sheds some light on and gives some insight to the behaviour of such systems; it gives tools for approximate analysis, and; it gives tools for their initial design.

The following observations regarding VCSWs systems could be made:

1. Under the assumptions made, a single parameter controls the response reduction of viscously coupled shear walls or wall-viscous frames w.r.t the corresponding uncoupled wall system.
2. In contrast to conventional coupled shear walls and wall-frame structures, the controlling parameter in the viscously coupled shear walls and wall-viscous frame systems, $\xi = 1/2\sqrt{c^2/mEI}$, does not depend on the height of the structure. This is a very important observation since it implies that this system can be efficient also for low rise buildings. Of course, depending on the aspect ratio, shear deformation of the wall system (not considered here) may play a role in those systems and should be accounted for.
3. The addition of damping can lead to a considerable reduction in most responses of interest (up to 60 % for cases where the relation given by CEN (1998) for the effect of damping on SDOF systems' response holds) in most responses of interest. These are: displacements, inter-story drifts, total accelerations, total and wall shear, overturning moment and wall bending moments. Based on the example, such a reduction is feasible using "off-the-shelf" dampers.

4. Increasing ξ further than 0.2 may lead to a slight further reduction in the responses listed in Item 2 above. This, however, will be accompanied by a large increase in the forces in the damping system. In turn, this would lead to a large increase in dampers' peak forces and in columns/walls axial forces. Hence, it is suggested using values of $\xi \leq 0.2$. Based on Eq. (13.4) and the derivations in this paper for $\xi = 0.2$ circa 50 % reduction in most response values can be attained. The force in the most loaded damper in this case would be circa $F_D = 0.19h/l \cdot \sqrt{d^2 + b^2}/d \cdot VB_{undamped}$ where $VB_{undamped}$ is the base shear of the undamped structure, and the damper stroke would approximately be $stroke = 0.5 \cdot l \cdot d/\sqrt{d^2 + b^2} \cdot ID_{undamped}$ where $ID_{undamped}$ is the inter-story drift of the undamped system (% of story height).
5. From Item 4 it can be concluded that for the same response reductions in two buildings having the same geometric properties *h*, *d*, *b*, and *l*, but with a different height, *H*, mass, *m*, and stiffness *EI*, the ratio of the force at the most loaded damper to the undamped base shear, would be similar. That is, the design damper forces are expected to be smaller in lower buildings, showing, again, the applicability of the concept to low raise buildings.

Acknowledgments Prof. Reinhorn has contributed significantly, and keeps contributing, to shaping me as a researcher, an engineer, an educator and a person. Some of his contributions, indirectly affected the work presented in this chapter. The main motivation for this work comes from recognizing the importance of insight into the behaviour of structural systems, their controlling parameters, the effect of these parameters on the responses of interest, and tools for initial design. This recognition was gained from studying Prof. Reinhorn's work presenting new concepts of structural systems, as well as from the stories he used to tell his students while discussing their own work. His work and those stories, I believe, encouraged his former students to aspire for contributions at the concept level rather than on fine tuning. In addition, a concept that was used in the present work utilises continuum models to describe the overall behaviour of structural systems. As both Prof. Reinhorn and I taught, at some point, classes related to the analysis and design of tall buildings, we had many discussions on such models, their advantages and limitations. Finally, this chapter uses dimensional analysis to enable a rigorous and efficient parametric study. Many past students of Prof. Reinhorn's "Experimental Methods" class are now leading professors and practitioners in the various fields of structural engineering. They probably remember his insistence on the theory of scaling, dimensional analysis and similitude as indispensable tools for such studies. Some of them also applied these tools in the context of experimental work as well as in the context of parametric studies, as was done in this chapter.

Support for this work was provided by the Israel Ministry of Housing and Construction through the National Building Research Institute at the Technion, Grant # 2013019. The author gratefully acknowledges this support.

References

Agrawal AK, Yang JN (1999) Design of passive energy dissipation systems based on LQR control methods. J Intell Mater Syst Struct 10(12):933–944

Attard TL (2007) Controlling all interstory displacements in highly nonlinear steel buildings using optimal viscous damping. J Struct Eng 133(9):1331–1340

Aydin E, Boduroglu MH et al (2007) Optimal damper distribution for seismic rehabilitation of planar building structures. Eng Struct 29:176–185

CEN. Eurocode (EC) 8: design of structures for earthquake resistance – Part 1 general rules, seismic actions and rules for buildings (prEN 1998-1), Brussels

Christopoulos C, Filiatrault A (2006) Principles of supplemental damping and seismic isolation. IUSS Press, Milan

Cimellaro GP, Reinhorn AM, Bruneau M, Rutenberg A (2006) Multi-dimensional fragility of structures: formulation and evaluation. Multidisciplinary Center for Earthquake Engineering Research, p 123

Constantinou MC, Symans MD (1992) Experimental and analytical investigation of seismic response of structures with supplemental fluid viscous dampers. N.C.f.E.E. Research, Buffalo

Constantinou MC, Soong TT, Dargush GF (1998) Passive energy dissipation systems for structural design and retrofit. Multidisciplinary Center for Earthquake Engineering Research, Buffalo

Dargush GF, Sant RS (2005) Evolutionary aseismic design and retrofit of structures with passive energy dissipation. Earthq Eng Struct Dyn 34(13):1601–1626

Fujita K, Moustafa A, Takewaki I (2010) Optimal placement of viscoelastic dampers and supporting members under variable critical excitations. Earthq Struct 1(1):43–67

Gluck J, Reinhorn A, Rutenberg A (1979) Dynamic torsional coupling in tall building structures. In ICE proceedings, vol 67, no 2, pp 411–424. Thomas Telford

Gluck N, Reinhorn AM et al (1996) Design of supplemental dampers for control of structures. J Struct Eng 122(12):1394–1399

Kanno Y (2013) Damper placement optimization in a shear building model with discrete design variables: a mixed-integer second-order cone programming approach. Earthq Eng Struct Dyn 42(11):1657–1676

Kusumastuti D, Reinhorn AM, Rutenberg A (2005) A versatile experimentation model for study of structures near collapse applied to seismic evaluation of irregular structures (No. 2). Multidisciplinary Center for Earthquake Engineering Research

Lavan O (2012) On the efficiency of viscous dampers in reducing various seismic responses of wall structures. Earthq Eng Struct Dyn 41(12):1673–1692

Lavan O, Avishur M (2013) Seismic behavior of viscously damped yielding frames under structural and damping uncertainties. Bull Earthq Eng 11(6):2309–2332

Lavan O, Dargush GF (2009) Multi-objective optimal seismic retrofitting of structures. J Earthq Eng 13:758–790

Lavan O, Levy R (2005) Optimal design of supplemental viscous dampers for irregular shear-frames in the presence of yielding. Earthq Eng Struct Dyn 34(8):889–907

Lavan O, Levy R (2006) Optimal peripheral drift control of 3D irregular framed structures using supplemental viscous dampers. J Earthq Eng 10(6):903–923

Lavan O, Cimellaro GP, Reinhorn AM (2008) Noniterative optimization procedure for seismic weakening and damping of inelastic structures. J Struct Eng ASCE 134(10):1638–1648

Levy R, Lavan O (2006) Fully stressed design of passive controllers in framed structures for seismic loadings. Struct Multidiscip Optim 32(6):485–498

Lopez Garcia D, Soong TT (2002) Efficiency of a simple approach to damper allocation in Mdof structures. J Struct Control 9(1):19–30

Madsen LPB, Thambiratnam DP, Perera NJ (2003) Seismic response of building structures with dampers in shear walls. Comput Struct 81:239–253

Makris N, Black CJ (2004a) Dimensional analysis of bilinear oscillators under pulse-type excitations. J Eng Mech 130(9):1019–1031

Makris N, Black CJ (2004b) Dimensional analysis of rigid-plastic and elastoplastic structures under pulse-type excitations. J Eng Mech 130(9):1006–1018

Marko J, Thambiratnam D, Perera N (2004) Influence of damping systems on building structures subject to seismic effects. Eng Struct 26:1939–1956

Rahimian A (2002) Coupled truss systems with damping for seismic protection of buildings. U.S. Patent 6,397,527, Assigned to the Cantor Seinuk Group, PC New York, NY

Reinhorn A, Rutenberg A, Glück J (1977) Dynamic torsional coupling in asymmetric building structures. Build Environ 12(4):251–261

Rutenberg A, Glück J, Reinhorn A (1978) On the dynamic properties of asymmetric wall-frame structures. Earthq Eng Struct Dyn 6(3):317–320
Shukla AK, Datta TK (1999) Optimal use of viscoelastic dampers in building frames for seismic force. J Struct Eng 125(4):401–409
Silvestri S, Trombetti T (2007) Physical and numerical approaches for the optimal insertion of seismic viscous dampers in shear-type structures. J Earthq Eng 11(5):787–828
Smith RJ, Willford MR (2007) The damped outrigger concept for tall buildings. Struct Des Tall Spec Build 16(4):501–517
Song J, Chu YL, Liang Z, Lee GC (2008) Modal analysis of generally damped linear structures subjected to seismic excitations. Multidisciplinary Center for Earthquake Engineering Research, State University of New York at Buffalo. Report no. MCEER-08-0005, New York
Sonmez M, Aydin E, Karabork T (2013) Using an artificial bee colony algorithm for the optimal placement of viscous dampers in planar building frames. Struct Multidiscip Optim 48(2):395–409
Soong TT, Dargush GF (1997) Passive energy dissipation systems in structural engineering. Wiley, Chichester
Takewaki I (1997) Optimal damper placement for minimum transfer functions. Earthq Eng Struct Dyn 26(11):1113–1124
Takewaki I (2000) Optimal damper placement for planar building frames using transfer functions. Struct Multidiscip Optim 20(4):280–287
Takewaki I (2009) Building control with passive dampers: optimal performance-based design for earthquakes. Wiley, Singapore
Taylor DP (2002) Mega brace dampers for the Torre Mayor project at Mexico City. http://www.taylordevices.eu/pdfs/Mega%20Brace%20Seismic%20Dampers-Taylor.pdf
Trombetti T, Silvestri S (2004) Added viscous dampers in shear-type structures: the effectiveness of mass proportional damping. J Earthq Eng 8(2):275–313
Trombetti T, Silvestri S (2006) On the modal damping ratios of shear-type structures equipped with Rayleigh damping systems. J Sound Vib 292:21–58
Wang YP, Reinhorn AM, Soong TT (1992) Development of design spectra for actively controlled wall-frame buildings. J Eng Mech 118(6):1201–1220
Whittle JK, Williams MS et al (2012) A comparison of viscous damper placement methods for improving seismic building design. J Earthq Eng 16(4):540–560
Zhang RH, Soong TT (1992) Seismic design of viscoelastic dampers for structural application. J Struct Eng 118(5):1375–1392

Chapter 14
An Energy-Based Method for Designing the Supplemental Damping of Inelastic Structures

R. Barron and A.G. Ayala

Abstract This paper presents a design method for calculating the supplemental damping of inelastic structures. This method is based on a balance of the mean energy dissipated per cycle by the structure and by the supplemental damping. The objective is to reduce the amplitude of the hysteretic cycles by adding the required amount of damping, so that the supplemental damping dissipates a large portion of the hysteretic energy with the reduced amplitude of vibration. The applicability of the method is shown in the design of the supplemental damping of an 8-story building highly irregular in plan and elevation. It is concluded that the use of a small amount of supplemental damping, optimally placed, considerably reduces the structural response. In this example, the added damping represents a damping ratio of 4.3 % and it was capable to dissipate more than the 80 % of the energy fed by the earthquake. The truss system of the supplemental damping took the majority of the inertial forces, realising the structure from these forces and reducing its overall response. For example, the mean reduction of the maximum interstory drifts and maximum story rotations were 46 % and 65 %, respectively. This showed that the energy-based method herein presented is a good method for designing the supplemental damping.

R. Barron (✉)
Facultad de Ingeniería, Universidad Autónoma de Zacatecas, Av. Ramón López Velarde No. 801, Zacatecas, Zac. 98000, Mexico
e-mail: raul.barron1@gmail.com

A.G. Ayala
Instituto de Ingeniería, Universidad Nacional Autónoma de México, Circuito Interior, Ciudad Universitaria, Delegación Coyoacán, México D.F., Mexico
e-mail: gayalam@iingen.unam.mx

G.P. Cimellaro et al. (eds.), *Computational Methods, Seismic Protection, Hybrid Testing and Resilience in Earthquake Engineering*, Geotechnical, Geological and Earthquake Engineering 33, DOI 10.1007/978-3-319-06394-2_14,
© Springer International Publishing Switzerland 2015

14.1 Introduction

The methodologies for aseismic design have evolved from a force-based philosophy to a displacement-based philosophy that relies on the ductility of the structure to dissipate the energy that an earthquake fed into the structure. However, with the advent of innovative systems of active and passive structural control, once more, the design methodologies are evolving. The objective of this work is to develop a method for designing the supplemental damping of structures using an energy-based philosophy.

The purpose of adding damping to a structure is to increase its capability to dissipate the energy fed by an earthquake. Without dampers, the input energy of a large earthquake usually exceeds the capacity of the structure to dissipate energy by its inherent damping. Once this capacity is surpassed, portions of the structure typically yield or crack, which may result in severe structural damage. However, with the use of the energy dissipation devices, which have as a main purpose the dissipation of the energy fed by the earthquake, the current design methodologies are been rethinking. As mentioned in Soong and Dargush (1997), "... a focus on energy as a design criterion is conceptually very appealing. With this approach, the engineer is concern, not so much with the resistance to lateral loads but rather with the need to dissipate the energy input into the structure from the seismic disturbance".

Housner was the first to propose an energy-based philosophy for aseismic design (1956). On his method of limit-design, the structure is designed to dissipate plastically the energy fed by the earthquake. He assumed that the maximum plastic strain energy is equal to the maximum kinetic energy minus the elastic strain energy, and suggested the use of elastic velocity response spectra to evaluate the maximum kinetic energy. On the other hand, the energy dissipation capacity of the structure is evaluated in a single-degree-of-freedom representation by multiplying the plastic strain by the yield stress. However the damage to a structure does not depend simply on the magnitude of the dissipated energy, but the local characteristics are also important to evaluate the energy flow into its structural components, as well the history of the inelastic incursions, which determine the level damage of its structural components. As noted in Soong and Dargush (1997), this is a serious shortcoming for the use of the energy concepts for limit design of traditional structures. On the other hand, with the use of passive dampers, which have stable hysteretic loops, the dampers are designed and placed in the areas of the structure where there is a great dissipation of the input energy. This reduces the structural response and, consequently, minimizes the damage to the primary structure.

Even though the great importance of dampers as protective devices, little work has been done in developing design methods aimed to determine the right amount and the placement that best reduces the structural response. In contrast, a significant amount of experimental research has been conducted to determine their characteristics and their effects on the structures (Reinhorn et al. 1995; Gluck et al. 1996; Goel and Booker 2001). A practical design method to determine the supplemental damping has to be simple and efficient. Simple for the practical engineers can use it in a routinely design, and efficient in determining the optimal amount and precise placement in order to use them efficiently.

Although there are some methods in the literature for designing the supplemental damping, some of them are too simple but not lead to an optimal design, or they are efficient but not ease to use in a routinely design. This paper describes the design method proposed in Barron and Ayala (2013) that has the desired characteristics mentioned in the former paragraph. A numerical example is provided to show its efficiency in determining the sizes and places of the dampers that reduce the structural response to the desired limit with a minimum amount of supplemental damping.

14.2 Procedure to Design a Structure with Supplemental Damping

As mentioned in the introductory paragraph, the purpose of adding damping to a structure is to increase its capability to dissipate the destructive energy that an earthquake fed to the structure. The dampers are characterised by having stable hysteresis loops that dissipate large amount of energy without a significant degradation of their strength and stiffness. Accordingly, the dampers are placed in the zones of the structure where a large portion of the input energy is dissipated by plastic strain (the energy related to the structural damage), so that they are able to dissipate that energy. But also it is effective to put the dampers at the places in the structure where the force of the damper have a large lever arm, making them more efficient in reducing the overall response of the structure. In that way, with a small number of dampers, the plastic strain energy of the whole structure can be reduced considerably. Guided by these considerations, the proposed procedure for designing the supplemental damping is based on energy concepts.

Before presenting the explanation of the design procedure, it is important to mention three key issues. (1) The use of diagonals and columns to support the dampers form trusses. This is an efficient structural system to resist lateral forces that takes a large portion of the inertial forces, which releases the structure from these forces. (2) In buildings, the better places to put the dampers are the exterior frames, where the lateral forces of the trusses that support the dampers have large lever arms. Consequently, the dampers are more effective in reducing the translational and torsional response of the structure. (3) Since the procedure for design the dampers requires locating the zones that dissipate more energy in the structure, a multi-degree-of-freedom model of the structure is required to evaluate the response of the different structural elements.

The starting point of the design procedure is the evaluation of the mean energies dissipated per cycle. The objective is to reduce the amplitude of the hysteretic cycles of plastic strain by adding the required amount of damping, so that each damper dissipates a large portion of the hysteretic energy with a reduced amplitude of vibration. Therefore, the objective is to reduce the mean amplitude of vibration, Um, to the target amplitude, U_0, at the places where a damper is required to reduce the large hysteretic cycles.

The mean amplitudes of vibration, Um, of the zones with the larger hysteretic cycles, can be evaluated by averaging the displacement of local maxima obtained from a deterministic analysis or can be evaluated directly from a stochastic analysis as shown in Barron and Ayala (2013). The stochastic approach is used in this work with the assumption of a stationary response for the strong segment of the response. This hypothesis is adequate if the structure does not suffer severe damage, which is possible with the help of the dampers that reduces the large displacements.

To obtain the equation of the dissipated energies, the equation of motion, Eq. 14.1 is required.

$$M\ddot{x} + C\dot{x} + f(x) + \Gamma(\dot{x}, x) = F(t) \tag{14.1}$$

where t is the time variable, x is the time-dependent displacement vector of the structure, M is the mass matrix, C is the damping matrix that represents the inherent damping of the structure as an equivalent viscous damping, $f(x)$ is a vector representing the nonlinear restoring forces of the structure, $\Gamma(\dot{x}, x)$ is a vector representing the forces produced by the added damping that, in general, can be a nonlinear function of the displacement and the velocity, and F is a vector that represents the wind or earthquake induced forces.

As shown in Soong and Dargush (1997), the equation of motion can be rewritten in the form of energy by multiplying it for the displacement $dx = \dot{x}dt$ and integrating over one cycle of vibration that is denoted by $\langle\cdot\rangle$, the equation Eq. 14.2 is obtained.

$$\langle M\ddot{x}\dot{x}\rangle + \langle C\dot{x}^2\rangle + \langle f(x)\dot{x}\rangle + \langle\Gamma(\dot{x}, x)\dot{x}\rangle = \langle F\dot{x}\rangle \tag{14.2}$$

For the steady-state response of a harmonic motion, or for a stationary random process, $\langle\ddot{x}\dot{x}\rangle = \langle\dot{x}x\rangle = 0$. Equation 14.2 is thus changed into the balance equation of the dissipated energies given in Eq. 14.3.

$$E_{ID} + E_{PS} + E_{AD} = E_I \tag{14.3}$$

where $E_{ID} = \langle C\dot{x}^2\rangle$ is the mean dissipated energy by the equivalent viscous damping in the structure; $E_{PS} = \langle f(x)\dot{x}\rangle$ is the mean plastic strain energy that is the non-recoverable energy of the restoring forces of the structure when it yields; $E_{AD} = \langle\Gamma(\dot{x}, x)\dot{x}\rangle$ is the mean energy dissipated by the added damping; and $E_I = \langle F\dot{x}\rangle$ is the mean input energy impose to the structure by the external force F.

The Eq. 14.3 represents the total mean energy dissipated per cycle by all structural elements. For the case of buildings, the energies are evaluated at the interstory of the frames where a damper is required to reduce the large hysteretic cycles. The increase of the amount of damping from $AD1$ to $AD2$ reduces the mean amplitude of vibration from U_m until it reaches the target displacement U_0. In this situation, the Eq. 14.4 represents the balance equation of the dissipated energies in a cycle of vibration. This equation equals the dissipated energy in a cycle of vibration at the mean amplitude U_m to the dissipated energy in a cycle of vibration of reduced amplitude U_0, in which the increased damping AD2 dissipates a larger portion of the energy.

$$[E_{ID} + E_{PS} + E_{AD1}]^{U_m} = [E_{ID} + E_{PS} + E_{AD2}]^{U_0} \tag{14.4}$$

The dissipated energy by the damping AD2 can be obtained by solving for E_{AD2} in Eq. 14.4.

$$E_{AD2} = [E_{ID} + E_{PS} + E_{AD1}]^{U_m} - [E_{ID} + E_{PS}]^{U_0} \tag{14.5}$$

This energy is of paramount importance, because no matter what the type of damper it is used (viscous, viscoelastic, friction, or metallic), it allow us to design the damper in the sense that given the amplitude of vibration U_0 in a harmonic test, its characteristics can be determined so it dissipates the required energy in Eq. 14.5.

For example, if a viscous damper is used, it is known that for the steady-state response of a harmonic motion, or for a stationary random process, the mean energy dissipated in a cycle of vibration at the frequency ω and amplitude U_0 is given in Eq. 14.6 (Clough and Penzien 1993; Chopra 1995).

$$E_{AD2} = c\pi\omega U_0^2 \tag{14.6}$$

From this equation, the required damping constant c of the viscous damper can be evaluated as given in Eq. 14.7. In an actual ground motion, ω is the response frequency (the central frequency in the power spectrum).

$$c = \pi\omega U_0^2 / E_{AD2} \tag{14.7}$$

It is important to mention that the procedure for designing the dampers follows an iterative scheme because the energy introduced by the earthquake depends on the amplitude of the response, which with the added damping is being reduced until it reaches the target displacement. Thus, at each step of the iterative process new values of the energy dissipated for each damper are evaluated using the equation Eq. 14.5 and new values for the damping constants c are obtained with the equation Eq. 14.7. The iterative process ends when the c values of all dampers does not change as we proceed with the iterations.

Usually, the most effective places to put the dampers are the exterior frames. Accordingly, at the start of the iterations, we assume that the dampers are placed in all the interstories of the selected frames. As the iterations proceed, some of the c values are too small meaning that there is no need to put a damper there, so we retired those locations from the iterative scheme. This increase the speed of convergence and improve the results because the dampers are being placed at the locations where they are more effective.

14.3 A Numerical Example

The procedure for designing the supplemental damping is shown in a numerical example in which the dampers of an 8-storey steel building are designed. The Fig. 14.1 shows an isometric view of the building generated with the structural analysis program SAP2000 (Computers and Structures Inc. 2012). From this Figure it can be seen the great asymmetries in plan and in elevation of the building.

The building was designed using the strong column-weak beam criterion, which ensures that the inelastic behavior of the structure appears only on the beams that developed plastic hinges. Only the columns of the first level developed plastic hinges at their base. Table 14.1 presents the story heights and masses, and the sections of beams and columns.

The columns base plates were considered semi rigid assuming a rigidity for the base plates five times the angular rigidity of the columns. The inherent damping is modeled as proportional damping using the 1st and 3rd modes of vibration, with a 2 % damping ratio.

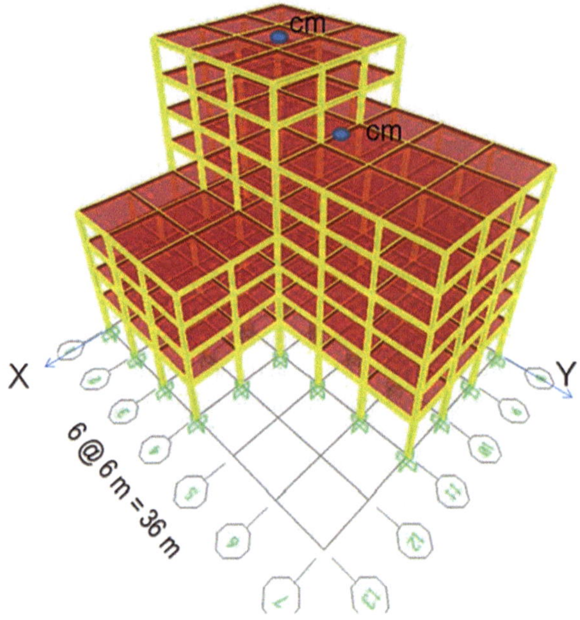

Fig. 14.1 Isometric view of the building generated with SAP-2000

Table 14.1 Story masses, their location and sections of beams and columns

Story	Height (m)	Mass (kg) × 10^3	Mom. of inertia (kg m^2) × 10^4	*X* cm (m)	*Y* cm (m)	Beams W section	Columns TS
1	4.0	628	11,570	12.75	15.75	18 × 60	16 × 16 × 5/8
2	3.8	622	11,461	12.75	15.75	16 × 50	16 × 16 × 5/8
3	3.8	615	11,327	12.75	15.75	16 × 45	16 × 16 × 1/2
4	3.8	601	11,185	12.5	15.9	16 × 45	16 × 16 × 1/2
5	3.8	459	6,185	9.0	18.0	16 × 40	16 × 16 × 1/2
6	3.8	438	5,903	9.0	17.6	16 × 40	16 × 16 × 3/8
7	3.8	229	1,235	9.0	9.0	16 × 36	16 × 16 × 3/8
8	3.8	208	1,120	9.0	9.0	16 × 36	16 × 16 × 3/8

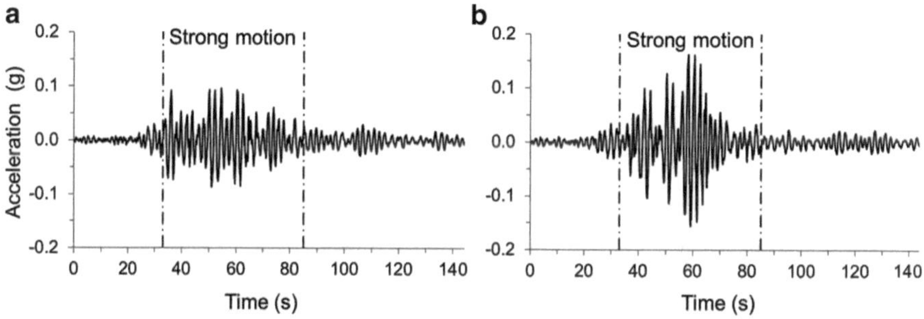

Fig. 14.2 Horizontal components of the ground acceleration recorded at the parking area of the building of Minister of Communications and Transportations (*SCT*), in Mexico City, generated by the earthquake occurred on September 19, 1985, in the cost of Michoacán, México

14.3.1 Modelling of the Ground Excitation

The ground accelerations applied to the building are the two components that were recorded at the parking area of the building of Minister of Communications and Transportations, in Mexico City, during the earthquake occurred on September 19, 1985, in the cost of Michoacán, México. The N00W component, in Fig. 14.2a, was applied to the *X* direction and the N90W components, in Fig. 14.2b, is applied to the *Y* direction.

For the stochastic analysis, the ground accelerations are considered as vector random process with components in the *X* and Y directions. These components are assumed zero-mean stationary gaussian processes. The unilateral auto- and cross-power spectral densities are shown in Fig. 14.3. These spectral densities were evaluated using the intense part of the records, with a duration of 50 s, defined between the times corresponding to the 5 and 95 % of the Arias intensity. To calculate these spectral densities, the "cpsd" function of MATLAB (Krauss et al. 1994) was used with a 30 s Hamming window with 50 % overlap.

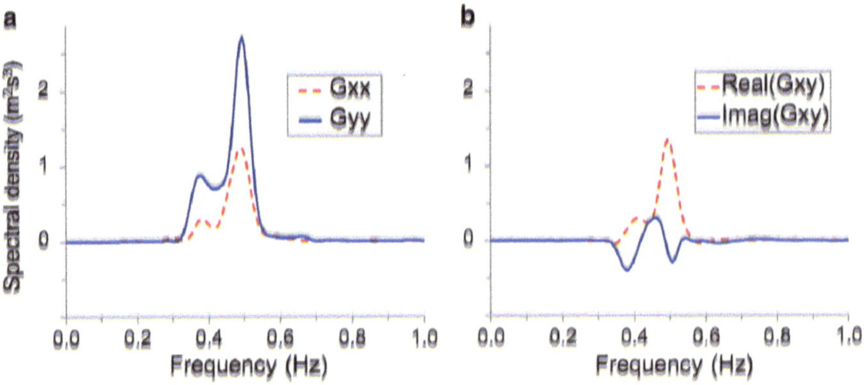

Fig. 14.3 The unilateral auto- and cross-power spectral densities of the two components of the ground acceleration recorded at the SCT building, in Mexico City, during the earthquake occurred on September 19, 1985, in the cost of Michoacán, México

Table 14.2 Story stiffness and yield shear-force of frames

	Stiffness (kN/cm)				Yielding shear force (kN)			
	Frames				Frames			
Interstory	1 2 3 4	5 6 7	8 9 10 11	12 13	1 2 3 4	5 6 7	8 9 10 11	12 13
1	480	310	550	310	1,800	1,250	2,900	1,250
2	340	210	390	200	1,560	1,070	1,960	1,090
3	280	170	320	160	1,350	910	1,570	990
4	240	165	290	180	1,060	830	1,400	800
5	170	140	290	–	920	650	1,280	–
6	135	120	250	–	760	550	900	–
7	130	–	150	–	680	–	680	–
8	100	–	105	–	500	–	500	–

14.3.2 *Modelling of the Building*

Since the design procedure requires several analyses to determine the optimal sizes and locations of the dampers, a simplified model of the building was used. Each frame is modeled as shear-type frame with a one shear-resisting element per interstory that takes into account the shear forces of the columns of the inter-story. The Bouc-Wen model (Bouc 1967; Wen 1976) with a smooth exponent of 3 and a 5 % post yielding stiffness, is used to represent the nonlinear behaviour of the shear-resisting elements. The story stiffnesses were calculated from a pushover analysis with lateral forces proportional to the first mode of vibration. Table 14.2 shows the stiffness and yield shear forces of the frames in the simplified model.

The yield shear-forces were evaluated considering that the plastic moments of the beams are distributed by half between the upper and the lower columns that connected them, dividing the sum of the ends moments of the columns by the

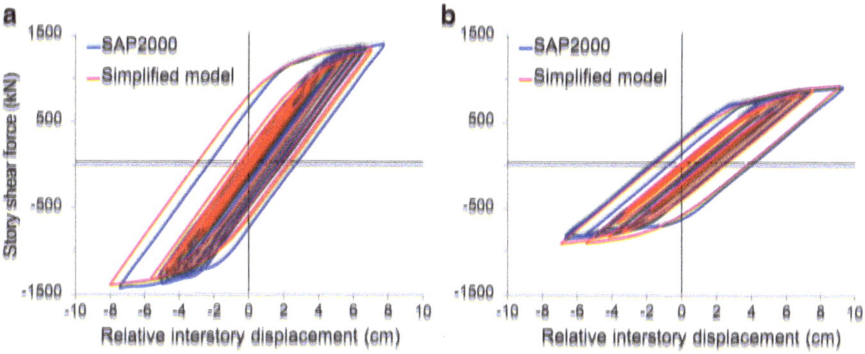

Fig. 14.4 Hysteresis loops of the simplified model vs. the SAP2000 model; (**a**) third story of frame 1 and (**b**) third story of frame 7

Table 14.3 Mean dissipated energy per cycle (kJoules) in all frames

Frame	1	2	3	4	5	6	7	8	9	10	11	12	13
Inherent damping	22	20	19	18	10	10	11	18	16	14	12	3	3
Plastic strain	40	34	29	27	11	12	13	20	14	10	8	1	1

story height gives the yield shear-forces. Some adjustment to the yield shear forces were made so the simplified model represents well the response of the more-precise SAP2000 model. Figure 14.4 presents a comparison of the hysteresis loops obtained with the simplified model and SAP2000, for the 3rd story of frames 1 and 7. From this figures it can be seen the good approximation of the simplified model.

The deterministic and stochastic analyses of the simplified model were performed with computer programs developed in MATLAB (2013). The analysis procedure for the stochastic analysis is briefly explained in Barron and Ayala (2013).

14.3.3 Design of the Dampers

The procedure for designing the dampers starts with the evaluation of the mean inter-story-displacement of local maxima, U_m, from which the dissipated plastic strain energy and the dissipated energy by the inherent damping are evaluated. Table 14.3 presents the dissipated energies on each frame. From this table it can be seen that the frames 1, 2, 3, 4 and 8 dissipate the larger amount of strain energy. At a first glance one can think that the best place to put the dampers are those frames. However, as it is shown next, the most effective locations to place the dampers are the exterior frames.

The procedure starts by selecting the frames were the dampers are going to be located. Usually, the most effective places are the exterior frames. Accordingly, the exterior frames 1, 7, 8 and 13 are selected to put the dampers, although the frames 7

Table 14.4 Calculated damping coefficients for the Design_1 and Design_2 obtained with the target displacements of 2.0 and 1.5 cm, respectively. Damping units in (kN-s/cm)

		Design 1				Design 2			
Frame		1	7	8	13	1	7	8	13
Interstory	7	18	–	0	–	0	–	0	–
	6	58	0	0	–	78	0	0	–
	5	82	0	16	–	105	0	54	–
	4	22	65	32	0	103	114	45	0
	3	123	113	111	26	194	160	171	73
	2	76	104	0	0	194	175	111	41
	1	0	0	0	0	0	0	0	0

and 13 does not dissipate much energy. At the start of the iterations it is assumed that the dampers are placed in all the interstories of the selected frames. As the iterations proceed, some of the c values are too small, meaning that there is no need to put a damper there, so we retired those locations from the iterative scheme. This increase the speed of convergence and improve the results because the dampers are being placed at the locations where they are more effective.

For purpose of comparison, two target displacements, U_0, are used to design the dampers. These displacements are 2.0 and 1.5 cm. The locations and sizes of the dampers obtained with these target displacements are named Design 1 and Design 2, respectively. Table 14.4 shows the calculated damping coefficients for these two designs. It can be observed that the 1st interstory did not required dampers in both designs, and that only the frame 1 required a damper in 7th interstory in the Design 1.

As will be shown in the next section, with a small amount of damping, the two damper configurations obtained produce a grate reduction in the displacements and the torsional response of the structure. The damping ratio of these designs were evaluated using the non-classical mode shapes. The complex characteristic values of the first mode are $\lambda_1 = -0.172 + 3.95i$ for the Design 1 and $\lambda_1 = -0.441 + 3.99i$ for the Design 2. The corresponding damping ratios are $\xi = 4.3\ \%$ and $\xi = 10.5\ \%$, respectively.

14.4 Presentation and Discussion of the Results of the Numerical Example

What follows presents a comparison of the seismic performance of the building in the numerical example with and without dampers, for the designs Design 1 and Design 2. The first response parameter to review is the mean interstory displacement of local maxima. Table 14.5 presents the mean interstory displacements of local maxima for the 1st and 3rd interstory of all frames. It can be seen that that these displacements never exceeded the target displacement of 2.0 and 1.5 cm established for the Design 1 and Design 2, respectively.

Table 14.5 Mean interstory displacements of local maxima for the 1st and 3rd interstory. Design 1 with a target displacement of 2.0 cm and Design 2 with a target displacement of 1.5 cm

Frames		1	2	3	4	5	6	7	8	9	10	11	12	13
Mean drift (cm)	No dampers	3.6	3.6	3.6	3.6	3.7	3.8	3.9	3.5	3.2	2.9	2.6	2.3	2.1
	Design 1	2.0	2.0	2.0	2.0	2.0	2.0	2.0	2.0	2.0	2.0	2.0	2.0	2.0
3rd interstory	Design 2	1.5	1.5	1.5	1.5	1.5	1.5	1.5	1.5	1.5	1.5	1.5	1.5	1.5
Mean drift (cm)	No dampers	2.5	2.4	2.4	2.4	2.5	2.5	2.6	2.3	2.1	1.9	1.7	1.6	1.4
	Design 1	1.6	1.6	1.6	1.6	1.6	1.6	1.7	1.6	1.5	1.5	1.4	1.3	1.2
1st interstory	Design 2	1.3	1.3	1.3	1.3	1.3	1.4	1.4	1.3	1.2	1.2	1.1	1.1	1.0

Table 14.6 Maximum interstory drifts (cm) of 3rd interstory of all frames

Frame	1	2	3	4	5	6	7	8	9	10	11	12	13
Without dampers	8.4	8.3	8.2	8.2	8.4	8.6	8.8	8.1	7.4	6.7	6.0	5.4	4.9
Design 1	4.5	4.5	4.5	4.5	4.5	4.5	4.5	4.6	4.5	4.5	4.5	4.5	4.6
Design 2	3.4	3.4	3.4	3.4	3.4	3.4	3.4	3.4	3.4	3.4	3.4	3.4	3.4

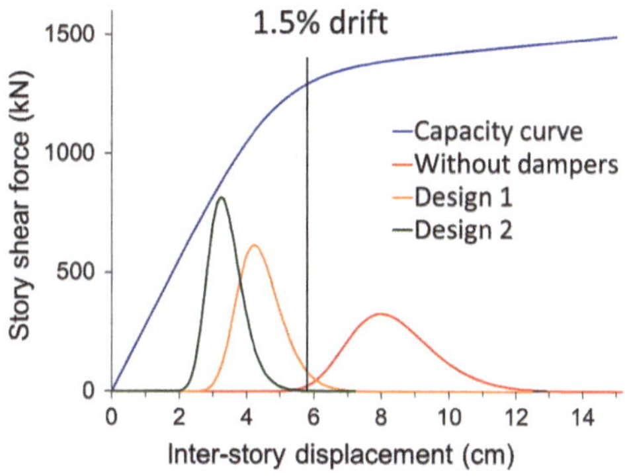

Fig. 14.5 Capacity curve and probability density function of the seismic demand (maximum drift of the 3rd interstory of frame 1) for the base structure and the structure with added damping denoted as Design 1 and Design 2

The second response parameter to review is the maximum inter-story-displacement. Table 14.6 presents the maximum interstory drifts for the 3rd interstory of all frames. It can be observed large displacements for the base structure, all of them are above 8.0 cm. The seismic norms establish limits for the interstory drift as a percentage of the height. For example, the FEMA P-750 (2009) establishes a drift limit ratio of 1.5 % for an occupancy category III (High occupancy). For the example building, with an interstory height of 380 cm, this limit is L = 5.7 cm. For the base structure, the drift limit is exceeded in most of the frames. On the other hand, the designs Design 1 and Design 2 comply with this limit. From Table 14.6 it can be seen that the larger interstory drifts got the larger reductions. For the Frames 1–7, which have the larger displacements, the mean reduction is 46 % for the Design 1 and 60 % for the Design 2.

A rational approach for the performance evaluation of a building has to consider the randomness of the ground motion. Figure 14.5 presents the probability density functions (pdf) of the maximum interstory drift of the 3rd interstory of frame 1, for

Table 14.7 Comparison of the dissipated energies (kJouls/cycle), base structure vs. Design 1 and Design 2

	Inherent	Added	Plastic strain	Total
Without dampers	176 (44 %)	–	220 (56 %)	396
Design 1	67 (16 %)	334 (81 %)	14 (3 %)	415
Design 2	40 (10 %)	359 (89 %)	4 (1 %)	403

the base structure and the structure with added damping, Design 1 and Design 2. These functions are scaled by 1,000 just for the purpose to see them at the scale of the shear force. From these pdf can be evaluated the probabilities that the peak interstory drift exceeds the limiting displacement L = 5.7 cm, these probabilities are 99.7 %, 4 % and 0 % for the base structure, Design 1 and Design 2, respectively.

The third response parameter to review are the dissipated energies in the structure. Table 14.7 presents a comparison of the energies dissipated in the base structure and the structure with added damping, Design 1 and Design 2. From this table it can be seen that both designs reduces considerably the energies dissipated by the inherent damping and the plastic strain. The dampers dissipating more than the 80 % of the energy fed by the earthquake. This reduces the forces in the primary structure because the dampers and the corresponding truss system that support them take the majority of the inertial forces. With a reduced force, the lateral and torsional responses of the building are diminished. For example, the mean reduction of the maximum interstory drifts and maximum story rotations were 46 % and 65 %, respectively. With these levels of reduction the structure has a behaviour practically linear (see Fig. 14.6b, c).

From the comparison of the level of damping in the Design 1 and Design 2, it can be seen that the Design 1 is more convenient because it gives a good and an acceptable level of reduction with a 40 % less damping than the Design 2, (4.3 %/10.5 %).

It is important to see a comparison of the effect of the added damping in the reduction of the hysteresis loops. Figure 14.6a–c present the hysteresis loops for the 3rd interstory of frame 1. From this figure it can be seen the reductions in the displacement and the force that is obtained with the added damping. The Design 1 reduces the maximum displacement from 8.2 cm, in the base structure, to 4.4 cm, which produces a response almost elastic. Whereas the Design 2, with more damping, reduces the maximum displacement to 3.9 cm, which produces an elastic structural response.

Figure 14.6d shows a comparison of the maximum displacement profile of frame 1 for the base structure vs. the Design 1 and the Design 2. It can be seen that the roof displacements are reduced 26 % for the Design 1 and 40 % for the Design 2. It is important to mention that a reduction of the roof displacement is no representative of the reductions of the interstory drifts because the roof displacement is the addition of all interstory displacements that are not reduced in the same proportion, the larger reductions are in the 2nd, 3rd, 4th and 5th interstory, with a mean reduction of 46 % and 59 % for the Design 1 and the Design 2, respectively.

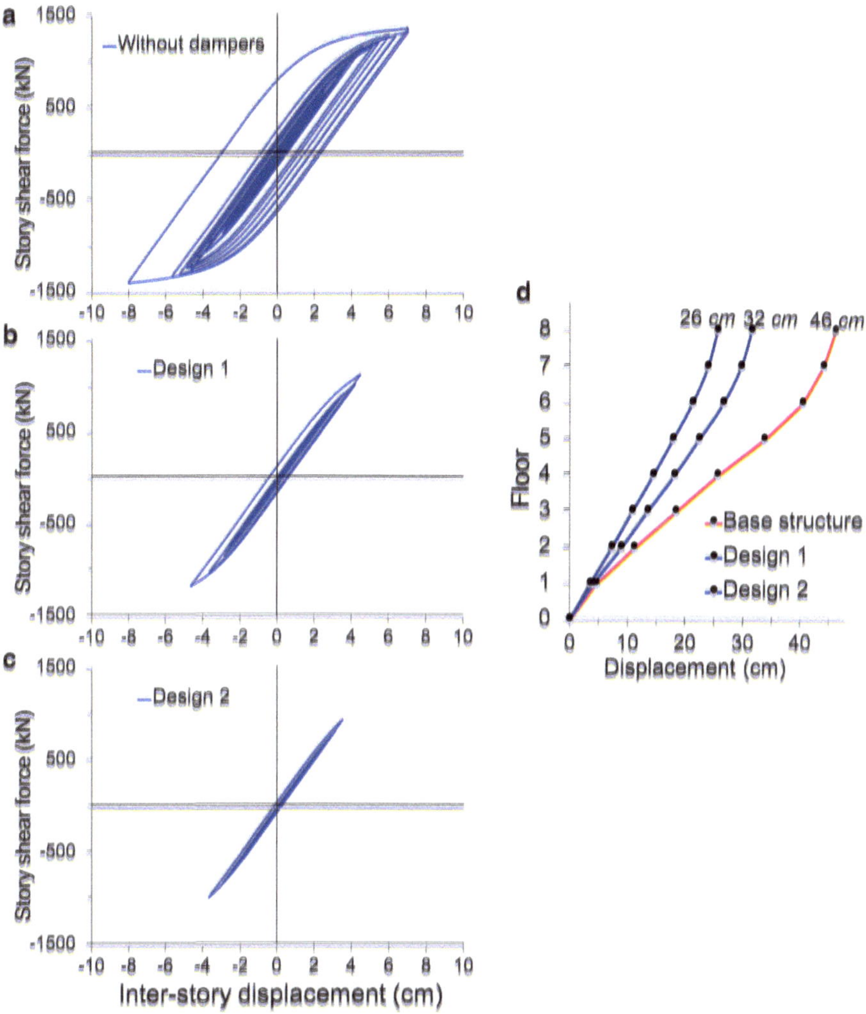

Fig. 14.6 Comparison of the hysteretic loops of the 3rd interstory of frame 1, and comparison of the displacement profiles of frame 1, base structure vs. the Design 1 and Design 2

The forth response parameter to review are the story rotations. Table 14.8 presents the maximum story rotations for the base structure and the structure with added damping, Design 1 and Design 2. From this table it can be seen that the reductions are almost uniform, and are more than 50 %. The mean values of reduction are 65 % for the Design 1 and 78 % for the Design 2.

To end this section, Table 14.9 summarizes the most relevant results of this numerical example that compares the effectiveness of the Design 1 and Design 2 in reducing the structural response. Table 14.9 shows the mean reductions, in percentage, of the maximum interstory drifts and maximum story rotations. From this table it can be seen that the Design 1 produces a good reduction of the structural

Table 14.8 Comparison of maximum story rotations, base structure vs. Design 1 and Design 2. (Rotations in degrees × 10^{-3})

Story	1	2	3	4	5	6	7	8
Without dampers	50	119	198	280	333	401	430	448
Design_1	21	42	59	96	106	136	156	163
Design_2	14	20	28	50	65	95	103	111

Table 14.9 Mean reductions, in percentage, of the maximum interstory drifts and maximum story rotations for the Design 1 and Design 2

	Design 1 ($\xi = 4.3$ %)	Design 2 ($\xi = 10.5$ %)
Interstory drift	46 %	60 %
Floor rotations	65 %	78 %

response with a 4.3 % added damping, and that an increase in damping to 10.5 % in the Design 2 does not increases substantially the reductions obtained with the Design 1. Therefore, the Design 1 gives good reduction of the translational and torsional response of the building with a less damping.

14.5 Conclusions

This paper presented a design method for calculating the supplemental damping of inelastic structures. This procedure is based on a balance of the mean dissipated energy per cycle by the structure and by the added damping. The objective is to reduce the amplitude of the hysteretic cycles of plastic strain by adding the required amount of damping, so that each damper dissipates a large portion of the hysteretic energy with the reduced amplitude of vibration. The dampers are also placed at the zones of the structure where they have a large lever arm, making them more efficient in reducing the overall response of the structure. In that way, with a small number of dampers, the plastic strain energy of the whole structure can be reduced considerably.

A numerical example was provided to illustrate the applicability of the design method. The example consisted on determining the supplemental damping of an 8-story building that is highly irregular in plan and elevation. The two horizontal components of the ground motion were considered. Two target displacements were established and the two levels of the calculated damping were named Design 1 and Design 2. From this example it was shown the effectiveness of the proposed design method in reducing the translational and torsional response of the building. From the analysis of the results of this example, the fallowing conclusions can be made.

- The use of a small amount of supplemental damping, optimally placed, reduces considerably the structural response. For the building example, the added damping represents a small damping ratio of 4.3 %. The supplemental damping was capable to dissipate more than the 80 % of the energy fed by the earthquake.

The truss system of the supplemental damping took the majority of the inertial forces, realising the structure from these forces and reducing its overall response. For example, the mean reduction of the maximum interstory drifts and maximum story rotations were 46 % and 65 %, respectively.

- It is confirmed that the energy-based method herein presented is a good method for designing the supplemental damping of a structure. This method determined the best size and placement of the dampers that produces a large reduction of the structural response with a small number of dampers.

Acknowledgments This is a joint research between the National Autonomous University of Mexico (UNAM) and the Autonomous University of Zacatecas (UAZ). Support for this research came from the Institute of Engineering of UNAM, General Directorate for Affairs of the Academic Personnel of UNAM and the Faculty of Engineering of UAZ; this support is gratefully acknowledged.

References

Barron R, Ayala AG (2013) An energy design method for buildings with supplemental damping and nonlinear behavior. In: Lavan O, De Stefano M (ed) Seismic behaviour and design of irregular and complex civil structures. Geotechnical, geological and earthquake engineering, vol 24. Springer Dordrecht Heidelberg New York London. Springer Science+Business Media Dordrecht 2013

Bouc R (1967) Forced vibration of mechanical systems with hysteresis. In: Proceedings of the 4th conference on non-linear oscillations, Prague

Chopra AK (1995) Dynamics of structures: theory and applications to earthquake engineering, 2nd edn. Prentice Hall, New Heaven

Clough RW, Penzien J (1993) Dynamics of structures, 2nd edn. McGraw Hill Co., New York

Computers and Structures Inc., SAP-2000 (2012) Static and dynamic finite element analysis of structures (version 15). Computers and Structures Inc., Berkeley

FEMA P-750 (2009) NEHRP recommended seismic provisions for new buildings and other structures. Building Seismic Safety Council of the National Institute of Building Sciences, Federal Emergency Management Agency

Gluck N, Reinhorn AM, Gluck J, Levy R (1996) Design of supplemental dampers for control of structures. J Struct Eng ASCE 122(12):1394–1399

Goel RK, Booker CA (2001) Effects of supplemental viscous damping on inelastic seismic response of asymmetric systems. Earthq Eng Struct Dyn 30:411–430

Housner GW (1956) Limit design of structures to resist earthquakes. In: Proceedings of the world conference on earthquake engineering. Earthquake Engineering Research Center, Berkeley, pp 5-1–5-13

Krauss TP, Shure L, Little JN (1994) Signal processing toolbox – for use with MATLAB. The MathWorks Inc., Natick

MATLAB (2013) The MathWorks Inc., Natick

Reinhorn AM, Li C, Constantinou MC (1995) Experimental and analytical investigation of seismic retrofit of structures with supplemental damping, Part 1: fluid viscous damping devices. Technical report NCEER-95-0001. National Center for Earthquake Engineering Research, Buffalo

Soong TT, Dargush GF (1997) Passive energy dissipation systems in structural engineering. Wiley, New York

Wen YK (1976) Method of random vibration of hysteretic systems. J Eng Mech ASCE 102(EM2):249–263

Chapter 15
Seismic Response and Stability of the Rocking Frame

Nicos Makris and Michalis F. Vassiliou

Abstract This paper investigates the planar rocking response of an array of free-standing columns capped with a freely supported rigid beam in an effort to explain the appreciable seismic stability of ancient free-standing columns which support heavy epistyles together with the even heavier frieze atop. Following a variational formulation the paper concludes to the remarkable result that the dynamic rocking response of an array of free-standing columns capped with a rigid beam is identical to the rocking response of a single free-standing column with the same slenderness; yet with larger size – that is a more stable configuration. Most importantly, the study shows that the heavier the freely supported cap-beam is (epistyles with frieze atop), the more stable is the rocking frame regardless the rise of the center of gravity of the cap-beam; concluding that top-heavy rocking frames are more stable than when they are top-light. This "counter intuitive" finding renders rocking isolation a most attractive alternative for the seismic protection of bridges with tall piers; while its potential implementation shall remove several of the concerns associated with the seismic connections of prefabricated bridges.

N. Makris (✉)
Department of Civil Engineering, Division of Structures, University of Patras, Patras GR-26500, Greece
e-mail: nmakris@upatras.gr

M.F. Vassiliou
ETH Zürich, IBK, Wolfgang-Pauli-Strasse 15, CH-8093 Zürich, Switzerland
e-mail: vassiliou@ibk.baug.ethz.ch

G.P. Cimellaro et al. (eds.), *Computational Methods, Seismic Protection, Hybrid Testing and Resilience in Earthquake Engineering*, Geotechnical, Geological and Earthquake Engineering 33, DOI 10.1007/978-3-319-06394-2_15,
© Springer International Publishing Switzerland 2015

15.1 Introduction

Under base shaking slender objects and tall rigid structures may enter into rocking motion that occasionally results in overturning. Early studies on the seismic response of a slender rigid block were presented by Milne (1885); however, it was Housner (1963) who uncovered a size-frequency scale effect which explained why: (a) the larger of two geometrically similar blocks can survive the excitation that will topple the smaller block; and (b) out of two same acceleration amplitude pulses, the one with the longer duration is more capable to induce overturning. Following Housner's seminal paper a number of studies have been presented to address the complex dynamics of one of the simplest man-made structures – the free standing rigid column.

Yim et al. (1980) conducted numerical studies by adopting a probabilistic approach, Aslam et al. (1980) confirmed with experimental studies that the rocking response of rigid blocks is sensitive to system parameters; while Psycharis and Jennings (1983) examined the uplift of rigid bodies supported on viscoelastic foundation. Subsequent studies by Spanos and Koh (1984) investigated the rocking response due to harmonic steady-state loading and identified "safe" and "unsafe" regions together with the fundamental and subharmonic modes of the system. Their study was extended by Hogan (1989, 1990) who further elucidated the mathematical structure of the problem by introducing the concepts of orbital stability and Poincare sections. The transient rocking response of free-standing rigid blocks was examined in depth by Zhang and Makris (2001) who showed that there exist two modes of overturning: (a) by exhibiting one or more impacts; and (b) without exhibiting any impact. The existence of the second mode of overturning results in a safe region that is located on the acceleration-frequency plane above the minimum overturning acceleration spectrum. The fundamental differences between the response of a rocking rigid column (inverted pendulum) and the response of the linear elastic oscillator (regular pendulum) led to the development of the rocking spectrum (Makris and Konstantinidis 2003). More recent studies pertinent to the rocking response of rigid columns have focused on more practical issues such as the representation of the impact (Prieto et al. 2004), the effect of the flexibility-yielding of the supporting base (Apostolou et al. 2007; Palmeri and Makris 2008), the effect of brittle and ductile restrainers (Makris and Zhang 2001) or the effect of seismic isolation on the rocking response of the free-standing column (Vassiliou and Makris 2012).

In this paper we investigate the planar rocking response of an array of free-standing columns capped with a freely supported rigid beam as shown schematically in Fig. 15.1. Herein we use the term "rocking frame" for the one degree of freedom structure shown in Fig. 15.1. Sliding does not occur either at the pivot points at the base or at the pivot points at the cap-beam. Our interest to this problem was partly motivated from the need to explain the remarkable seismic stability of ancient free-standing columns which support heavy free-standing epistyles together with the even heavier frieze atop. As an example, Fig. 15.2 shows the entrance view of the

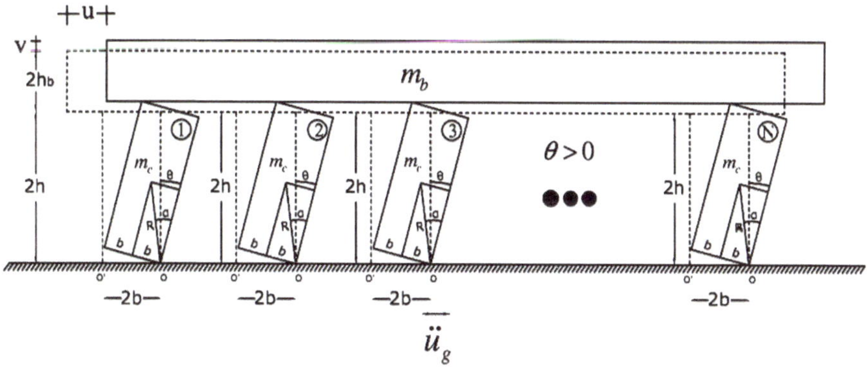

Fig. 15.1 Rocking array of free-standing columns capped with a freely supported rigid beam

Fig. 15.2 View of the Temple of Aphaia, in Aegina, Greece. Its monolithic, free-standing columns support massive epistyles and the frieze atop and the entire rocking frame remains standing for more than 2,500 years in a region with high seismicity

late archaic temple of Aphaia in the island of Aegina nearby Athens, Greece. Dates ranging from 510BC to 470BC have been proposed for this temple. All but three of the 32 outer columns of the temple are monolithic and they have been supporting for 2.5 millennia the front and back epistyles together with the heavy frieze (triglyph and metope).

The understanding of the rocking response and stability of the configuration shown in Fig. 15.1 is also pertinent to the growing precast bridge construction technology where bridge piers supporting heavy decks are allowed to rock atop their foundation in order to achieve re-centering of the bridge bent after a seismic event.

15.2 A Notable Limitation of the Equivalent Static Lateral Force Analysis

15.2.1 Seismic Resistance of Free-Standing Columns Under "Equivalent Static" Lateral Loads

Consider a free-standing rigid column with size $R = \sqrt{b^2 + h^2}$ and slenderness $b/h = \tan\alpha$ as shown in Fig. 15.3 (left). Let us first assume that the base of the column is moving (say to the left) with a "slowly" increasing acceleration, $\ddot{u}_g$ (say a very long-duration acceleration pulse which allows for an equivalent static analysis). Uplift of the block (hinge formation) happens when the seismic demand (overturning moment) $= m\ddot{u}_g h$ reaches the seismic resistance (recentering moment) $= mgb$. When uplifting is imminent, "static" moment equilibrium of the block about the pivoting point O gives

$$\underbrace{m\ddot{u}_g h}_{\text{demand}} = \underbrace{mgb}_{\text{resistance}} \quad \text{or} \quad \underbrace{\ddot{u}_g}_{\text{demand}} = g\frac{b}{h} = \underbrace{g\tan\alpha}_{\text{resistance}} \tag{15.1}$$

Equation (15.1) shows that the block <*b*, *h*> will uplift when $\ddot{u}_g \geq g\tan\alpha$. Now, given that this is a "quasistatic" lateral inertial loading, the inertia moment due to

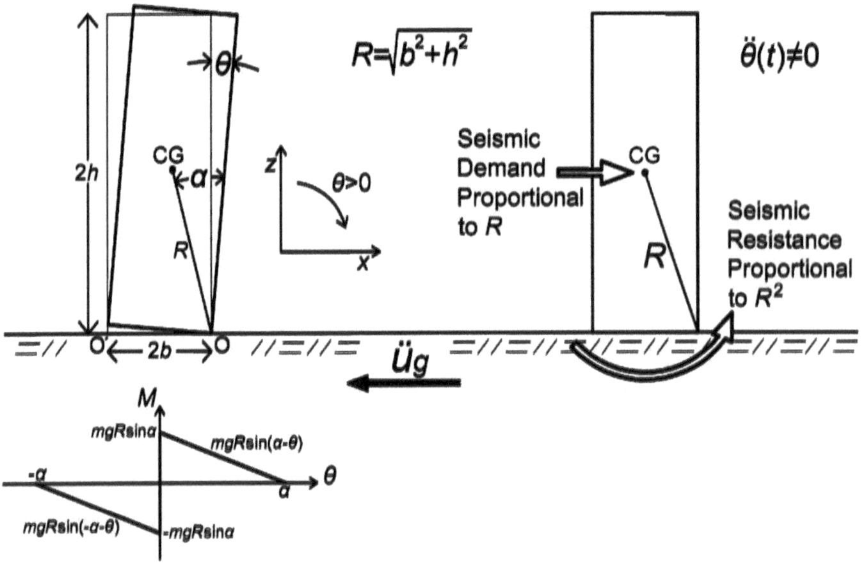

Fig. 15.3 *Left*: Geometric characteristics of a free-standing rocking column together with its moment rotation diagram. *Right*: During rocking motion $\left(\ddot{\theta}(t) \neq 0\right)$ the seismic resistance is proportional to R^2; while, the seismic demand is proportional to R

the nearly zero rotational accelerations of the blocks is negligible $\left(\ddot{\theta}(t) = 0\right)$. Upon uplift has occurred, the rocking block experiences a positive rotation, $\theta(t)$; therefore, the seismic moment demand is $m\ddot{u}_g R\cos(\alpha - \theta(t))$; while the seismic resistance is merely $mgR\sin(\alpha - \theta(t))$, since $\ddot{\theta}(t) = 0$. For $\theta > 0$, the resistance of the rocking block upon uplifting under quasistatic lateral loading is $\tan(\alpha - \theta(t))$ which is smaller than $\tan\alpha$. Accordingly; once the block uplifts, it will also overturn. From this analysis one concludes that under quasistatic lateral loading the stability of a free-standing column depends solely on its slenderness ($g\tan\alpha$) and is independent to the size $\left(R = \sqrt{b^2 + h^2}\right)$.

15.2.2 Seismic Resistance of Free-Standing Columns Subjected to Dynamic Loads

In reality, earthquake shaking, $\ddot{u}_g$, is not a quasistatic loading and upon uplifting has occurred the block will experience a finite rotational acceleration $\left(\ddot{\theta}(t) \neq 0\right)$. In this case, dynamic moment equilibrium gives

$$\underbrace{-m\ddot{u}_g(t)R\cos\left[\alpha = \theta(t)\right]}_{\text{seismic demand}} = \underbrace{I_o\ddot{\theta}(t) + mg\ R\sin\left[\alpha - \theta(t)\right]}_{\text{seismic resistance}} \quad \theta > 0 \qquad (15.2)$$

where I_o is the rotational moment of inertia of the column about the pivot point at the base – a quantity that is proportional to the square of the size of the column R. As an example for rectangular columns, $I_o = \frac{4}{3}mR^2$, and Eq. (15.2) simplifies to

$$\underbrace{-\ddot{u}_g(t)R\cos\left[\alpha = \theta(t)\right]}_{\text{seismic demand}} = \underbrace{\frac{4}{3}R^2\ddot{\theta}(t) + gR\sin\left[\alpha - \theta(t)\right]}_{\text{seismic resistance}} \quad \theta > 0 \qquad (15.3)$$

Equation (15.3) indicates that when a slender free-standing column is set into rocking motion the seismic demand (overturning seismic moment) is proportional to R (first power of the size); whereas, the seismic resistance (opposition to rocking) is proportional to R^2 (second power of the size). Consequently, Eq. (15.3) dictates that regardless how slender a block is (small α) and how intense the ground shaking, $\ddot{u}_g$, is (seismic demand is a first power of R), when a rotating block $\left(\ddot{\theta}(t) = \text{finite}\right)$ is large enough, the second power of R in the right-hand-side (seismic resistance term) can always ensures stability. Simply stated, Housner's (1963) size effect uncovered in 1963 is merely a reminder that a quadratic term eventually dominates over a linear term regardless the values of their individual coefficients.

Figure 15.3 (right) shows schematically the relations with the size R of the seismic demand (linear relation) and the seismic resistance (quadratic relation). From its very conception the "equivalent static lateral force analysis" is not meant to deal with any rotational acceleration term; therefore, its notable failure to capture the ample seismic stability (resistance) of tall free-standing structures. Consequently, it is worth noting that while in modern earthquake resistance (capacity) design we follow essentially a "static" approach based on the concepts of strength, stiffness and ductility, the earthquake resistance design of the ancient temples as the one shown in Fig. 15.2, is based entirely on a truly dynamic approach based on the concept of rotational inertia; while, the three basic quantities that we use in modern capacity design are essentially immaterial (strength = $g\tan\alpha$, stiffness = negative, ductility = nonexistent).

15.3 Review of the Dynamics of the Solitary Rocking Column

With reference to Fig. 15.3 (left) and assuming that the coefficient of friction is large enough so that there is no sliding, the equation of motion of a free standing column for negative rotation ($\theta(t) < 0$) is (Yim et al. 1980; Hogan 1989; Makris and Roussos 2000; Zhang and Makris 2001, among others).

$$I_o\ddot{\theta}(t) + mgR\sin\left[-\alpha - \theta(t)\right] = -m\ddot{u}_g(t)R\cos\left[-\alpha - \theta(t)\right] \quad \theta < 0 \tag{15.4}$$

Equations (15.2) and (15.4) are valid for arbitrary values of the slenderness angle $\alpha = \tan^{-1}(b/h)$ and can be expressed in the compact form

$$\ddot{\theta}(t) = -p^2\left\{\sin\left[\alpha\ \mathrm{sgn}\left[\theta(t)\right] - \theta(t)\right] + \frac{\ddot{u}_g}{g}\cos\left[\alpha\ \mathrm{sgn}\left[\theta(t)\right] - \theta(t)\right]\right\} \tag{15.5}$$

The oscillation frequency of a rigid block under free vibration is not constant, because it strongly depends on the vibration amplitude (Housner 1963). Nevertheless, the quantity $p = \sqrt{\frac{mRg}{I_o}}$ is a measure of the dynamic characteristics of the block. For the 5.3 m × 1.0 m free-standing column of the Temple of Aphaia in Aegina, $p = 1.67$ *rad/s*; whereas, and for a household brick, $p \approx 8$ *rad/s*.

Figure 15.3 (bottom) shows the moment-rotation relationship during the rocking motion of a free-standing block. The system has infinite stiffness until the magnitude of the applied moment reaches the value $mgR\sin\alpha$, and once the block is rocking, its restoring force decreases monotonically, reaching zero when $\theta = \alpha$. This negative stiffness, which is inherent in rocking systems, is most attractive in earthquake engineering given that such systems do not resonate.

During the oscillatory rocking motion of a free-standing rigid column, the moment-rotation curve follows the curve shown in Fig. 15.3 (bottom) without enclosing any area. Energy is lost only during impact, when the angle of rotation reverses. When the angle of rotation reverses, it is assumed that the rotation continues smoothly from points O to O′ and that the impact force is concentrated at the new pivot point, O′. With this idealization, the impact force applies no moment around O′, hence the angular momentum around O′ is conserved. Conservation of angular momentum about point O′ just before the impact and right after the impact gives

$$I_o\dot{\theta}_1 - m\dot{\theta}_1 2bR\sin(\alpha) = I_o\dot{\theta}_2 \tag{15.6}$$

where $\dot{\theta}_1$ = angular velocity just prior to the impact; and $\dot{\theta}_2$ = angular velocity right after the impact. The ratio of kinetic energy after and before the impact is

$$r = \frac{\dot{\theta}_2^2}{\dot{\theta}_1^2} \tag{15.7}$$

which means that the angular velocity after the impact is only $\sqrt{r}$ times the velocity before the impact. Substitution of (15.6) into (15.7) gives

$$r = \left[1 - \frac{3}{2}\sin^2\alpha\right]^2 \tag{15.8}$$

The value of the coefficient of restitution given by (15.8) is the maximum value of r under which a free-standing rigid block with slenderness α will undergo rocking motion. Consequently, in order to observe rocking motion, the impact has to be inelastic. The less slender a block (larger α), the more plastic is the impact, and for the value of $\alpha = \sin^{-1}\sqrt{2/3} = 54.73°$, the impact is perfectly plastic. During the rocking motion of slender blocks, if additional energy is lost due to the inelastic behavior at the instant of impact, the value of the true coefficient of restitution r will be less than the one computed from Eq. (15.8).

15.4 Equation of Motion of the Rocking Frame

The free standing rocking frame shown in Fig. 15.1 is a single-degree-of-freedom structure with identical columns, having size $R = \sqrt{b^2 + h^2}$ and slenderness $\alpha = \text{atan}(b/h)$. The only other parameter that influences the dynamics of the rocking frame is the ratio of the mass of the cap-beam, m_b, to the mass of all the N rocking columns, m_c, $\gamma = \frac{m_b}{Nm_c}$. For the temple of Aphaia in Aegina where the frieze γ is as low as 0.6; whereas in prefabricated bridges $\gamma > 4$. As in the case of the single rocking column, the coefficient of friction is large enough so that sliding does not occur at the pivot point at the base and at the cap-beam. Accordingly, the horizontal translation displacement $u(t)$ and the vertical lift $v(t)$ of the cap-beam are functions of the single degree of freedom $\theta(t)$. For a positive horizontal ground acceleration (the ground is accelerating to the right), the rocking frame will initially rock to the left ($\theta(t) < 0$). Assuming that the rocking frame will not topple, it will recenter, impacts will happen at the pivot points (at the base and at the cap-beam) and subsequently it will rock to the right ($\theta(t) > 0$). During rocking the dependant variables $u(t)$, $v(t)$ and their time derivatives are given for $\theta(t) < 0$ and $\theta(t) > 0$ by the following expressions

$$u = 2R\left(\sin\alpha - \sin\left(\alpha \pm \theta\right)\right) \tag{15.9}$$

$$\dot{u} = \mp 2R\cos\left(\alpha \pm \theta\right)\dot{\theta} \tag{15.10}$$

$$\ddot{u} = 2R\left(\sin\left(\alpha \pm \theta\right)\left(\dot{\theta}\right)^2 - \cos\left(\alpha \pm \theta\right)\ddot{\theta}\right) \tag{15.11}$$

and

$$v = 2R\left(\cos\left(\alpha \pm \theta\right) - \cos\alpha\right) \tag{15.12}$$

$$\dot{v} = \mp 2R\sin\left(\alpha \pm \theta\right)\dot{\theta} \tag{15.13}$$

$$\ddot{v} = 2R\left(-\cos\left(\alpha \pm \theta\right)\left(\dot{\theta}\right)^2 \mp \sin\left(\alpha \pm \theta\right)\ddot{\theta}\right) \tag{15.14}$$

In the equations above, whenever there is a double sign (say $\pm$) the top sign is for $\theta(t) < 0$ and the bottom sign is for $\theta(t) > 0$.

During rocking motion Langrange's equation must be satisfied,

$$\frac{d}{dt}\left(\frac{dT}{d\dot{\theta}}\right) - \frac{dT}{d\theta} = Q \tag{15.15}$$

In Eq. (15.15), T is the kinetic energy of the system and Q is the generalized force acting on the system

$$Q = \frac{dW}{d\theta} \tag{15.16}$$

in which W is the work done by the external field forces acting on the rocking frame during an admissible rotation $\delta\theta$. During this admissible rotation $\delta\theta$, the variation of work is

$$\delta W = \frac{dW}{d\theta}\delta\theta \tag{15.17}$$

In either case were $\theta(t) < 0$ or $\theta(t) > 0$ the kinetic energy of the system is

$$T = N\frac{1}{2}I_o\left(\dot{\theta}\right)^2 + \frac{1}{2}m_b\left((\dot{u})^2 + (\dot{v})^2\right) \tag{15.18}$$

Using Eqs. (15.10) and (15.13), Eq. (15.18) reduces to

$$T = \left(\frac{N}{2}I_o + 2m_b R^2\right)\left(\dot{\theta}\right)^2 \tag{15.19}$$

Our analysis proceeds by first investigating the rocking motion of a free-standing frame subjected to a horizontal ground acceleration $\ddot{u}_g(t)$ when $\theta(t) < 0$. During this segment of the motion the variation of the work, W, is

$$\delta W = \left(m_b + \frac{N}{2}m_c\right)\left(\ddot{u}_g \delta u - g\delta v\right) \tag{15.20}$$

or

$$\delta W = \left(m_b + \frac{N}{2}m_c\right)\left(\ddot{u}_g \frac{du}{d\theta} - g\frac{dv}{d\theta}\right)\delta\theta \tag{15.21}$$

The combination of Eqs. (15.17) and (15.21) gives

$$\frac{dW}{d\theta} = \left(m_b + \frac{N}{2}m_c\right)\left(\ddot{u}_g \frac{du}{d\theta} - g\frac{dv}{d\theta}\right) \tag{15.22}$$

which simplifies to

$$\frac{dW}{d\theta} = 2R\left(m_b + \frac{N}{2}m_c\right)\left(-\ddot{u}_g \cos\left(\alpha + \theta\right) + g \sin\left(\alpha + \theta\right)\right) \tag{15.23}$$

after using the expression given by (15.9) and (15.12).

The substitution of Eqs. (15.19) and (15.23) into Lagrange's equation given by (15.15) results to the equation of motion of the rocking frame for $\theta(t) < 0$

$$\left(\frac{\frac{I_o}{2m_c R} + 2\gamma R}{\left(\gamma + \frac{1}{2}\right) g}\right)\ddot{\theta} = \sin\left(\alpha + \theta\right) - \frac{\ddot{u}_g}{g}\cos\left(\alpha + \theta\right) \tag{15.24}$$

where $\gamma = \frac{m_b}{Nm_c}$ is the ratio of the mass of the cap-beam (epistyle), m_b, to the mass of all the N columns $= Nm_c$.

For the case where the rotation is positive $\theta(t) > 0$ the variation of the work is

$$\delta W = -\left(m_b + \frac{N}{2}m_c\right)\left(\ddot{u}_g \delta u - g\delta v\right) \tag{15.25}$$

and Eq. (15.16) takes the form

$$\frac{dW}{d\theta} = -2R\left(m_b + \frac{N}{2}m_c\right)\left(\ddot{u}_g \cos\left(\alpha + \theta\right) + g \sin\left(\alpha + \theta\right)\right) \tag{15.26}$$

The substitution of Eqs. (15.19) and (15.26) into Lagrange's equation given by (15.15) offers the equation of motion of the rocking frame for $\theta(t) > 0$

$$\left(\frac{\frac{I_o}{2m_c R} + 2\gamma R}{\left(\gamma + \frac{1}{2}\right) g}\right)\ddot{\theta} = -\sin(\alpha - \theta) - \frac{\ddot{u}_g}{g}\cos(\alpha - \theta) \tag{15.27}$$

For rectangular columns $I_o = (4/3)mR^2$; and Eqs. (15.24) and (15.27) can be expressed in a single compact form

$$\ddot{\theta}(t) = -\frac{1+2\gamma}{1+3\gamma}p^2\left(\sin\left[\alpha\,\mathrm{sgn}\left[\theta(t)\right] - \theta(t)\right] + \frac{\ddot{u}_g(t)}{g}\cos\left[\alpha\,\mathrm{sgn}\left[\theta(t)\right] - \theta(t)\right]\right) \tag{15.28}$$

Equation (15.28) which describes the planar motion of the free-standing rocking frame is precisely the same as Eq. (15.3) which describes the planar rocking motion of a single free-standing rigid column with the same slenderness α, except that in the rocking frame the term p^2 is multiplied with the factor $\frac{1+2\gamma}{1+3\gamma}$. Accordingly, the frequency parameter of the rocking frame, $\widehat{p}$, is

$$\widehat{p} = \sqrt{\frac{1+2\gamma}{1+3\gamma}}\,p \tag{15.29}$$

where $p = \sqrt{\frac{3g}{4R}}$ is the frequency parameter of the rocking column and $\gamma = \frac{m_b}{Nm_c}$ is the mass of the cap-beam to the mass of all N columns.

For a light cap-beam $\left(\gamma = \frac{m_b}{Nm_c} \to 0\right)$, the multiplication factor $\frac{1+2\gamma}{1+3\gamma} \to 1$ and the array of free standing columns coupled with a light epistyle exhibit precisely the dynamic rocking response of the solitary free-standing column. On the other hand, as the mass of the epistyle increases

$$\lim_{\gamma\to\infty}\frac{1+2\gamma}{1+3\gamma} = \frac{2}{3} \tag{15.30}$$

Accordingly, the dynamic behavior of a rocking frame with a very heavy cap-beam supported on columns with slenderness α and frequency parameter, $p = \sqrt{\frac{3g}{4R}}$, is identical to the dynamic rocking response of a single rigid column with slenderness α and frequency parameter $\widehat{p} = \sqrt{\frac{2}{3}}p$ – that is a smaller frequency parameter; therefore a larger, more stable column.

This remarkable result offered by Eq. (15.28) – that the heavier the cap-beam is, the more stable is the free standing rocking frame despite the rise of the center of gravity of the cap-beam – has been also confirmed by obtaining Eq. (15.28) for a pair of columns with the algebraically-intense direct formulation after deriving the equations of motion of the two-column frame through dynamic equilibrium

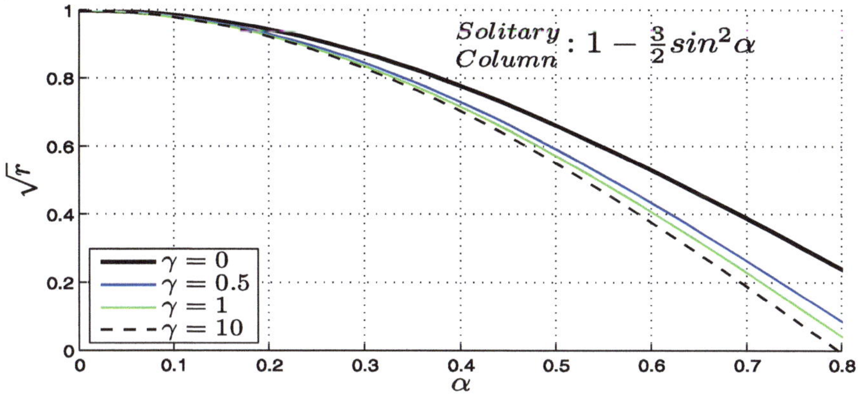

Fig. 15.4 Values of the maximum coefficient of restitution as a function of the slenderness, α of the columns of the rocking frame for different values of the mass ratio $\gamma = \frac{m_b}{Nm_c}$

(Makris and Vassiliou 2014). Furthermore, numerical studies with the discrete element method by Papaloizou and Komodromos (2009) concluded to the same result – that the planar response of free-standing columns supporting epistyles is more stable than the response of the solitary, free-standing column.

According to Eq. (15.28) the rocking response and stability analysis of the free-standing rocking frame with columns having slenderness, α, and size R, is described by all the past published work on the rocking response of the free-standing single block (Housner 1963; Yim et al. 1980; Aslam et al. 1980; Spanos and Koh 1984; Zhang and Makris 2001; Makris and Konstantinidis 2003; Vassiliou and Makris 2012; Makris and Roussos 2000, among others), where the block has the same slenderness, α, and a larger size $\widehat{R}$ given by

$$\widehat{R} = \frac{1+3\gamma}{1+2\gamma}R = \left(1+\frac{\gamma}{1+2\gamma}\right)R \tag{15.31}$$

During rocking motion of a free-standing frame, the moment-rotation curve follows the curve shown in Fig. 15.3 (bottom) where R is replaced by $\widehat{R}$. Energy is lost during impact when the angle of rotation reverses. At this instant it is assumed that the rotation continues smoothly and that the impact forces are concentrated at the new pivot points. Application of the angular momentum-impulsive theorem in association with the change of the linear momentum of the cap-beam (Makris and Vassiliou 2013) offers the ratio of the kinetic energy of the rocking frame after and before impact

$$r = \left(\frac{\dot{\theta}_2}{\dot{\theta}_1}\right)^2 = \left(\frac{1-\frac{3}{2}\sin^2\alpha + 3\gamma\cos 2a}{1+3\gamma}\right)^2 \tag{15.32}$$

Equation (15.32) indicates that the angular velocity of the rocking frame after the impact is only $\sqrt{r}$ times the velocity before the impact. Figure 15.4 plots the value of the minimum coefficient of restitution $\sqrt{r} = \frac{1-\frac{3}{2}\sin^2\alpha+3\gamma\cos 2a}{1+3\gamma}$ as a function of the slenderness α for different values of the mass ratio $\gamma = \frac{m_b}{Nm_c}$.

Figure 15.4 indicates that the maximum coefficient of restitution $\sqrt{r}$ of the rocking frame is always smaller than the maximum coefficient of restitution of the solitary column $= 1 - \frac{3}{2}\sin^2\alpha$, indicating that when a free-standing frame engages into rocking motion, it dissipates more energy than the equal slenderness equivalent solitary free-standing column with size $\widehat{R}$ due to the additional impacts that happen between the columns and the cap-beam (epistyles and frieze).

15.5 Overturning Spectra – Self Similar Response

The relative simple form yet destructive potential of near source ground motions has motivated the development of various closed form expressions which approximate their dominant kinematic characteristics. The early work of Veletsos et al. (1965) was followed by the papers of Hall et al. (1995), Makris (1997), Makris and Chang (2000), Alavi and Krawinkler (2001) and more recently by the papers of Mavroeidis and Papageorgiou (2003) and Vassiliou and Makris (2011). Physically realizable pulses can adequately describe the impulsive character of near-fault ground motions both qualitatively and quantitatively. The minimum number of parameters of the mathematical pulse is two, which are the acceleration amplitude, a_p and the duration T_p. The more sophisticated model of Mavroeidis and Papageorgiou (2003) involves four parameters which are the pulse period, the pulse amplitude, the pulse phase and the number of half cycles. Recently, Vassiliou and Makris (2011) used the Mavroeidis and Papageorgiou model (2003) in association with wavelet analysis to develop a mathematically formal and objective procedure to extract the time scale and length scale of strong ground motions.

The current established methodologies for estimating the pulse characteristics of a wide class of records are of unique value, since the product, $a_pT_p{}^2 = L_p$ is a characteristic length scale of the ground excitation and is a measure of the persistence of the most energetic pulse to generate inelastic deformation (Makris and Black 2004a, b). It is emphasized that the persistence of the pulse, $a_pT_p{}^2 = L_p$, is a different characteristic than the strength of the pulse which is measured with the peak pulse acceleration, a_p. The reader may recall that among two pulses with different acceleration amplitude (say $a_{p1} > a_{p2}$) and different pulse durations (say $T_{p1} < T_{p2}$) the inelastic deformation does not scale with the peak pulse acceleration (most intense pulse) but with the strongest length scale (larger $a_pT_p{}^2 =$ most persistent pulse) (Makris and Black 2004a, b; Karavasilis et al. 2010).

The heavy dark line in Fig. 15.5 (Top) which approximates the long-period acceleration pulse of the NS component of the 1992 Erzincan, Turkey, record is a scaled expression of the second derivative of the Gaussian distribution, $e^{-t^2/2}$, known in the seismological literature as the symmetric Ricker wavelet (Ricker 1943, 1944).

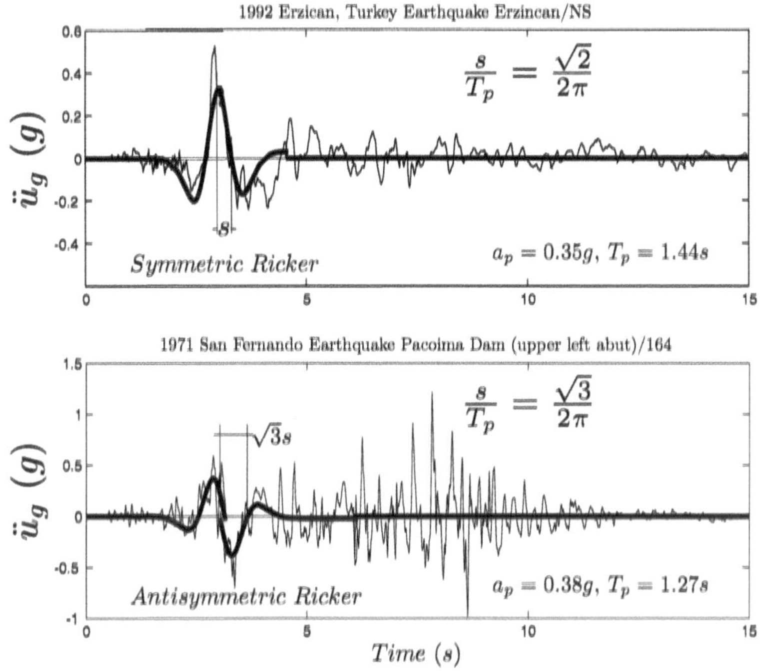

Fig. 15.5 *Top*: North-South components of the acceleration time history recorded during the 1992 Erzican, Turkey earthquake together with a symmetric Ricker wavelet. *Bottom*: Fault-normal component of the acceleration time history recorded during the 1971 San Fernando earthquake, together with an antisymmetric Ricker wavelet

$$\psi(t) = a_p\left(1 - \frac{2\pi^2 t^2}{T_p^2}\right)e^{-\frac{1}{2}\frac{2\pi^2 t^2}{T_p^2}} \tag{15.33}$$

The value of $T_p = \frac{2\pi}{\omega_p}$, is the period that maximizes the Fourier spectrum of the symmetric Ricker wavelet. Similarly, the heavy dark line in Fig. 15.5 (Bottom) which approximates the long-period acceleration pulse of the Pacoima Dam motion recorded during the February 9, 1971 San Fernando, California earthquake is a scaled expression of the third derivative of the Gaussian distribution $e^{-t^2/2}$

$$\psi(t) = \frac{a_p}{\beta}\left(\frac{4\pi^2 t^2}{3T_p^2} - 3\right)te^{-\frac{1}{2}\frac{4\pi^2 t^2}{3T_p^2}} \tag{15.34}$$

in which β is a factor equal to 1.3801 that enforces the above function to have a maximum equal to a_p.

The choice of the specific functional expression to approximate the main pulse of pulse–type ground motions has limited significance in this work. What is important to recognize is that several strong ground motions contain a distinguishable acceleration pulse which is responsible for most of the inelastic deformation of structures (Hall et al. 1995; Makris and Chang 2000; Alavi and Krawinkler 2001; Makris and Psychogios 2006; Karavasilis et al. 2010, among others). A mathematically rigorous and easily reproducible methodology based on wavelet analysis to construct the best matching wavelet has been recently proposed by Vassiliou and Makris (2011).

Consider the free-standing rocking frame shown in Fig. 15.1 that is subjected to an acceleration pulse (like those shown in Fig. 15.5) with acceleration amplitude a_p and pulse duration, $T_p = \frac{2\pi}{\omega_p}$. From Eq. (15.28) it results that the response of a free-standing rocking frame subjected to an acceleration pulse is a function of six variables

$$\theta(t) = f\left(p, \alpha, \gamma, g, a_p, \omega_p\right) \tag{15.35}$$

The seven variables appearing in Eq. (15.35) involve only two reference dimensions; that of length [L] and time [T]. According to Buckingham's Π-Theorem, the number of dimensionless products with which the problem can be completely described is equal to [number of variables = 7] – [number of reference dimensions = 2] = 5. Herein we select as repeating variables the characteristics of the pulse-excitation, a_p and ω_p and the five independent Π-products are: $\Pi_\theta = \theta$, $\Pi_\omega = \omega_p/p$, $\Pi_\alpha = \tan\alpha$, $\Pi_\gamma = \gamma$ and $\Pi_g = a_p/g$. With these five dimensionless Π-products, Eq. (15.35) reduces to

$$\theta(t) = \phi\left(\frac{\omega_p}{p}, \tan\alpha, \gamma, \frac{a_p}{g}\right) \tag{15.36}$$

The rocking response of the free-standing frame shown in Fig. 15.1 when subjected to a horizontal base acceleration history $\ddot{u}_g(t)$ is computed by solving Eq. (15.28) in association with the minimum energy loss expression given by Eq. (15.32) which takes place at every impact.

Figure 15.6 shows the minimum overturning acceleration spectra of a free-standing rocking frame when subjected to a symmetric Ricker pulse (left) and an antisymmetric Ricker pulse (right) for different values of the mass ratio $\gamma = \frac{m_b}{Nm_c}$. The top plots are for values of the column slenderness $\alpha = 10°$ and the bottom plots are for $\alpha = 14°$.

In constructing Fig. 15.6, the frequency parameter p is the frequency parameter of the columns of the frame (not $\widehat{p}$) and the enhanced stability of the rocking frame due to (a) the corresponding larger size, $\widehat{p} = \sqrt{\frac{1+2\gamma}{1+3\gamma}}\,p$, and (b) the reduced coefficient of restitution (see Eq. (15.32)) is given by the curves for each given value of γ.

Figure 15.6 indicates that up to values of $\omega_p/p = 4$ the additional stability of the rocking frame versus the stability of the equal slenderness solitary column is marginal. For values of $\omega_p/p > 4$ (larger columns or shorter period pulses) the minimum acceleration overturning spectra of the rocking frame are higher than

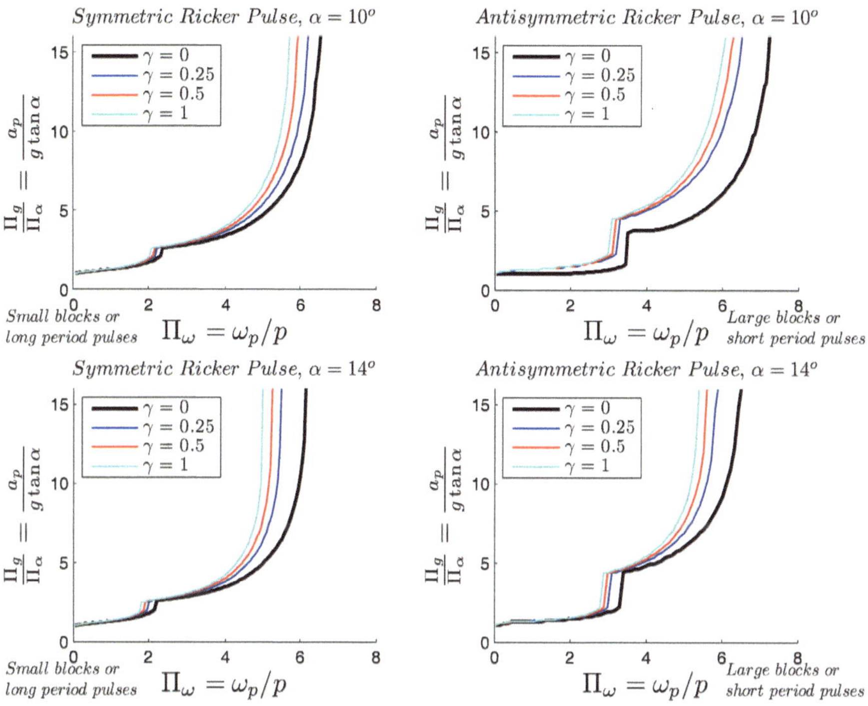

Fig. 15.6 Minimum overturning acceleration spectra of the free-standing rocking frame shown in Fig. 15.1 when subjected to a symmetric Ricker pulse (*left*) and an antisymmetric Ricker pulse (*right*) for different values of the mass ratio $\gamma = \frac{m_b}{N m_c}$. *Top*: $\alpha = 10°$, *Bottom* $\alpha = 14°$. The values of the coefficient of restitution are given by Eq. (15.32)

the corresponding spectrum of the solitary rocking column showing the enhanced seismic stability of the top-heavy rocking frame. This enhanced seismic stability is indifferent to the height of the center of gravity of the cap-beam.

15.6 Seismic Stability of Ancient Columns Supporting Epistyles and the Frieze Atop

In ancient Greek temples the epistyles are positioned from the vertical axis of one column to the vertical axis of the neighboring column; therefore, the joint of the epistyles are along the vertical axis of the column (see Fig. 15.2). With this configuration during lateral loading of the peristyle of the temple, each epistyle in addition to the horizontal translation, u, shown in Fig. 15.1 it will also experience a small rotation. Nevertheless, the tendency of the epistyle to rotate is partially prevented from the presence of the neighboring high-profile epistyle and the heavy stone of the frieze atop which goes over the joint of the epistyles (see Fig. 15.2).

According to this construction pattern with very tight joints the ancient builders constructed a nearly continuous and massive structure atop the columns which according to this study enhanced appreciably the seismic-rocking stability of the peristyle of the temples.

Two of the strongest ground motions recorded in Greece are the 1973 Lefkada record and the 1995 Aigion record (Vassiliou and Makris 2011). Both records exhibit distinguishable acceleration pulses with durations $T_p \approx 0.6$ s. We concentrate on the Temple of Apollo in Corinth where its 7.5 m × 1.8 m monolithic columns remain standing since 540BC in an area with high seismicity. The dimensions of its columns yield a frequency parameter $p = \sqrt{\frac{3g}{4R}} = 1.4 rad/s$ and a slenderness $\alpha = \tan^{-1}(b/h) = 13.5^{\circ}$. By taking the pulse duration $T_p = 0.6$ s of the nearby Aigion record, the dimensionless term Π_ω assumes the value $\frac{\omega_p}{p} = \frac{2\pi}{pT_p} = 7.5$. For such large value of $\omega_p/p \approx 7.5$ the bottom plots of Fig. 15.6 give for the solitary free-standing column ($\gamma = 0$ line) an overturning ground acceleration $a_p > 15 g\tan\alpha = 15\ g \times 0.24 = 3.6\, g$ – which is an unrealistically high acceleration. Consider now the extreme situation for Greece, where the predominant pulse of the ground shaking exhibits a period $T_p = 0.9$ s. A pulse period $T_p = 0.9$ s may be a rare event for the fault size and earthquake magnitude that prevail in Greece; nevertheless, it helps one understanding the appreciable seismic stability of rocking structures. With $T_p = 0.9\, s$ and $p = 1.4\, rad/s$, $\omega_p/p = 5$ and according to the bottom plots of Fig. 15.6 which are for slenderness $\alpha = 14^{\circ}$, the minimum overturning acceleration of a rocking frame with $\gamma = 0.25$ exceeds the value of $a_p \approx 5\ g \times 0.24 = 1.2$ g. This analysis shows that the free-standing peristyles of ancient temples can survive acceleration pulses as long as 0.9 s and as intense as 1.2 g. While this is a physically realizable pulse (Loh et al. 2000), it is an unlikely strong shaking for the seismicity of Greece that apparently never happened over the 2,500 years of the lifespan of the temple.

15.7 The Emerging Concept of Rocking Isolation for Bridges

The concept of allowing the piers of bridges to rock is not new. For instance, the beneficial effects that derive from uplifting and rocking have been implemented since the early 1970s in the South Rangitikei bridge in New Zealand (Beck and Skinner 1973). During the last decade, the benefits/challenges associated with the rocking of bridge piers have been receiving increasing attention partly because of growing interest in the prefabricated bridge technology (Wacker et al. 2005; Pang et al. 2008; Cohagen et al. 2008, and references reported therein) and partly because of the need for the bridge structure to recenter after a strong seismic event (Sakai et al. 2006; Cheng 2008, among others).

In the prefabricated bridge technology the bridge piers and the deck are not free standing. The structural system is essentially a hybrid system (see Wacker et al. 2005; Cheng 2008) where the bridge pier is connected to its foundation

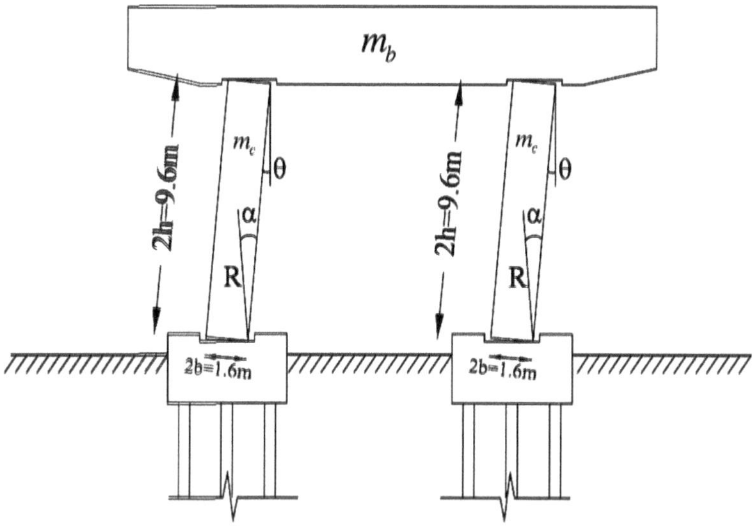

Fig. 15.7 Free Standing rocking bridge bent. Potential sliding during impact is prevented with the recess shown. No vertical post-tensioning, no continuation of the longitudinal reinforcement of the columns through the rocking interfaces at the pile-caps and the cap-beam

and at the deck with a post-tensioned tendon that passes through the axis of the column together with longitudinal mild steel reinforcement running near the circumference of the column. During earthquake loading the majority of deformation is concentrated at the pier-foundation and pier-cap-beam interfaces and the overall deformation pattern of the post-tensioned pier-cap-beam system resembles the deformation pattern of the free-standing rocking frame that is under investigation in this study. Nevertheless, the post-tensioning tendons and the mild-steel longitudinal reinforcement that extends into the foundation and the cap-beam contributes appreciably to the lateral moment capacity of the system and in most prefabricated bridge applications, the moment-rotation curve of the hybrid systems follows a positive slope. Furthermore, during seismic shaking the post-tensioned, prefabricated bridge frame may be subject to resonance due to its overall positive stiffness; while, the longitudinal mild-steel reinforcement that extends through the seismic connection is subject to buckling and fracture.

Within the context of a proof-of-concept, in this study we present the planar rocking response of a free-standing two-column bridge bent where its moment rotation curve follows a negative slope given that the frame is entirely free to rock (see Fig. 15.3-bottom). Figure 15.7 shows schematically the free-standing two-column bridge bent of interest in its deformed configuration. Sliding at the pivot point during impact is prevented with a recess at the pile-cap and the cap-beam as shown in Fig. 15.7. In this numerical application the cylindrical piers of the free-standing bridge bent are 9.6 m tall with a diameter $d = 2b = 1.6$ m. These are typical dimensions of bridge piers for highway overpasses and other bridges in Europe and USA. Taller bridge piers will result to even more stable configurations. With $2h = 9.6$ m and $2b = 1.6$ m, the slenderness of the bridge pier is $\tan\alpha = 1/6 = 0.166$ and its frequency parameter $p = 1.23$.

Depending on the length of the adjacent spans and the per-length weight of the deck, the mass ratio $\gamma = \frac{m_b}{Nm_c}$ assumes values from four and above ($\gamma > 4$). The larger the value of γ (heavier deck), the more stable is the free standing rocking frame (see Fig. 15.6). The seismic response analysis of the rocking frame has been studied until this section by using as ground excitation acceleration pulses described either by symmetric or the antisymmetric Ricker wavelets. The acceleration amplitude, a_p, and the duration, T_p, of any distinct acceleration pulse allow the use of the dimensional analysis presented in this work and the derivation of the associated Π-products which improve the understanding of the physics that govern the problem together with the organization and presentation of the response. Nevertheless, in order to stress the main finding of this study – that top-heavy free-standing rocking frames enjoy ample seismic stability – we examine the planar seismic response of the free standing two column bridge bent shown in Fig. 15.7 when subjected to six strong-motion historic records listed in Table 15.1. The values of the acceleration amplitude, a_p, and pulse period, T_p, shown in last two columns of Table 15.1 have been determined with the extended wavelet transform (Vassiliou and Makris 2011).

Figure 15.8 plots the time histories of the normalized rotation, θ/α, together with the vertical uplift, $v(t)$, and the horizontal drift, $u(t)$, of the free standing rocking bridge bent shown in Fig. 15.7 with $\gamma = \frac{m_b}{Nm_c} = 4$. Note that for all six strong ground motions selected in this analysis the frame rotation, θ, is less than one-third of the slenderness, α, of the columns ($\theta/\alpha < 0.33$); therefore the free-standing rocking frame exhibits ample seismic stability.

The peak horizontal displacement $u_{\max}$ ranges from 20 to 50 cm; while the vertical uplift is as high as 5 cm. The evaluation of these response quantities shall be conducted in association with the equivalent response quantities from vertically post-tensioned hybrid frames (Wacker et al. 2005; Pang et al. 2008; Cheng 2008) and seismically isolated decks (Constantinou et al. 1998; Makris and Zhang 2004; Buckle et al. 2006, among others) after considering the effects of the end-conditions of the deck at the abutments. This comparison/evaluation is the subject of an ongoing study which also examines other practical issues such as the effect of the crushing of the pivoting points of the columns (Roh and Reinhorn 2010a, b) and the accommodation of the deck uplift at the end-abutments.

The main conclusion of this study is that heavy decks freely supported on free-standing piers exhibit ample seismic stability and that the heavier is the deck (even if the center of gravity rises) the more stable is the rocking frame. This conclusion may eventually lead to the implementation of the free-standing rocking frame – a structural configuration where all the issues associated with seismic connections such as buckling and fracture of the longitudinal reinforcing bars or spalling of the concrete cover (Wacker et al. 2005; Pang et al. 2008; Cohagen et al. 2008; Sakai et al. 2006; Cheng 2008) are removed as they are not an issue in the ancient temple shown in Fig. 15.2.

Table 15.1 Earthquake records used for the seismic response analysis of the free-standing rocking bridge bent

Earthquake	Record	Magnitude (Mw)	Epicentral distance (km)	PGA (g)	PGV (m/s)	a_p (g)	T_p (s)
1966 Parkfield	CO2/065	6.1	0.1	0.48	0.75	0.41	0.6
1971 San Fernando	Pacoima Dam/164	6.6	11.9	1.23	1.13	0.38	1.27
1986 San Salvador	Geotech Investig. Center	5.4	4.3	0.48	0.48	0.34	0.8
1992 Erzican,	Erzincan/EW	6.9	13	0.50	0.64	0.34	0.9
1994 Northridge	Jensen Filter Plant/022	6.7	6.2	0.57	0.76	0.39	0.5
1995 Kobe	Takarazuka/000	6.9	1.2	0.69	0.69	0.50	1.1

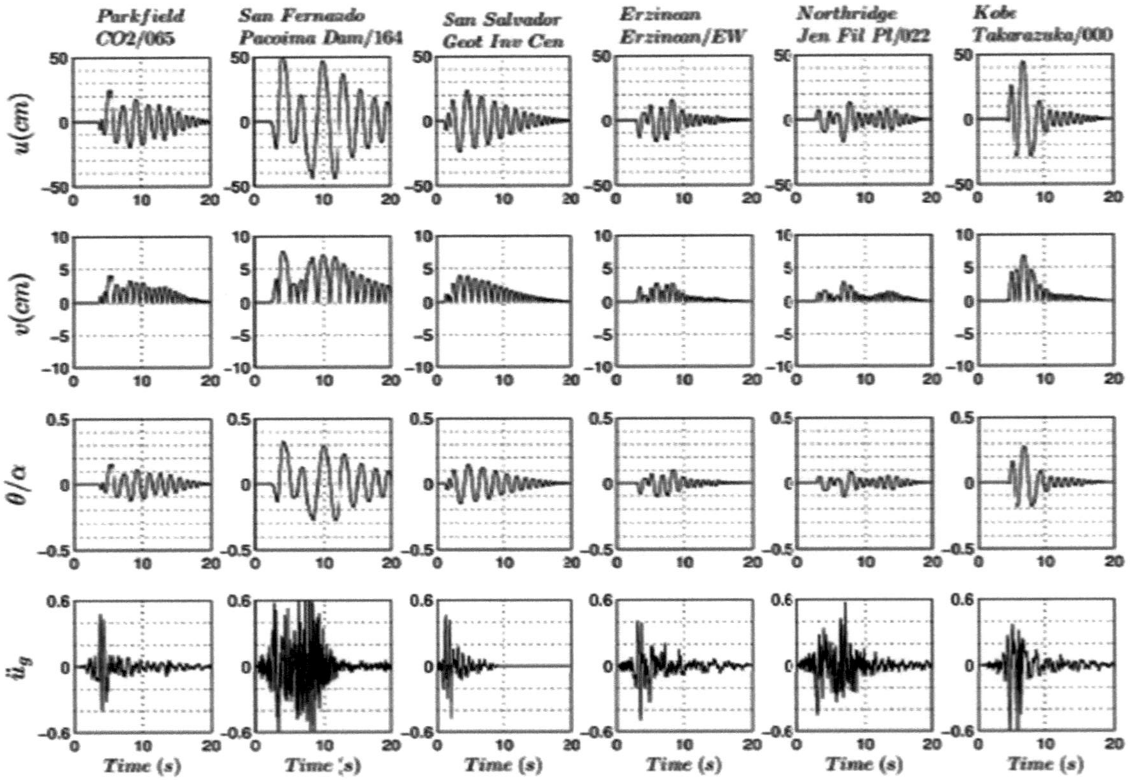

Fig. 15.8 Rotation, vertical and horizontal displacement histories of the free standing rocking frame shown in Fig. 15.7 ($p = 1.23$, $\tan\alpha = 1/6$, $\gamma = 4$) when subjected to the recorded ground motions listed in Table 15.1

Despite the successful design of the South Rangitikei bridge and the ample dynamic stability of the rocking frame as documented in Makris and Vassiliou (2012, 2013, 2014), most modern tall bridges (with tall slender piers) are protected from seismic action via base (shear) isolation of the deck, rather than from (the most natural) rocking isolation. Part of the motivation of this work is to show in the simplest possible way that in the event that a rocking system is selected, the heavy deck atop the tall slender columns not only does not harm the stability of the columns but in contrast enhances the stability of the entire system as shown by Eq. (15.28).

Our work comes to support the emerging design concept (mainly advanced by the prefabricated bridge technology) of concentrating the inelastic deformations of bridge frame at the locations where the bridge-piers meet the foundation and the deck (Mander and Cheng 1997; Sakai and Mahin 2004; Wacker et al. 2005; Mahin et al. 2006; Cheng 2008; Cohagen et al. 2008; Yamashita and Sanders 2009, among others). It shall however be stressed that in the prefabricated bridge technologies, the bridge piers and the deck are not free-standing, therefore, the structural system is essentially a hybrid system in-between the rocking frame examined in this work and a traditional ductile moment-resisting frame.

At present, the equivalent static lateral force procedure is deeply rooted in the design philosophy of the structural engineering community which is primarily preoccupied on how to improve the ductility and performance of the seismic connections; while the ample dynamic rocking stability that derives from the beneficial coexistence of negative stiffness and gravity as described by Eq. (15.28) is ignored. At the same time, it shall be recognized that during the last decade there have been several publications which have voiced the need to go beyond the elastic response spectrum and the associated equivalent static lateral force procedure (Makris and Konstantinidis 2003; Lagomarsino et al. 2004; Apostolou et al. 2007; Resemini et al. 2008; Garini et al. 2011, among others). The time is therefore ripe for the development of new, physically motivated response/design curves which are relevant (in a technically sound way) with the response/design of large, slender structures. Part of the motivation for this paper is to bring forward the ample seismic stability associated with the free rocking of large, slender structures and the corresponding rocking frame.

To this end Fig. 15.9 shows evidence of hinging of the top frame of a bell-tower in Gaio, Italy which remained standing due to the beyond limit-state rocking mechanism that developed (Lagomarsino, Personal communication, 2008). The main objective of this paper is through the derivation of Eq. (15.28) to accept and establish the hinging mechanism (inherent in the rocking frame) not just as a limit-state mechanism but as an operational-state (seismic protection) mechanism for large slender structures as was accepted more than 2.5 millennia ago by the builders of ancient temples as shown in Fig. 15.2.

Fig. 15.9 Evidence of hinging of the top of a bell tower in Gaio, Italy which remained standing after earthquake shaking due to the beyond limit-state stability of the developed rocking frame (Lagomarsino Personal communication, 2008)

15.8 Conclusions

This paper investigated the planar rocking response of an array of free-standing columns capped with a freely supported rigid beam. Following a variational formulation, the paper concludes to the remarkable result that the dynamic rocking response of an array of free standing columns capped with a rigid beam is identical to the rocking response of a single free-standing column with the same slenderness as the slenderness of the columns of the rocking frame; yet with larger size and more energy loss during impacts. A larger size rocking column corresponds to a more stable configuration; therefore, the presence of the freely supported cap-beam renders the rocking frame more stable despite the rise of the center of gravity.

Most importantly, the study shows that the heavier the freely supported cap-beam is, the more stable is the rocking frame concluding that top-heavy rocking frames are more stable than when they are top-light. The stability of the rocking frame is independent to the number of columns and depends only on the ratio of the weight that is transferred to the column to the weight of the column together with the size and the slenderness of the columns.

The acceleration needed to create uplift of the rocking frame is independent to the mass and the height of the center of gravity of the cap-beam and depends only on the slenderness, α, of the columns ($u_g^{up} = g \tan \alpha$).

The findings above render rocking isolation a most attractive alternative for the seismic protection of bridges given that the heavier is the deck the more stable is the rocking bridge. The future implementation of a truly rocking frame where there is neither post-tensioning nor continuation of the longitudinal reinforcement through the rocking interfaces shall remove several of the concerns associated with the seismic connections of prefabricated bridges such as buckling and fracture of the longitudinal reinforcing bars or spalling of the concrete cover.

Acknowledgments Financial Support for this study has been provided by the action "Aristeia" of the "OPERATIONAL PROGRAMME EDUCATION AND LIFELONG LEARNING" and is co-funded by the European Social Fund (ESF) and National Resources of Greece.

References

Alavi B, Krawinkler H (2001) Effects of near-source ground motions on frame-structures. Technical report no. 138. The John A. Blume Earthquake Engineering Center, Stanford University

Apostolou M, Gazetas G, Garini E (2007) Seismic response of slender rigid structures with foundation uplift. Soil Dyn Earthq Eng 27(7):642–654

Aslam M, Scalise DT, Godden WG (1980) Earthquake rocking response of rigid bodies. J Struct Div ASCE 106(2):377–392

Beck JL, Skinner RI (1973) The seismic response of a reinforced concrete bridge pier designed to step. Earthq Eng Struct Dyn 2(4):343–358

Buckle IG, Constantinou MC, Diceli M, Ghasemi H (2006) Seismic isolation of highway bridges. Research report MCEER-06-SP07. MCEER, University of Buffalo

Cheng C-T (2008) Shaking table tests of a self-centering designed bridge substructure. Eng Struct 30(12):3426–3433

Cohagen LS, Pang JBK, Stanton JF, Eberhard MO (2008) A precast concrete bridge bent designed to recenter after an earthquake. Research report, Federal Highway Administration

Constantinou MC, Soong TT, Dargush GF (1998) Passive energy dissipation systems for structural design and retrofit. Monograph series. MCEER, University of Buffalo

Garini E, Gazetas G, Anastasopoulos I (2011) Asymmetric 'Newmark' sliding caused by motions containing severe 'directivity' and 'fling' pulses. Geotechnique 61(9):733–756

Hall JF, Heaton TH, Halling MW, Wald DJ (1995) Near-source ground motion and its effects on flexible buildings. Earthquake Spectra 11(4):569–605

Hogan SJ (1989) On the dynamics of rigid-block motion under harmonic forcing. Proc R Soc Lond A 425(1869):441–476

Hogan SJ (1990) The many steady state responses of a rigid block under harmonic forcing. Earthq Eng Struct Dyn 19(7):1057–1071

Housner GW (1963) The behaviour of inverted pendulum structures during earthquakes. Bull Seismol Soc Am 53(2):404–417

Karavasilis TL, Makris N, Bazeos N, Beskos DE (2010) Dimensional response analysis of multistory regular steel MRF subjected to pulselike earthquake ground motions. J Struct Eng 136(8):921–932

Lagomarsino S, Podestà S, Resemini S, Curti E, Parodi S (2004) Mechanical models for the seismic vulnerability assessment of churches. In: Proceedings of IV SAHC, Padova, Italy, vol 2. A.A. Balkema, London, pp 1091–1101

Loh C-H, Lee Z-K, Wu T-C, Peng S-Y (2000) Ground motion characteristics of the Chi-Chi earthquake of 21 September 1999. Earthq Eng Struct Dyn 29(6):867–897

Mahin S, Sakai J, Jeong H (2006) Use of partially prestressed reinforced concrete columns to reduce post-earthquake residual displacements of bridges. In: Proceedings of the 5th national seismic conference on Bridges & Highways, San Francisco, CA, 18–20 September, paper no. B25

Makris N (1997) Rigidity–plasticity–viscosity: can electrorheological dampers protect base-isolated structures from near-source ground motions? Earthq Eng Struct Dyn 26(5):571–591
Makris N, Black CJ (2004a) Dimensional analysis of rigid-plastic and elastoplastic structures under pulse-type excitations. J Eng Mech ASCE 130(9):1006–1018
Makris N, Black CJ (2004b) Dimensional analysis of bilinear oscillators under pulse-type excitations. J Eng Mech ASCE 130(9):1019–1031
Makris N, Chang S-P (2000) Effect of viscous, viscoplastic and friction damping on the response of seismic isolated structures. Earthq Eng Struct Dyn 29(1):85–107
Makris N, Konstantinidis D (2003) The rocking spectrum and the limitations of practical design methodologies. Earthq Eng Struct Dyn 32(2):265–289
Makris N, Psychogios T (2006) Dimensional response analysis of yielding structures with first-mode dominated response. Earthq Eng Struct Dyn 35(10):1203–1224
Makris N, Roussos YS (2000) Rocking response of rigid blocks under near-source ground motions. Geotechnique 50(3):243–262
Makris N, Vassiliou MF (2012) Sizing the slenderness of free-standing rocking columns to withstand earthquake shaking. Arch Appl Mech 82(10–11):1497–1511
Makris N, Vassiliou MF (2013) Planar rocking response and stability analysis of an array of free-standing columns capped with a freely supported rigid beam. Earthq Eng Struct Dyn 42(3): 431–449
Makris N, Vassiliou MF (2014) Are some top-heavy structures more stable? J Struct Eng ASCE 140(5):06014001-1–06014001-5
Makris N, Zhang J (2001) Rocking response of anchored blocks under pulse-type motions. J Eng Mech 127(5):484–493
Makris N, Zhang J (2004) Seismic response analysis of a highway overcrossing equipped with elastomeric bearings and fluid dampers. J Struct Eng 130(6):830–845
Mander JB, Cheng C-T (1997) Seismic resistance of bridge piers based on damage avoidance design. Tech. rep. no. NCEER- 97-0014. National Center for Earthquake Engineering Research, Dept. of Civil and Environmental Engineering, State Univ. of New York, Buffalo
Mavroeidis GP, Papageorgiou AS (2003) A mathematical representation of near-fault ground motions. Bull Seismol Soc Am 93(3):1099–1131
Milne J (1885) Seismic experiments. Trans Seismol Soc Jpn 8:1–82
Palmeri A, Makris N (2008) Response analysis of rigid structures rocking on viscoelastic foundation. Earthq Eng Struct Dyn 37(7):1039–1063
Pang JBK, Steuck KP, Cohagen LS, Stanton JF, Eberhard MO (2008) Rapidly constructible large-bar precast bridge-bent seismic connection, Research report WA-RD684.2. Washington State Department of Transportation
Papaloizou L, Komodromos K (2009) Planar investigation of the seismic response of ancient columns and colonnades with epistyles using a custom-made software. Soil Dyn Earthq Eng 29(11–12):1437–1454
Prieto F, Lourenço PB, Oliveira CS (2004) Impulsive Dirac-delta forces in the rocking motion. Earthq Eng Struct Dyn 33(7):839–857
Psycharis IN, Jennings PC (1983) Rocking of slender rigid bodies allowed to uplift. Earthq Eng Struct Dyn 11(1):57–76
Resemini S, Lagomarsino S, Cauzzi C (2008) Dynamic response of rocking masonry elements to long period strong ground motion. In: Proceedings of the 14th world conference on earthquake engineering, 12–17 October, Beijing
Ricker N (1943) Further developments in the wavelet theory of seismogram structure. Bull Seismol Soc Am 33(3):197–228
Ricker N (1944) Wavelet functions and their polynomials. Geophysics 9(3):314–323
Roh H, Reinhorn A (2010a) Nonlinear static analysis of structures with rocking columns. J Struct Eng ASCE 136(5):532–542
Roh H, Reinhorn A (2010b) Modeling and seismic response of structures with concrete rocking columns and viscous dampers. Eng Struct 32(8):2096–2107

Sakai J, Mahin S (2004) Analytical investigations of new methods for reducing residual displacements of reinforced concrete bridge columns, PEER report 2004/02. Pacific Earthquake Engineering Research Center, University of California, Berkeley

Sakai J, Hyungil J, Mahin SA (2006) Reinforced concrete bridge columns that re-center following earthquakes. In: Proceedings of the 8th US national conference on earthquake engineering, 18–22 April 2006, San Francisco, CA, USA, paper no. 1421

Spanos PD, Koh AS (1984) Rocking of rigid blocks due to harmonic shaking. J Eng Mech ASCE 110(11):1627–1642

Vassiliou MF, Makris N (2011) Estimating time scales and length scales in pulse-like earthquake acceleration records with wavelet analysis. Bull Seismol Soc Am 101(2):596–618

Vassiliou MF, Makris N (2012) Analysis of the rocking response of rigid blocks standing free on a seismically isolated base. Earthq Eng Struct Dyn 41(2):177–196

Veletsos AS, Newmark NM, Chelapati CV (1965) Deformation spectra for elastic and elastoplastic systems subjected to ground shock and earthquake motions. In: Proceedings of the 3rd world conference on earthquake engineering, vol II, Wellington, pp 663–682

Wacker JM, Hieber DG, Stanton JF, Eberhard MO (2005) Design of precast concrete piers for rapid bridge construction in seismic regions. Research report, Federal Highway Administration

Yamashita R, Sanders D (2009) Seismic performance of precast unbonded prestressed concrete columns. ACI Struct J 106(6):821–830

Yim CS, Chopra AK, Penzien J (1980) Rocking response of rigid blocks to earthquakes. Earthq Eng Struct Dyn 8(6):565–587

Zhang J, Makris N (2001) Rocking response of free-standing blocks under cycloidal pulses. J Eng Mech ASCE 127(5):473–483

Chapter 16
The Dispersion of Concrete Compressive Strength of Existing Buildings

M.T. Cristofaro, R. Pucinotti, M. Tanganelli, and M. De Stefano

Abstract The correct estimation of the compressive concrete strength plays a key role in the evaluation of the structural performance of existing RC buildings. Both Italian (NTC 2008) and European (EC8) Standards define different levels of knowledge according to the number of tests carried out on a building. They indicate a reduced value to assume in the analysis, defined as the *mean* value of the compressive strength, divided by a confidence factor. However, such a procedure completely neglects the dispersion of the test data, as represented by the high values of the *coefficient of variation*. Instead, this aspect is treated by FEMA 356 where a limit to the *coefficient of variation* was introduced. In this paper, with reference to a significant number of existing buildings located in Tuscany, the *coefficient of variation* (*cov*) of concrete strength is evaluated and the frequency of high *cov* values is determined. The dispersion of compressive strength, obtained by SonReb method, using correlation curves calibrated ad hoc on single building, shows that increasing the number of data for each building the *coefficient of variation* does not necessary decrease. Moreover, the strength value considered by EC8 in the analysis for a single building, i.e. the mean value of compressive strength, is often not conservative, while the approach provided by FEMA 356 is safer since it dependent on the *cov* itself.

M.T. Cristofaro (✉) • M. Tanganelli • M. De Stefano
Department of Architecture (DiDA), University of Florence, Piazza Brunelleschi, 6, 50121 Florence, Italy
e-mail: crismt@alice.it; marco.tanganelli@unifi.it; mario.destefano@unifi.it

R. Pucinotti
Dipartimento di Patrimonio, Architettura e Urbanistica, Università Mediterranea di Reggio Calabria, Salita Melissari – Feo di Vito, Reggio Calabria, Italy
e-mail: raffaele.pucinotti@unirc.it

G.P. Cimellaro et al. (eds.), *Computational Methods, Seismic Protection, Hybrid Testing and Resilience in Earthquake Engineering*, Geotechnical, Geological and Earthquake Engineering 33, DOI 10.1007/978-3-319-06394-2_16,
© Springer International Publishing Switzerland 2015

16.1 Introduction

The assessment of existing buildings is an important issue that involves researchers and engineers in many countries. Estimation of concrete strength in existing structures is essential to evaluate their seismic capacity and to design seismic strengthening.

Recent innovations in codes (CEN. EN 1998–3 2005; Ministero delle Infrastrutture 2008, 2009; FEMA 356 2000) provide clear rules to assess the safety and to conduct the static strengthening on existing buildings. Type and number of *in situ* tests depend on the level of knowledge, which can be classified as Limited (KL1), Normal (KL2) and Full (KL3). The evaluation of seismic performance is influenced by many factors; one of these is the concrete compressive strength, as shown in Pucinotti and Tripodo (2009), De Stefano et al. (2013a, b), D'Ambrisi et al. (2009) and Pucinotti (2013). The assessment of the *in situ* concrete compressive strength is influenced by the methodology used (destructive or non-destructive tests), but also from the strong heterogeneity of the concrete (Pucinotti and Modaferri 2010; Pucinotti 2013).

The Italian code NTC 2008 (Ministero delle Infrastrutture 2008) specifies that the mean value of the *in situ* strength (structural resistance) is usually lower than the mean value of the strength obtained on cast specimens that are prepared during the construction phases, stored and cured in laboratory (potential strength). The mean value of the structural resistance is acceptable if it is at least equal to 85 % the design value (Pucinotti 2013). The NTC08 and its instructions (Ministero delle Infrastrutture 2008, 2009) do not take into account the dispersion of the data in the definition of the *in situ* mean compressive strength of the concrete; in fact they introduce coefficients of confidence which depend on the level of knowledge (KL) only.

FEMA 356 (2000) considers instead a statistical parameter, the *coefficient of variation* (*cov*), which depends on the scattering data. If the *cov* of the sample is lower than 14 %, the *mean* strength can be used in the analysis. Instead, for the *cov* greater than 14 %, the concrete strength value to use in analyses is defined as the *mean* minus one standard deviation. As alternative, if the *cov* exceeds 14 %, additional tests shall be performed until a *cov* lower than 14 % is achieved. The paper, with references of a large number of RC structures built between the years 1950 and 1980, highlights the large dispersion of the *in situ* concrete strength. It also shows that often the *mean* value of strength (f_{core_mean}) is not representative of the actual value. Data were collected by the Regional Seismic Department of Tuscany Region, by a series of destructive and non-destructive tests.

16.2 Description of Compressive Strength Population

Since the 1990s the Regional Seismic Department of Tuscany, operating in the framework of national and regional programmes on seismic prevention (VSCA 2004), has started campaigns of destructive and non-destructive investigations on

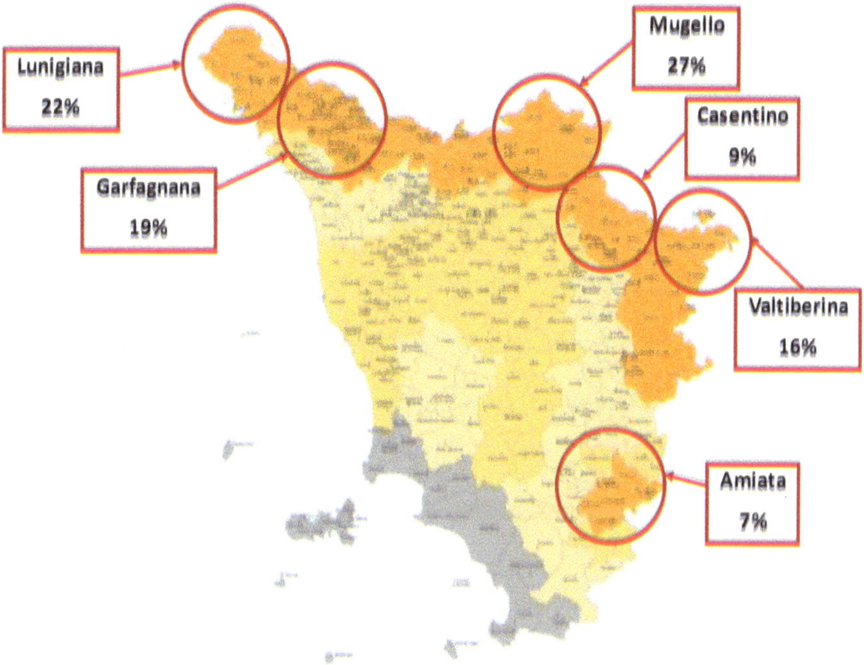

Fig. 16.1 Tuscany region: seismic geographic areas investigated

Table 16.1 Number of buildings and samples investigated

	1950s	1960s	1970s	1980s
Buildings	7	25	38	20
Samples	65	266	302	170
Mean samples for a single building	9	10	9	8
f_{core_mean} (MPa)	11.20	13.90	18.90	23.50

RC existing buildings built in Tuscany between the 1950s and the 1980s. These data have been organized into a data-base (Cristofaro 2009). Figure 16.1 shows the assumed geographic classification of Tuscany region: Lunigiana, Garfagnana, Mugello, Casentino, Valtiberina, Amiata.

The buildings have undergone both non-destructive tests and destructive tests. In detail, destructive tests were conducted on cores extracted from 803 structural elements of 90 buildings. Cores were cut and leveled, capped and subjected to compressive tests in laboratory. The experimental tests were conducted on cores with ratio length over diameter (l/d) equal to 2. Moreover non-destructive tests were conducted on 3,162 structural building elements. A such significant number of available samples has allowed a relevant statistical analysis.

Table 16.1 shows, for all decades, the total number of buildings, the total number of samples, the mean number of samples for a single building and, lastly, the mean value of the core concrete strength (f_{core_mean}).

16.3 Statistical Analysis

For each building the mean value of core concrete strength (f_{core_mean}), the standard deviation and the *cov* were valued. In the case of f_{core_mean} and *cov* a normal distributions and a log-normal distributions were considered for all decades.

Figure 16.2 shows the comparison between the histograms representing the analyzed samples, for each considered decade, and the two assumed distributions, respectively normal and lognormal. As can be seen in Table 16.1, the considered samples have different size, and consequently even the obtained histograms are differently detailed. In almost all cases, anyway, the histograms presents a substantial asymmetry, and, therefore, the lognormal distribution seems to better represent the considered samples.

Below, the *error of the estimate*, calculated as the ratio of the confidence interval at 95 % (confidence) and the *mean* value so as to obtain an *error of the estimate* in function of number of samples, is presented. These estimates were performed on four values of the *cov* varying between 14 and 56 %. In detail, Fig. 16.3 shows the results of the *error of the estimate* of mean rebound index value (I_{r_mean}) for each building of the 1950s, 1960s, 1970s and 1980s. It appears that, independently from construction period of the building, the *cov* remains low (around 14 %) when the number of samples increases.

Figure 16.4 shows instead the results of the *error of the estimate* of mean ultrasonic velocity (V_{us_mean}) for each building of the 1950s, 1960s, 1970s and 1980s. In this case the *cov* not systematically decreases as the number of samples increase, and it ranges between 14 and 28 %.

In both cases (see Figs. 16.3 and 16.4) the dispersion of data on a single building, and the consequent *cov*, are small. This evidence does not occur for of destructive testing.

Figure 16.5 shows the results of the *error of the estimate* of the f_{core_mean}. It is possible to see that at the increase of data does not correspond a significant reduction in the error *of the estimate*. Also it is possible to see as the *cov*, in some cases, is close to 50 %.

The influence of the number of data on the error of estimate and *cov* has been evaluated. For each building a formulation ad hoc (Cristofaro 2009) by SonReb method has been calibrated by using the regression analysis. Such analysis provides a mathematical relationship between a dependent variable and one or more independent variables. In the case of the combined SonReb method the dependent variable is the f_{core}, while the independent variables are the rebound index I_r and the ultrasonic velocity V_{us}. The obtained correlation formula is the following:

$$f_{core} = a \cdot I_r^b \cdot V_{us}^c \qquad (16.1)$$

where a, b, and c are the parameters to be determinate to calibrate the correlation law for individual building.

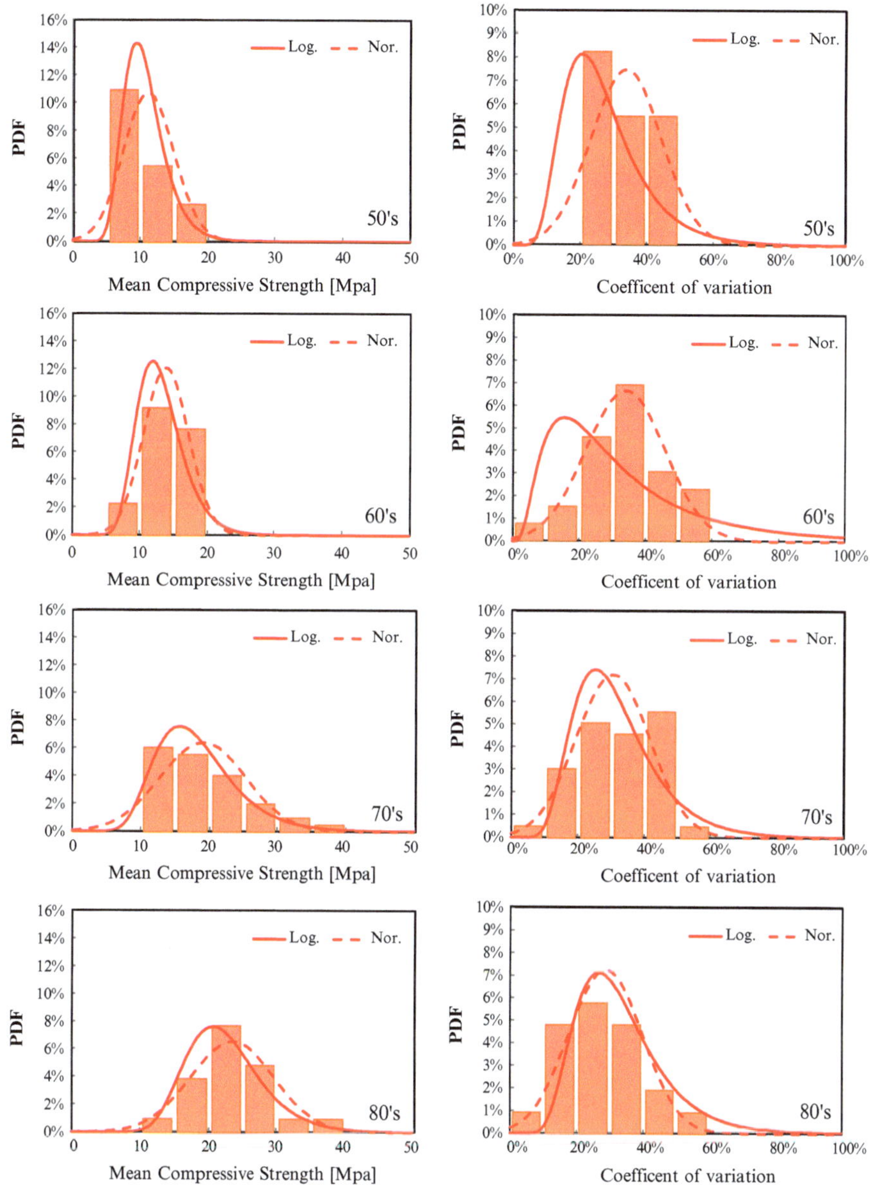

Fig. 16.2 Normal and log-normal distributions of both mean compressive strength and *cov*

Figure 16.6 shows the evaluation of the *error of the estimate* refers to the mean value of concrete compressive strength (f_{SonReb_mean}), obtained using the Eq. (16.1), for a single building. As can be seen, the increase of number of data induces a decrease of the *error of the estimate*. Also, for relevant number of data, values of *cov* are not necessary low; in fact in some cases the is over *cov* 42 %.

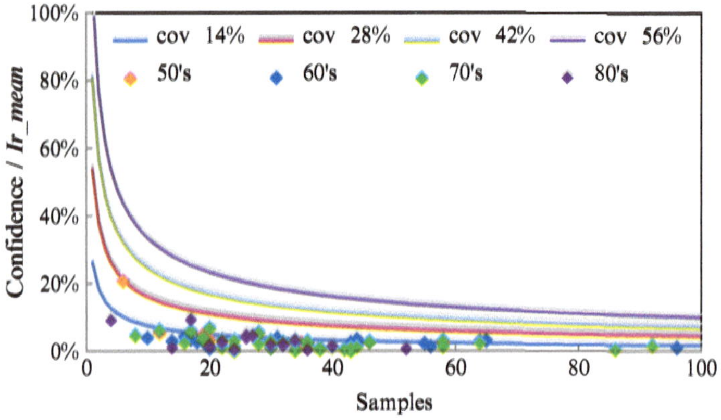

Fig. 16.3 Estimate error of I_{r_mean} vs. *cov*

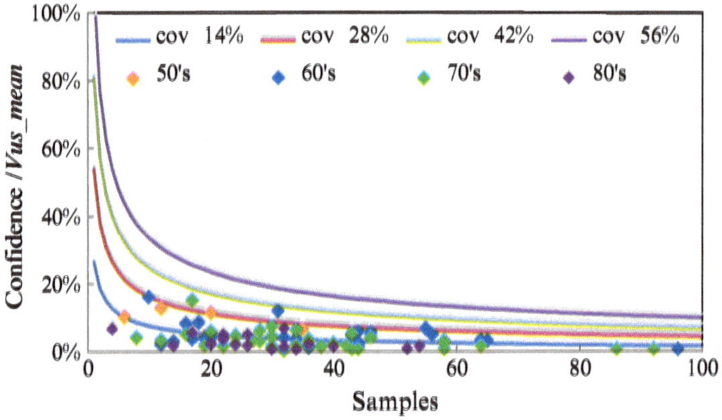

Fig. 16.4 Estimate error of V_{us_mean} vs. *cov*

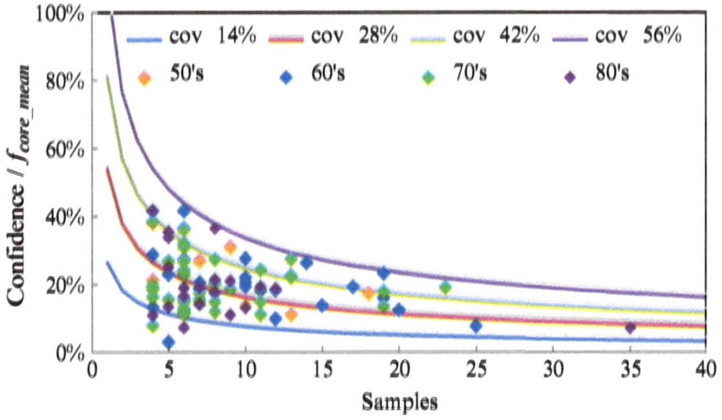

Fig. 16.5 Estimate error of f_{core_mean} vs. *cov*

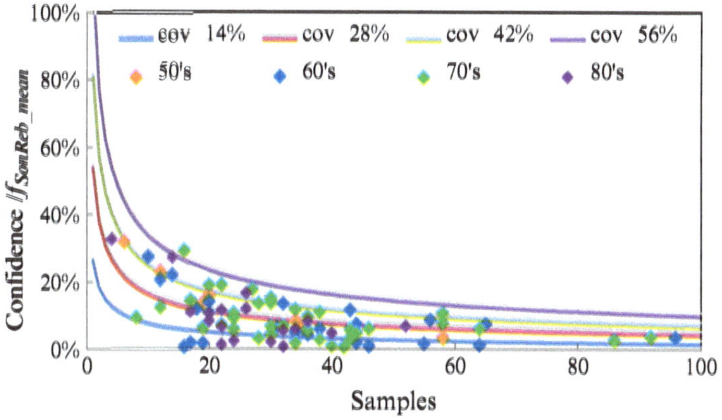

Fig. 16.6 Estimate error of f_{SonReb_mean} vs. *cov*

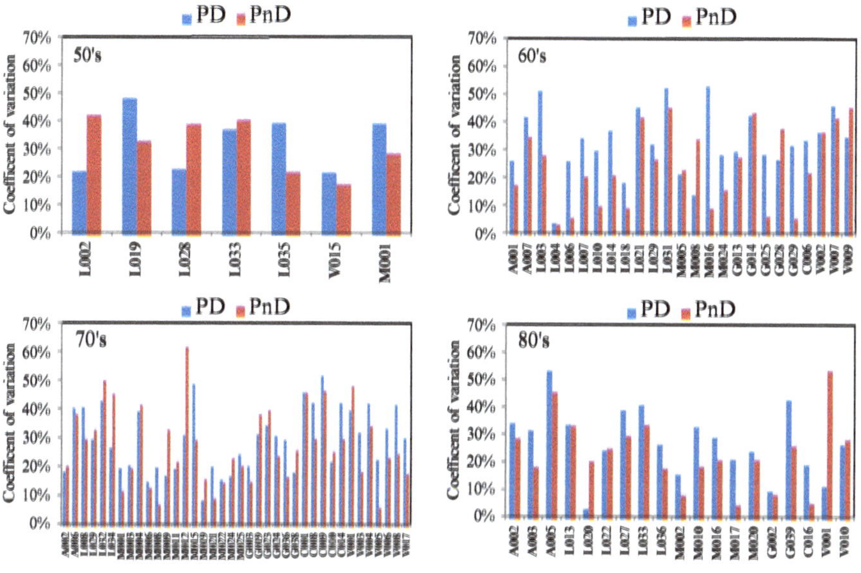

Fig. 16.7 Comparison between the *cov* relating to destructive and non-destructive testing

Figure 16.7 shows, for each building, the results of the *coefficient of variation* of the compressive strength carried out on cores (PD) and those obtained by the SonReb method (PnD). The figure shows as the PnD method does not provides a significant reduction in the *cov*, but in some cases a increase of it.

Figure 16.8 shows the results of *error of the estimate* related to concrete compressive strength obtained with destructive and non-destructive (SonReb) methods, found for each building. Moreover it shows, for all considered decades and for almost all buildings, that SonReb method provides a relevant reduction of the *error of the estimate*.

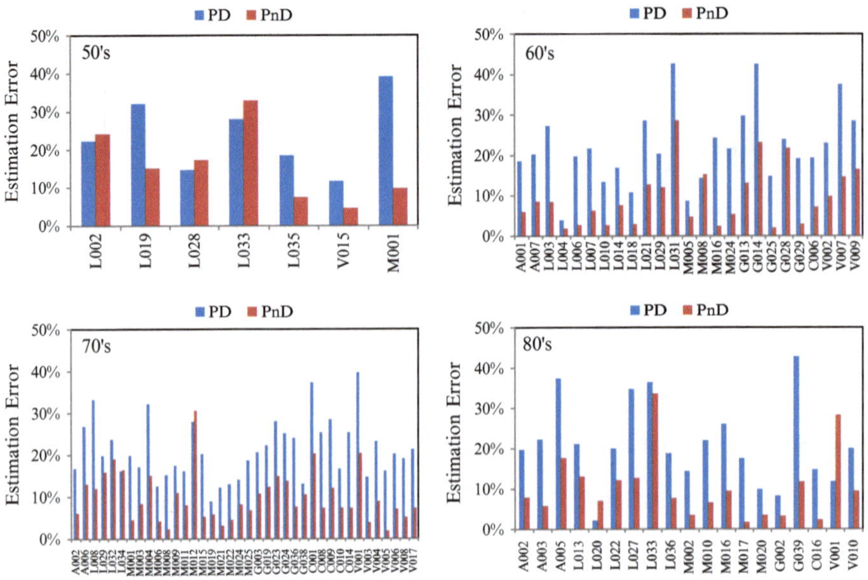

Fig. 16.8 Comparison between the error of evaluation relating to destructive and non-destructive testing

16.4 Comparison Between EC8 and FEMA 356

European Standards (EC8) define different levels of knowledge according to the number of tests carried out on a building; therefore the value of the concrete compressive strength assumed in the analysis is defined as the *mean* value appropriately reduced by a confidence factor. However, such procedure neglects the dispersion of the test data, i.e. the amount of the *coefficient of variation*. Instead, this aspect is treated by FEMA 356 where a limit to the *coefficient of variation* of 14 % is introduced.

Figure 16.9 shows, for each building, the results of the f_{core} calculated with both EC8 and FEMA 356 (CEN. EN 1998–3 2005; FEMA 356 2000). It is possible to see as the FEMA 356 previsions are safer respect the EC8 previsions also in the case of level of knowledge limited (KL1).

Finally, Fig. 16.10 shows the percentage difference between the FEMA 356 and EC8 predictions for different levels of knowledge (i.e. LC1, LC2 and LC3).

In detail, it is possible to see that for both the limited and normal knowledge (LC2 and LC3), EC8 overestimates the strength values compared to FEMA 356.

When the level of knowledge LC1 is considered the estimated values are lower than those calculated with application of the FEMA 356. In this case, according to EC8, only the lateral force analysis is applicable.

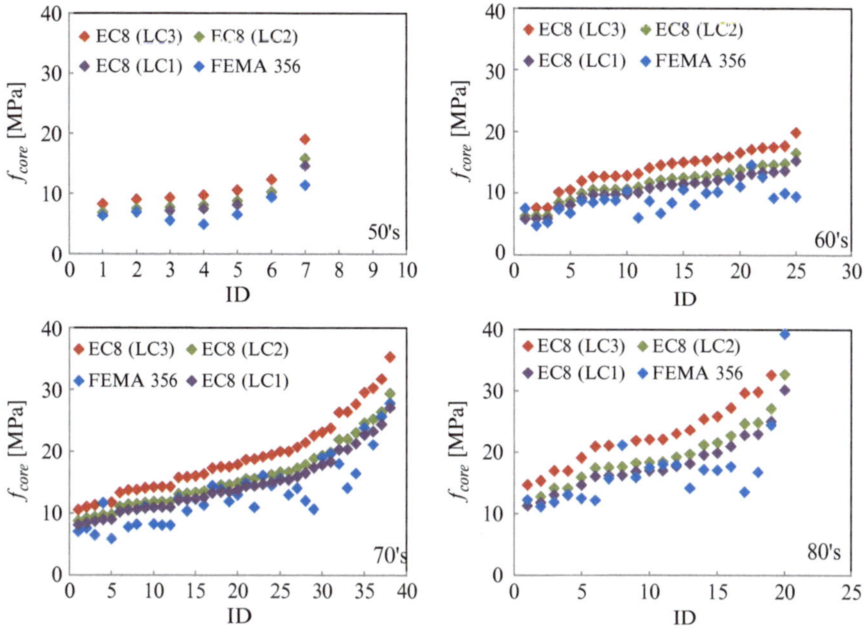

Fig. 16.9 f_{core} calculated with EC8 and FEMA 356

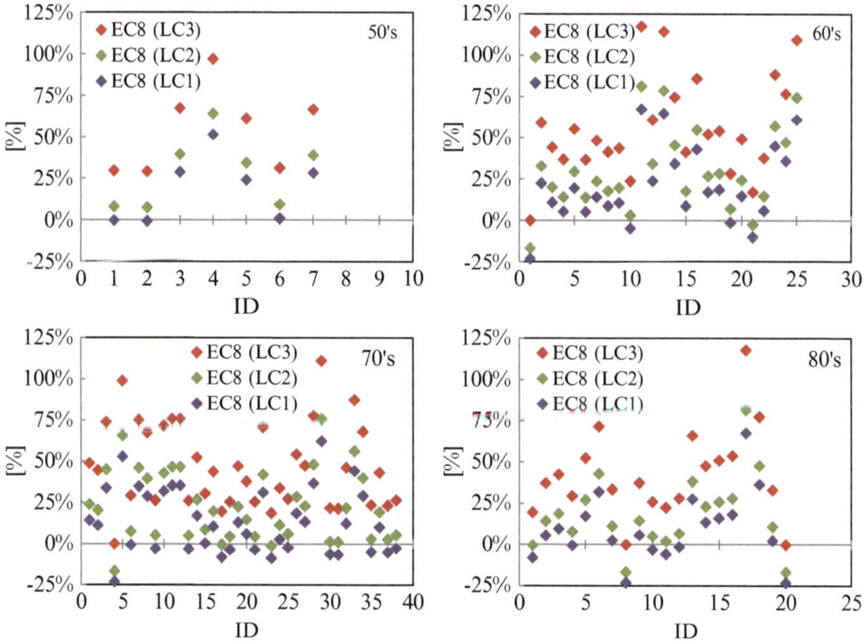

Fig. 16.10 f_{core} calculated with EC8 and FEMA 356

16.5 Conclusions

In this paper a series of destructive and non-destructive tests were performed on 90 RC existing buildings built in Tuscany between the 1950s and the 1980s with the purpose to estimate both the dispersion of compressive strength and the *coefficient of variation*. Nondestructive test results showed a low dispersion of the data, with consequent low values of *error of estimate* and *coefficient of variation*. Destructive tests carried out on cores, for all considered periods, showed *coefficient of variation* more than 30 %, with a peak, in some cases, of 60 %. Even the *error of the estimate* is high for all the decades under consideration, achieving in some cases 40 %. In general, destructive tests showed an *error of the estimate* and a *coefficient of variation* much higher than those of the SonReb method. In the evaluation of the mechanical properties of the concrete of existing buildings, it is therefore important to combine destructive tests with an adequate numbers of non-destructive tests.

The assessment of the concrete compressive strength obtained from the Eurocode 8 returns values significantly different from those of the FEMA 356. It would be desirable to determine the *mean* value of the compressive strength as a function of the amount of dispersion of data, as prescribed by FEMA356. Alternatively, the considered experimental data should be increase, including non-destructive testing calibrated on destructive test results.

In conclusion, results demonstrate as the assessment of the *in situ* concrete compressive strength is the most delicate stage in the evaluation of assessment of existing buildings and therefore must be carefully performed.

Acknowledgments The authors thank the Regional Seismic Department of Tuscany for making available the data-base of experimental data.

References

CEN. EN 1998–3 (2005). Eurocode 8 – design of structures for earthquake resistance – Part 3: assessment and retrofitting of buildings, Brussels

Cristofaro MT (2009) Metodi di valutazione della resistenza a compressione del calcestruzzo di strutture in c.a. esistenti. Tesi di Dottorato, Università degli Studi di Firenze, Firenze, Italia (in Italian)

D'Ambrisi A, De Stefano M, Tanganelli M (2009) Use of pushover analysis for predicting seismic response of irregular buildings: a case study. J Earthq Eng 13:1089–1100. doi:10.1080/13632460902898308

De Stefano M, Tanganelli M, Viti S (2013a) Effect of the variability in plan of concrete mechanical proprieties on the seismic response of existing RC framed structures. Bull Earthq Eng 11:1049–1060. doi:10.1007/s10518-012-9412-5

De Stefano M, Tanganelli M, Viti S (2013b) On the variability of concrete strength as a source of irregularity in elevation for existing RC buildings: a case study. Bull Earthq Eng 11:1711–1726. doi:10.1007/s10518-013-9463-2

FEMA 356 (2000) Federal emergency management agency. Prestandard and commentary for the seismic rehabilitation of buildings, Reston, Va, 2000

Ministero delle Infrastrutture. DM 14 gennaio 2008 (2008) Norme tecniche per le costruzioni (NTC 2008). Suppl Ord. n.30 alla G.U. n.29 del 4/2/2008

Ministero delle Infrastrutture. Circolare 2 febbraio 2009 (2009) Istruzioni per l'applicazione delle Nuove norme tecniche per le costruzioni. Supplement or. n.27 alla G.U. n.47 del 26/2/2009

Pucinotti R (2013) Assessment of in situ characteristic concrete strength. Construct Build Mater 44:63–73. doi:10.1016/j.conbuildmat. 2013.02.041

Pucinotti R, Modaferri G (2010) Variability of the compressive strength of concrete cores within the same drilling. In: Proceedings of the first workshop "Le nuove frontiere del calcestruzzo strutturale" – the new boundaries of structural concrete editors: A. Grimani, G. Plizzari, R. Realfonzo. ACI, Salreno, 22–23 Aprile 2010. ISBN: 978-88-985028-55-2

Pucinotti R, Tripodo M (2009) The fiumarella bridge: concrete characterization and deterioration assessment by non-destructive testing. Int J Microstruct Mater Prop 4(1):128–139. doi:10.1504/IJMMP.2009.028438

VSCA, Regione Toscana (2004) Delibera di G.R.T. n. 4301 del 21 luglio 2004, Istruzioni Tecniche con i criteri per lo svolgimento di indagini diagnostiche in edifici esistenti in cemento armato(VSCA) e successivi aggiornamenti, Firenze (in Italian)

Chapter 17
Concrete Strength Variability as a Source of Irregularity for Existing RC Structures

M. De Stefano, M. Tanganelli, and S. Viti

Abstract Mechanical properties of concrete can consistently affect the seismic performance of RC buildings. A proper determination of the concrete strength is therefore essential for a reliable modeling of the structure. The current European Technical Code, Eurocode 8, provides a criterion for the strength assumption related to the knowledge level of the structure, which does not take into account the variability of the strength. If the concrete strength is affected by a large variability, the conventional value suggested by Eurocode 8 can be not conservative, since it does not consider the increase in the demand due to a not homogeneous strength distribution and the reduced capacity of the weaker members. In this paper the concrete strength variability is investigated as a source of in-plan and in-elevation irregularity. The effects of the strength variability on the seismic response and performance are evaluated on a case study, that is a 4-storeys 3D framed building. The seismic response of the case study has been represented by performing a nonlinear static analysis, while the seismic performance has been measured in terms of chord rotation. Results obtained from the analysis have been compared with the Eurocode 8 previsions.

17.1 Introduction

A correct evaluation of the concrete strength appears as an important issue of the seismic analysis of existing RC buildings. In many cases RC buildings made in the 1960s and 1970s are built with poor materials, due to the scarce control during

M. De Stefano (✉) • M. Tanganelli • S. Viti
Department of Architecture (DiDA), University of Florence, Piazza Brunelleschi, 6, 50121 Florence, Italy
e-mail: mario.destefano@unifi.it; marco.tanganelli@unifi.it; stefania.viti@unifi.it

G.P. Cimellaro et al. (eds.), *Computational Methods, Seismic Protection, Hybrid Testing and Resilience in Earthquake Engineering*, Geotechnical, Geological and Earthquake Engineering 33, DOI 10.1007/978-3-319-06394-2_17,
© Springer International Publishing Switzerland 2015

the construction phases. Concrete usually has a low strength, and it is affected by a very large variability, even within a single building.

In order to properly model an existing structure, a correct characterization of the concrete strength and a conservative assumption of its distribution have to be made.

Due to concrete strength variability, in fact, buildings can experience an irregular seismic response, both in plan and in elevation (De Stefano et al. 2006, 2012, 2013; D'Ambrisi et al. 2013a, b; Magliulo et al. 2012), with a consequent decrease in their seismic performance. Strength variability, in fact, induces both an irregular strength and stiffness distribution, since the concrete Young modulus is related to the effective strength; as a consequence, the structure can experience an irregular seismic response, with an increase of the seismic demand, together with a reduction of the capacity in the weakest members.

In these last years, many Authors have made investigations about concrete strength characterization, focusing their attention both on the topic parameters to assume in the characterization and on the variability of the strength, with the consequent assumption of a value for analysis (Masi and Vona 2009; Franchin et al. 2007, 2009; Fardis 2009; Masi et al. 2008; Rajeev et al. 2010; Cosenza and Monti 2009; Marano et al. 2008; Monti et al. 2007; Jalayer et al. 2008).

In this paper the strength variability as a source of in-plan and in-elevation irregularity is investigated. The paper is organized in three different parts.

In the first part the work is introduced, and the main choices made to perform the analysis are presented. The strength distribution has been introduced on the basis of a large database provided by Regione Toscana, including buildings made in different decades (ranging between 1950 and 1980) in a homogeneous area. From this database, the most significant statistic parameters have been extracted to describe the strength distribution. The case-study is described, as well as the assumed strength distribution, the limit values assumed to check the seismic performance and the numerical model.

In the second part the strength variability has been considered as source of in-plan irregularity. This research follows on from previous works (De Stefano et al. 2012, 2014) made on the same case study. In De Stefano et al. (2012) the strength variability was introduced at each storey of the building, and at all storeys together. The seismic performance proved to be largely affected by the introduced variability. The reduction in the seismic performance came out to be maximum when the strength variability was at the first storey. Therefore in this paper the strength variability has been assumed to involve the first storey only, and different values of *cov*, ranging between 10 and 35 %, have been assumed for the concrete strength, in order to investigate the relationship between the amount of the maximum reduction in seismic performance and the strength variability.

The third part of the analysis focuses on the effects of the in-elevation strength variability. It is introduced by assuming different values of strength from floor to floor (De Stefano et al. 2013), according to the statistical parameters obtained from the available database. The response domains found as a consequence of the assumed strength samples were compared with the results obtained by following the EC8 provisions (EC 8-3 2005).

17.2 The Analysis

17.2.1 Concrete Strength Characterization

According to EC8 (EC 8-3 2005), existing RC buildings have to be represented assuming the *mean* value of concrete strength in the analysis, and reducing such value by a proper reduction factor, (Confidence Factor, *CF*), for verification. The Confidence Factor ranges between 1.00 and 1.35 (Italian National Annex), depending on the knowledge level of the structure. When a fully satisfactory knowledge of the structure is achieved, the *mean* value of the strength domain can therefore be adopted both in the structural analysis and in the verification ($CF = 1.00$). EC8 implicitly assumes that the high concrete strength variability in existing buildings depends on the unsatisfactory size of the tested sample. In fact it recommends to extend the *in situ* tests to an adequate number of samples, in order to reduce the *CF*. The adoption of a conventional (reduced) value of f_c in the verification is supposed to compensate the variability of the strength.

In this work the strength distribution has been assumed on the basis of a database (Cristofaro 2009) collected by Regione Toscana, which encompasses about 300 buildings, and over 1,000 destructive and not destructive tests. All the tested buildings are located in Tuscany in a not negligible seismic area. In this work only results relative to buildings made in the 1950s, 1960s and 1970s have been considered. Statistical indexes (Cristofaro et al. 2012) were evaluated for each building with a minimum of three test results. The *cov* of specimens taken from each single building ranges in a large interval, reaching 50 % for the three decades. In the analysis, the f_c population is assumed to follow a Gaussian distribution with a *mean* equal to 19.36 MPa and different *cov* values, ranging between 10 and 35 %. Figure 17.1 shows the assumed strength distribution; it consists in five values, respectively obtained from the *mean* by adding and subtracting once and twice the *standard deviation*.

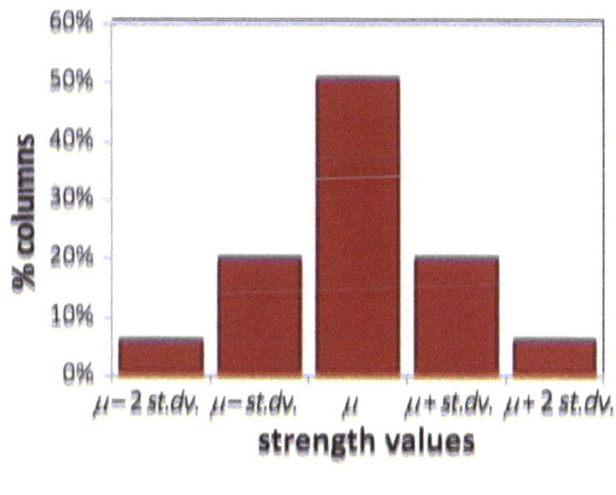

Fig. 17.1 The assumed strength distribution

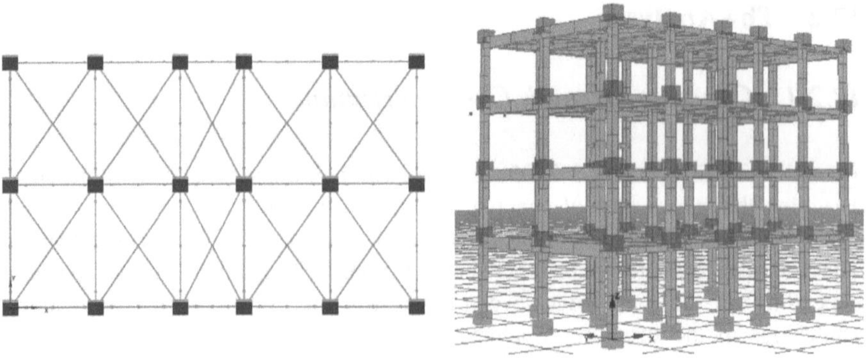

Fig. 17.2 Case-study

17.2.2 Sample Structure

The sample structure, shown in Fig. 17.2, is a 4 story 3D reinforced concrete frame with two bays, 4.5 m long, in the *y*-direction and five bays in the *x*-direction, four 3.2 m long and one 2.4 m long. The building is symmetric both along *x* and *y* axes. The concrete strength of the columns has been described according to the distribution model shown in the previous section, while the strength of the beams has been assumed equal to the *mean* value ($f_c = 19.36$ MPa). The reinforcement is assumed to have the same mechanical properties of the Italian FeB38k steel (yield stress over 375 N/mm^2, ultimate stress over 430 N/mm^2), since it has compatible characteristics with the steel used in the 1970s. The building is designed for vertical loads only, ignoring seismic loads. Vertical loads consist of dead loads and live loads equal to 2 KN/m^2.

17.2.3 The Structural Model

The analysis has been performed by using the program Seismostruct (Seismosoft 2006). A fiber model has been adopted to describe the cross section, and each member has been subdivided into four segments. The Mander et al. (1988) model has been assumed for the core concrete, a three-linear model has been assumed for the unconfined concrete, and a bilinear model has been assumed for the reinforcement steel. The damping has been assumed according to the Rayleigh model, whose parameters were obtained by imposing 5 % damping to the first two vibration modes, in terms of participating mass, of the structure.

Diaphragmatic action of floor slabs has been considered by introducing two diagonal struts in each floor bay.

17.2.4 The Assumed Performance Limits

In this work only ductile, flexure-controlled failure modes were considered. Shear failure has not be considered since it is strongly affected by reinforcement details, joint features and typology of secondary components which are not included in the modeling.

The structural response has been evaluated in terms of interstorey drifts, which have been compared with the ultimate limit values provided by Eurocode 8. EC8 considers two ultimate Limit States: *Severe Damage*, *SD* and *Near Collapse*, *NC*. Their limit chord rotations have been found for each column as a function of the ultimate rotation (see eq. A.1 par. A.3.1.1 in EC 8-3 2005). Each column has a different limit value, since the chord rotation depends on the cross section features and on the assumed f_c value.

17.3 The Strength Variability as Source of In-Plan Irregularity

17.3.1 The Assumed Strength Distribution Layouts

The current research follows on from a previous work (De Stefano et al. 2012) made on the same case study. From the previous investigation the effects of the strength variability resulted to affect the first storey much more than the other ones. In this work, therefore, the in-plan strength variability has been introduced at the first storey only. The columns of the other storeys, therefore, have been assumed to have the mean strength value, as well as the beams.

Since the seismic analysis has been performed in the *y*-direction only, the variability has been given mostly along the *x*-direction. Figure 17.3 shows the considered models assumed for strength distribution in the columns of the first storey. Two different models having the same *mean* and *cov* ranging between 10 and 35 % have been considered. The first model is one-way asymmetric (*1-way asym*, Fig. 17.3a), and it has the lower strengths all located at the left side of building plane, while the second model is two-way asymmetric (*2-way asym*, Fig. 17.3b), and it has the weaker columns towards the upper left side.

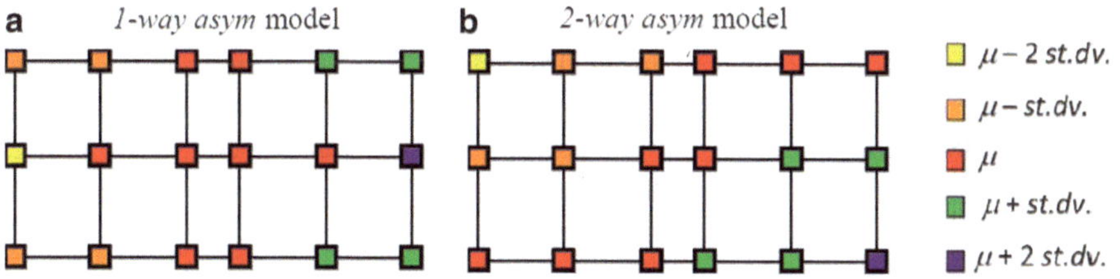

Fig. 17.3 Considered strength distributions in the columns of the first storey

The assumed plan strength layouts represent extreme situations, as they have the most skewed distributions which can be expected from the assumed strength domains. The effects related to the assumed strength layouts, both in terms of increase in the seismic demand and reduction in the seismic performance, have therefore to be considered as the maximum possible consequence of the strength variability.

17.3.2 The Seismic Analysis

Many recent studies (D'Ambrisi et al. 2009; Bosco et al. 2009, 2013a, b, c) have proved that the seismic response of irregular buildings can be found by performing nonlinear static analyses. In this study the N2 method, according to the approach suggested by Eurocodes (EC 8-3 2005), has been adopted. Two different heightwise pattern distributions, both of them applied along the *y*-direction, have been assumed to perform the analysis, respectively proportional to the product of the masses and their distance to the foundation plan (inverse triangular distribution) and to the masses only (mass-proportional distribution). The elastic spectrum provided by EC8 for a soil-type *B* has been assumed as seismic input. Different values of PGA ranging between 0.05 and 0.30g, with a step equal to 0.05g, have been considered.

17.3.3 The Effects of the Strength Variability on the Seismic Response

Due to an unfavorable distribution of the concrete strength, the building can experience unexpected torsional effects, with a consequent increase in the seismic response. The maximum seismic demand found by considering the strength variability has been compared with the one coming from the EC8 conventional approach. In Fig. 17.4 the maximum drifts at the first storey obtained with the two assumed models (*1-way asym* and *2-way asym*) are shown and compared with the values provided by EC8. As should be noted, increasing the assumed PGA, the introduced strength variability induces an increase in the maximum drift, whose amount depends on the considered *cov*.

When the nonlinear static procedure is adopted to find the seismic response of buildings, EC8 recommends to introduce a mass-eccentricity of ±5 %, in order to consider the possible sources of irregularity. If the 5 % eccentricity is introduced in the analysis, the consequent torsional effects are larger than those related to the strength variability in all the considered cases, as can be seen in the graphs in Figs. 17.5 and 17.6. It must be underlined, anyway, that the 5 % eccentricity provided by EC8 is aimed to cover all the possible plan irregularities, and not only those related to the strength variability.

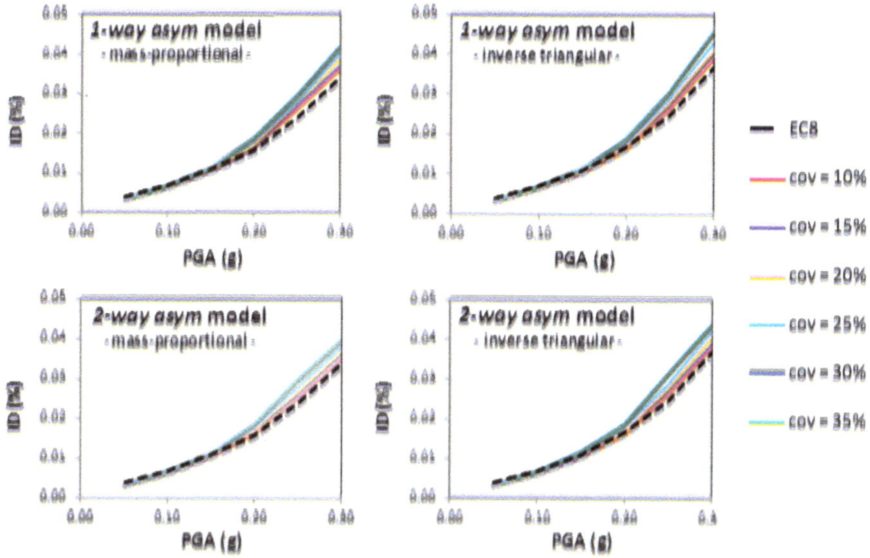

Fig. 17.4 Maximum ID at the first storey obtained by considering the strength variability and EC8 approach

Figures 17.5 and 17.6 show the normalized Top Displacement (*TD*) along the *x*-direction column-lines obtained with the two models, respectively. In each figure results obtained by considering both the two assumed heightwise pattern distributions have been shown.

17.3.4 The Effects of the Strength Variability on the Seismic Performance

The strength variability can affect the seismic performance of the structure even by reducing the capacity in the weakest members. When the concrete strength is affected by a large variability, in fact, some members can present a resistance much smaller than the expected one, so inducing a reduction in the seismic performance (demand vs. capacity) of the overall structure. Figure 17.7 shows the ratio between demand and capacity for the considered Ultimate Limit States, *Severe Damage* (*SD*) and *Near Collapse* (*NC*), at the increasing of the seismic input for all the considered values of *cov*. When the ratio demand/capacity exceeds the unity, the seismic performance is not satisfactory according to the EC8 prescriptions.

The performance found by applying the EC8 approach is larger than the performance related to the unfavorable strength distributions assumed in the analysis. Therefore, the EC8 evaluation appears to be unconservative, despite the adoption of the 5 % eccentricity, when compared with the correspondent results found with

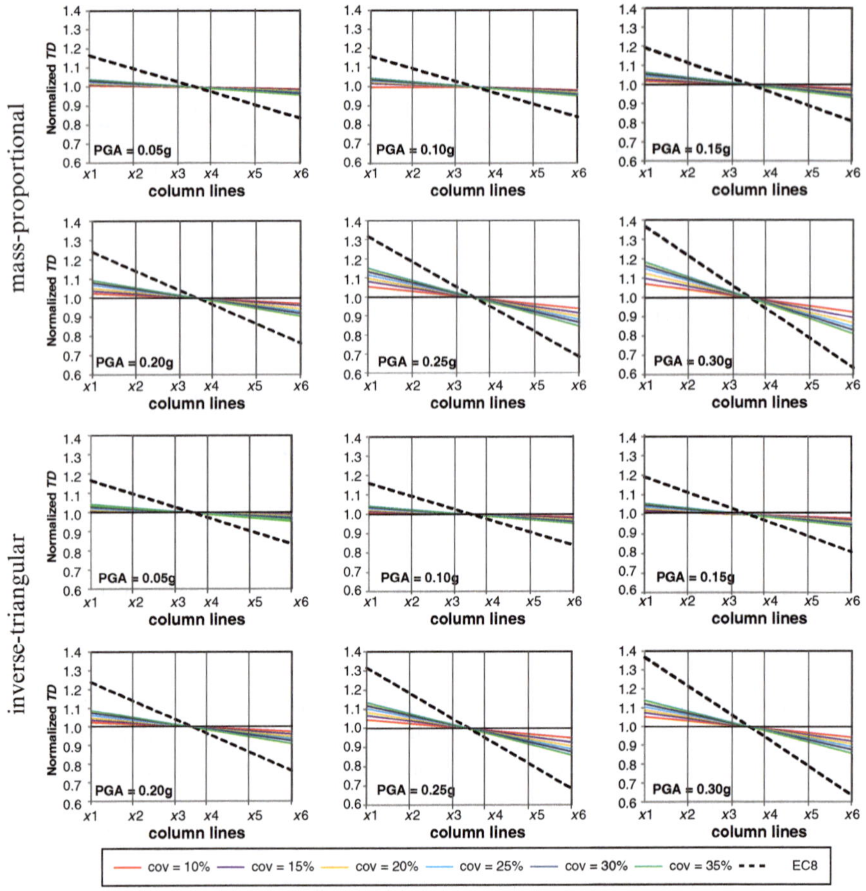

Fig. 17.5 Normalized Top Displacement (*TD*) at the column lines along the *x*-direction: *1-way asym* model

the assumed strength distributions. As regards the *SD* limit state, the performance related to the lower f_c value provided by EC8 approach ($CF = 1.35$) is smaller (more conservative) than the one coming from analysis for *cov* below 25 % (*1-way asym* model) and 30 % (*2-way asym* model), respectively. If the higher value of f_c is adopted ($CF = 1.00$), the seismic performance provided by the EC8 approach is conservative only for *cov* below 15 % when compared with the considered strength distributions. The results obtained for the seismic performance don't seem to be affected by the assumed heightwise pattern.

As regards the *NC* limit state, the safest estimation of the seismic performance by EC8 (5 % eccentricity and $CF = 1.35$), covers the results found by introducing the strength variability for *cov* below 30 %. Even in this case, when a $CF = 1.00$ is adopted, the EC8 approach overestimates the performance, comparing to the

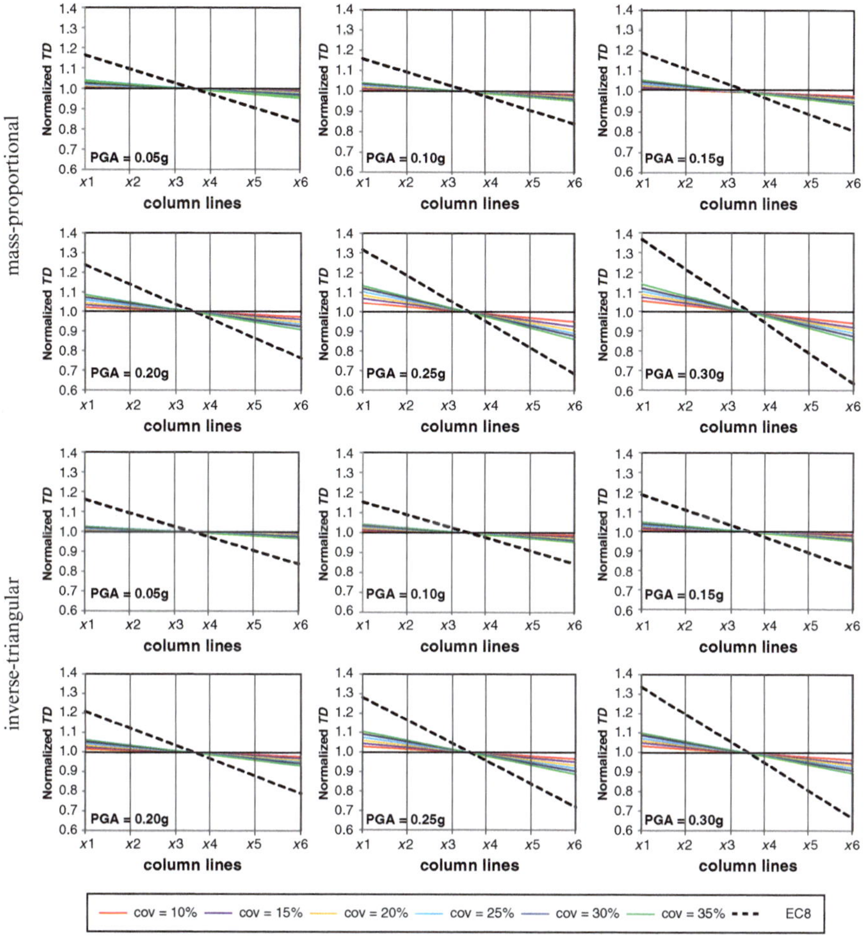

Fig. 17.6 Normalized Top Displacement (*TD*) at the column lines along the *x*-direction: *2-way asym* model

analysis, for *cov* over 25 %. It has to be noted that, when the 5 % eccentricity is not considered, the seismic performance provided by EC8 is not conservative even when compared to the smaller *cov* values.

Figure 17.8 shows the comparison between the seismic performance (*SD* and *NC*, respectively), found with the EC8 approach and with the two assumed models (*1-way asym*, *2-way asym*) for a specific value of PGA (PGA = 0.25g) at the varying of the *cov*. The ranges of the ratio demand/capacity obtained by applying the EC8 approach are evidenced and compared with the trend of the same ratio found with the two strength distribution models. It can be noted that the two models provide different trends of the demand/capacity ratio, especially for *cov* between 20 and 30 %.

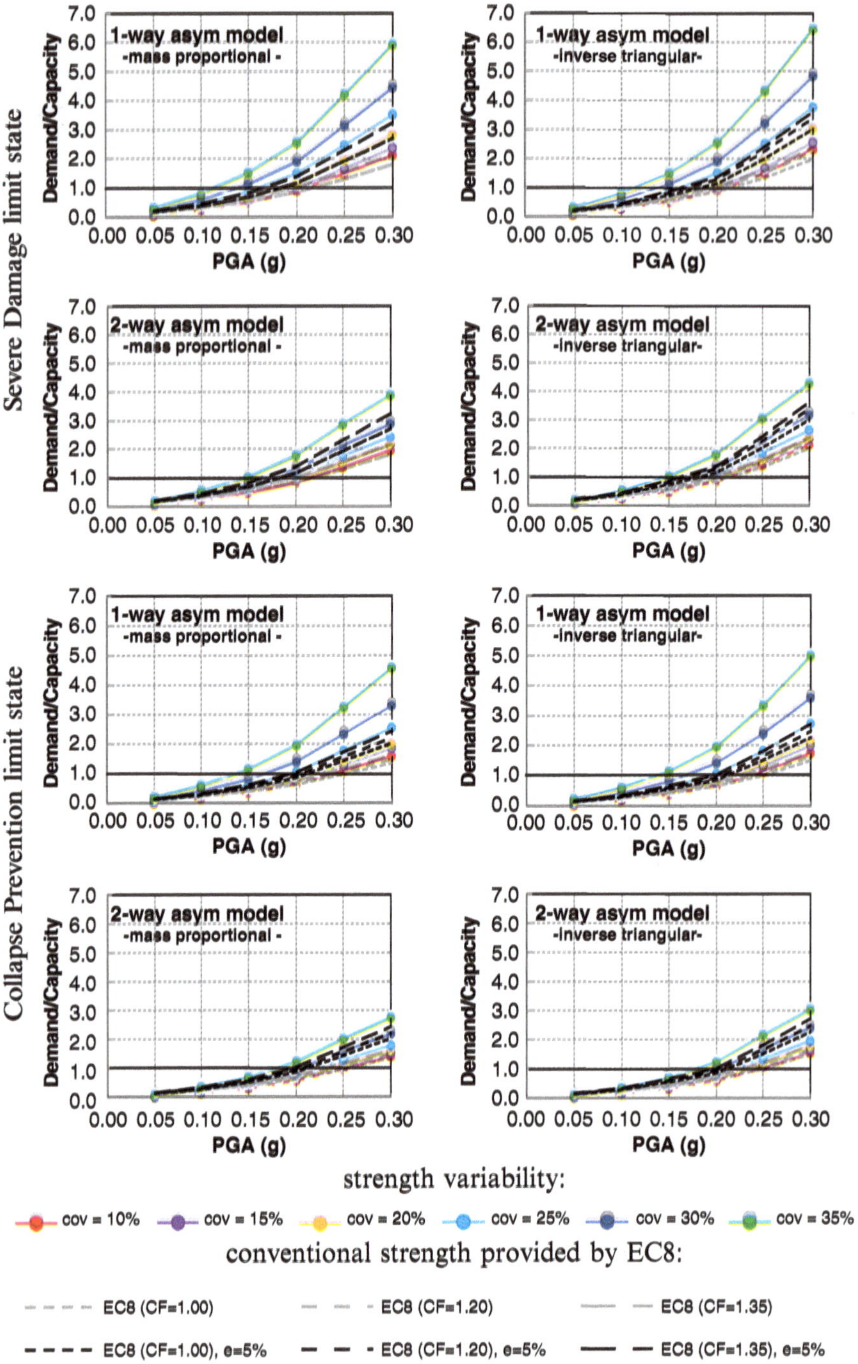

Fig. 17.7 Seismic performance: *SD* and *CP* limit states

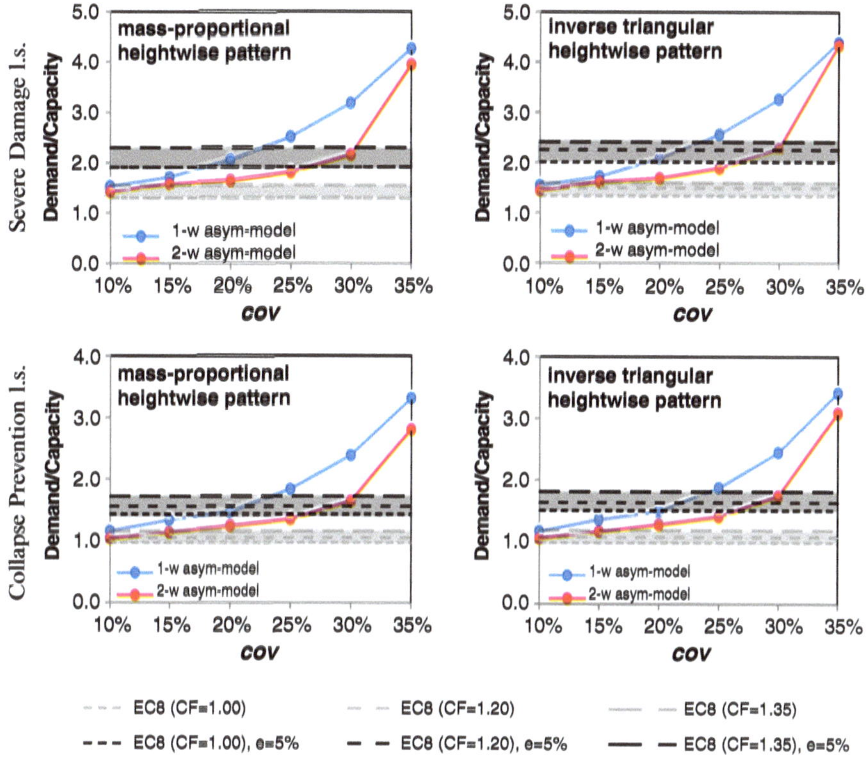

Fig. 17.8 Seismic performance: *SD* and *CP* limit states (PGA = 0.25g)

17.4 The Strength Variability as Source of In-Elevation Irregularity

17.4.1 The Assumed Strength Models

In order to investigate the effects of the in-elevation strength variability on the response of the case-study, a different strength value has been given to each storey. To this purpose, the five values constituting the strength domain have been combined to achieve a different mean strength at each level. The strength distribution at each storey, shown in Fig. 17.9, is symmetric along both the main directions, in order to avoid torsional effects, and therefore to get results easy to understand.

By combining the five f_c samples, four different mean strength values, respectively equal to 14.2, 16.7, 21.9 and 24.5 MPa have been obtained. The possible combinations of these four storey strength are represented by the 24 models shown in Fig. 17.10.

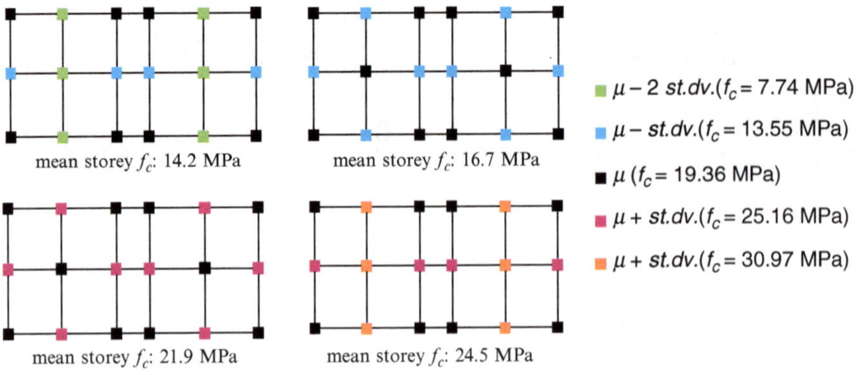

Fig. 17.9 Strength distribution in the columns

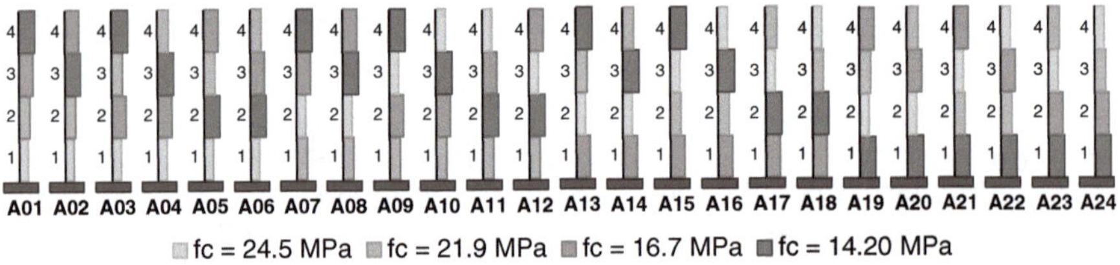

Fig. 17.10 Strength distribution in elevation

17.4.2 The Seismic Analysis

Both a nonlinear static (i.e. pushover) and a dynamic (i.e. time-history) analyses were performed to determine the seismic response of the case study. As regards the nonlinear static analysis, the N2 method was applied and two different horizontal patterns, respectively mass-proportional and inverse triangular, were assumed, like in the previous section.

The seismic input is represented through a set of seven ground motions with an average spectrum closely approaching that provided by Eurocode 8, assuming soil type B. The assumed set of ground motions was provided by Working Group Itaca (2008), on the basis of a PGA equal to 0.25g, a nominal life of the structure of 50 years and a magnitude between 5.5 and 6.5. The seven ground motions have been selected in order to match with the elastic design spectrum for a period range of between 0.5 and 1.0 s. Table 17.1 lists the main information for the adopted ground motions, together with their elastic spectra compared to that from EC8.

Table 17.1 Ground motions data

Name	Location	Date dd/mm/yyyy	PGA (g)	Duration (s)
STRNSC	Sturno	23/11/1980	0.225	70.75
CLTWEC	Calitri	23/11/1980	0.174	85.99
AQGWEC	L'Aquila	06/04/2009	0.446	100.00
AQKNSC	L'Aquila	06/04/2009	0.353	100.00
AQKWEC	L'Aquila	06/04/2009	0.330	100.00
AQVNSC	L'Aquila	06/04/2009	0.545	100.00
AQVWEC	L'Aquila	06/04/2009	0.657	100.00

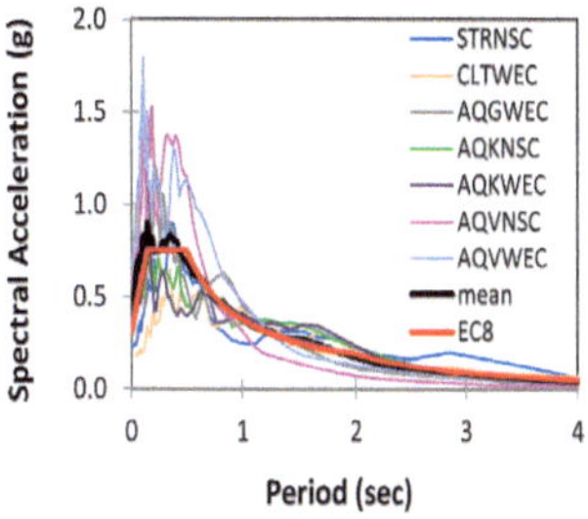

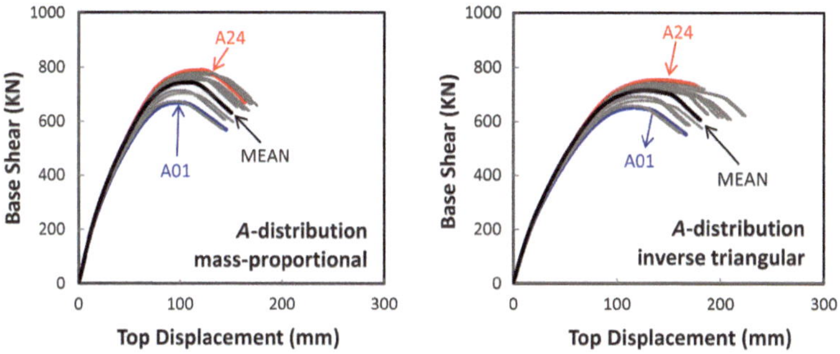

Fig. 17.11 Capacity curves of the case study with the assumed strength distributions

17.4.3 The Effects of the Strength Variability on the Seismic Response

17.4.3.1 Nonlinear Static Analysis

Each response quantity was found to be the mean value of the correspondent response domain, made by seven samples. Figure 17.11 shows the capacity curves of the two considered distributions under the two assumed heightwise patterns. It can be noted that the most favorable and unfavorable capacities occur, in all cases, for the same distribution models. In fact the A01 strength distribution (weakest storey at the bottom, strongest storey at the top) provides the lowest capacity, while the A24 one (strongest storey at the bottom, weakest storey at the top) provides the highest capacity.

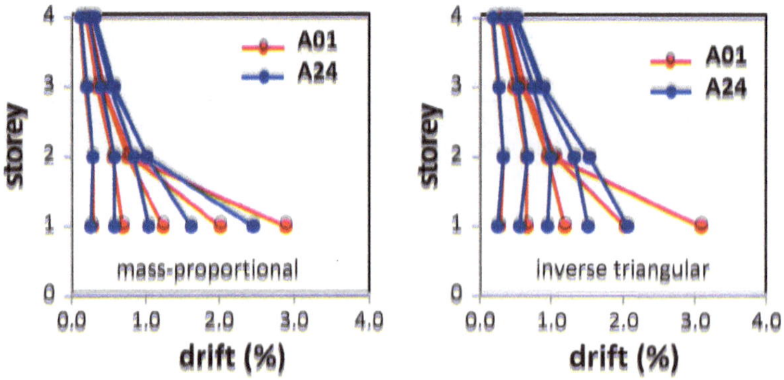

Fig. 17.12 Drift distribution along the building height

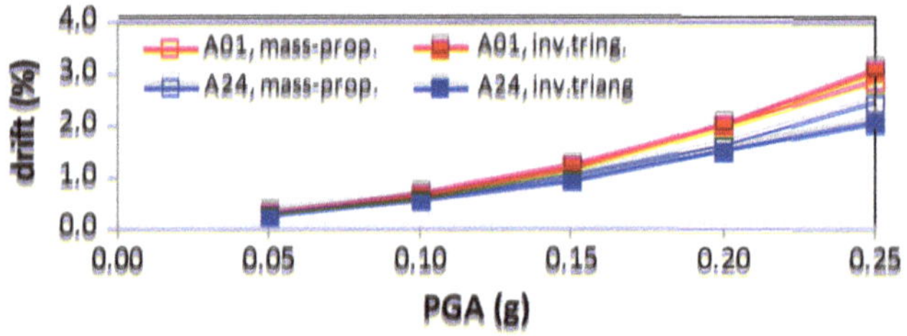

Fig. 17.13 Max drift at the first storey

Below, therefore, only the results obtained for these extreme strength distributions (A01, A24) have been shown, as they represent the maximum and the minimum seismic responses for the two considered distributions, so embracing all the other results.

Figure 17.12 shows the maximum drift profiles (averaged from the seven ground motions) for the most favourable and unfavourable models, by considering the two assumed heightwise patterns and the five seismic intensities. When the most unfavourable strength distribution is adopted, the maximum drift occurs at the first storey in all considered cases. When the most favourable distribution model is considered together with the inverse triangular heightwise pattern, instead, both the first and the second storeys experience large drifts, and the maximum can occur at any of the two levels.

In Fig. 17.13 the drift values at the first storey found by adopting the two alternative heightwise patterns have been compared. As can be seen from the diagrams, the results are very similar. In the following paragraphs, therefore, only the results obtained from the inverse triangular heightwise pattern will be shown.

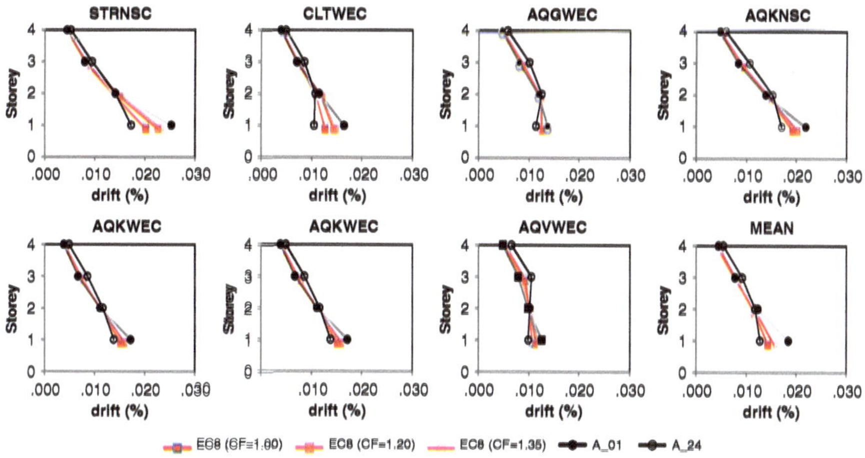

Fig. 17.14 Maximum drift profile found with all the considered f_c values (PGA = 0.25g)

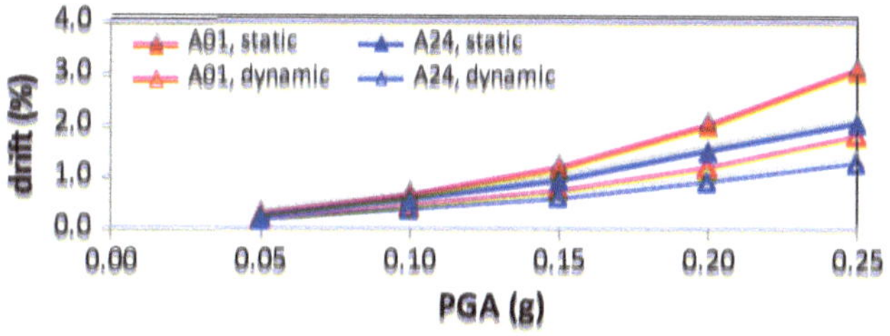

Fig. 17.15 Max drift at the first storey

17.4.3.2 Time-History Analysis

Figure 17.14 shows the effects of the strength variability on the obtained drift profiles. For sake of brevity, only the results obtained for the highest PGA (PGA = 0.25g) have been shown. The scatter in the drift domains due to the assumed concrete strength is maximum at the first storey, and it is sensitive to the considered ground motion. In fact the dynamic response of the case study appears to be largely affected by the assumed concrete strength value for some ground motions (AQKNSC, STRSNC, CLTWEC) more than for the others.

Figure 17.15 shows the maximum interstory drifts at the first storey as a function of the PGA. The trend of the drift as the PGA increases is the same in all cases, but the results provided by the nonlinear static analysis are higher than those from the time-history analysis. This difference between the two analytical approaches is not unexpected, since the static analysis is usually conservative (Magliulo et al. 2007) compared to the dynamic one. The maximum drifts obtained from the two strength distributions were compared with those obtained by the strength values suggested by EC8 (see Fig. 17.16).

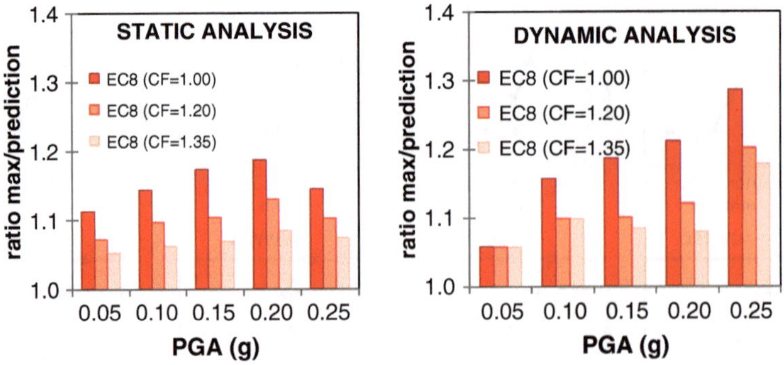

Fig. 17.16 Ratio between the maximum value of the response domains and EC8 predictions at the first storey

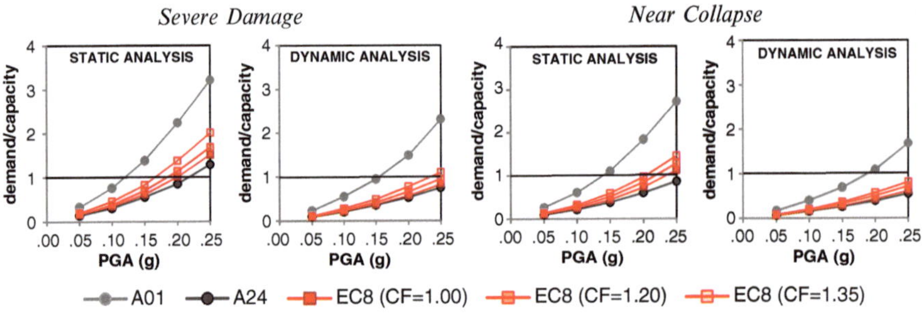

Fig. 17.17 Seismic performance of the case study (first storey)

17.4.4 The Effects of the Strength Variability on the Seismic Performance

The diagrams represented in Fig. 17.17 express the maximum values of the ratio between the seismic response (averaged on the seven ground motions) and the correspondent limit, as provided by EC8 for the two considered limit states: *Severe Damage* and *Collapse Prevention.*

In the *Severe Damage* limit state the structural demand found by the static analysis for the unfavorable strength distributions (A01) exceeds the correspondent capacity for PGA over 0.10g (0.15g for dynamic analysis), and the demand, for PGA equal to 0.25g, is three times the capacity (A01, static analysis). When the conventional strength values are adopted, the seismic performance is acceptable for PGA below 0.25g when the dynamic analysis is performed, while it overcomes the limit values already for PGA over 0.15 when the static analysis is applied.

In the *Near Collapse* limit state the structural demand provided by the time-history analysis exceeds the structural capacity only when the most unfavorable case (A01) is considered. For PGA equal to 0.25g the ratio demand/capacity found by the model A01 is twice (with both the two analysis types) than the same ratio found by the EC8 approach, when a CF = 1.00 is adopted.

17.5 Conclusive Remarks

17.5.1 Strength Variability as Source of In-Plan Irregularity

The strength variability has proved to affect the seismic response both in the demand, by inducing torsional effects, and in the capacity, by reducing the strength of the weakest members.

The maximum demand found by introducing the strength variability has been compared to the one provided by the EC8 procedure, that resulted to be unconservative. Anyway, when the accidental eccentricity of 5 % is introduced in the analysis, according to the EC8 prescriptions, the torsional effects provided by EC8 approach largely cover those induced by the assumed strength variability.

The seismic performance proved to be even more sensitive to the strength variability than the seismic demand. In fact, even when the safest conventional value of f_c ($CF = 1.35$) is adopted in the verification, the ratio between demand and capacity found according to EC8 prescriptions is smaller (i.e. less conservative) than the ratio obtained by considering the strength variability for *cov* over 20 %.

The two alternative strength distributions, respectively asymmetric along one of the main axes (*1-way asym* model) and along both axes (*2-way asym*-model) provide significant differences in the estimated seismic performance of the case study, especially for *cov* varying between 20 and 30 %. The two heightwise patterns used to perform the nonlinear static analysis, respectively mass-proportional and inverse triangular, provide instead very similar results, proving not to affect the analysis.

17.5.2 Strength Variability as Source of In-Elevation Irregularity

The strength variability proved to be an in-elevation irregularity source, inducing relevant effects both on the seismic demand and on the performance of the case study.

When the variability of the concrete strength is included in the analysis, the case-study experiences a large variation in the seismic response, with a *coefficient of variation* in the maximum drift of 17 % for PGA equal to 0.25g. The maximum drift due to the strength variability largely exceeds the response evaluated by assuming the concrete strength as suggested by codes provisions. EC8, in fact, provides maximum drifts up to 35 % lower than those from the most unfavorable assumed distribution.

The effect of f_c variability affects the seismic response even more substantially in terms of chord rotation. The maximum ratio between demand and capacity found by considering the strength variability is much higher than the ratio found by following the Codes' provisions. The demand/capacity ratio obtained with the considered f_c domains exceeds (for PGA = 0.25g, static analysis) the value of 3.0, while the same quantity, when calculated with the EC8, never exceeds 2.0.

Since all the analyses were performed using both static (pushover) and dynamic (time-history) analyses, the effect of the performed analysis on the obtained results can be observed. The nonlinear static analysis, even in this work, has proved to be conservative comparing to the time-history one providing larger values (up to 80 %) of the response parameters. Instead, the difference in results related to the adopted heightwise pattern (mass-proportional and inverse triangular) in the nonlinear static analysis is negligible.

17.5.3 Conclusions

The strength variability proved to be a source of in-plan and in-elevation irregularity, inducing a relevant increase both in the seismic demand and in the seismic performance of the case-study. When the concrete strength has an intrinsic high variability, the current European Technical Code, EC8, does not provide a reliable approach for concrete strength characterization, since it completely neglects the effects of the inherent variability.

Since the subject has proved to be relevant in defining the seismic performance of existing RC buildings, further analyses should be done in order to obtain more general results and to suggest possible improvements to the current European Codes.

Acknowledgments This work has been supported by the project DPC-ReLUIS 2010/2013 Task 1.1.2. "Strutture in cemento armato ordinarie e prefabbricate".

References

Bosco M, Ghersi A, Marino EM (2009) On the evaluation of seismic response of structures by nonlinear static methods. Earthq Eng Struct Dyn 38(13):1465–1482, John Wiley & Sons, Ltd

Bosco M, Marino E, Rossi PP (2013a) An analytical method for the evaluation of the in-plan irregularity of non-regularly asymmetric buildings. Bull Earthq Eng. doi:10.1007/s10518-013-9438-3

Bosco M, Ghersi A, Marino EM, Rossi PP (2013b) Comparison of nonlinear static methods for the assessment of asymmetric buildings. Bull Earth Eng. doi:10.1007/s10518-013-9516-6

Bosco M, Ghersi A, Marino EM, Rossi PP (2013c) Corrective eccentricities for assessment by the nonlinear static method of 3D structures subjected to bidirectional ground motions. Earthq Eng Struct Dyn 41:1751–1773

Cosenza E, Monti G (2009) Assessment and reduction of the vulnerability of existing reinforced concrete buildings. Atti Convegno finale ReLuis, pp 51–110

Cristofaro MT (2009) Metodi di valutazione della resistenza a compressione del calcestruzzo di strutture in c.a. esistenti. PhD dissertation, Università di Firenze

Cristofaro MT, D'Ambrisi A, De Stefano M, Pucinotti R, Tanganelli M (2012) Studio sulla dispersione dei valori di resistenza a compressione del calcestruzzo di edifici esistenti. Il giornale delle prove non distruttive, monitoraggio, diagnostica, vol 2/2012, pp 32–39, ISSN: 1721–7075

D'Ambrisi A, De Stefano M, Tanganelli M (2009) Use of pushover analysis for predicting seismic response of irregular buildings: a case study. J Earthq Eng 13:1089–1100, Taylor & Francis

D'Ambrisi A, De Stefano M, Tanganelli M, Viti S (2013a) Sensitivity of seismic performance of existing framed RC structures to irregular mechanical properties. In: Lavan O, De Stefano M (eds) Seismic behaviour and design of irregular and complex civil structures. Geotechnical, geological and earthquake engineering 24. Springer Science + Business Media, Dordrecht. doi:10.1007/978-94-007-5377-8_5

D'Ambrisi A, De Stefano M, Tanganelli M, Viti S (2013b) The effect of common irregularities on the seismic performance of existing RC framed buildings. In: Lavan O, De Stefano M (eds) Seismic behaviour and design of irregular and complex civil structures. Geotechnical, geological and earthquake engineering 24. Springer Science + Business Media, Dordrecht. doi:10.1007/978-94-007-5377-8_4

De Stefano M, Marino EM, Rossi PP (2006) Effect of overstrength on the seismic behaviour of multi-storey regularly asymmetric buildings. Bull Earthq Eng 4:23–42, Springer, Ltd

De Stefano M, Tanganelli M, Viti S (2012) Effect of the variability in plan of concrete mechanical properties on the seismic response of existing RC framed structures. Bull Earthq Eng. doi:10.1007/s10518-012-9412-5

De Stefano M, Tanganelli M, Viti S (2013) On the variability of concrete strength as a source of irregularity in elevation for existing RC buildings: a case study. Bull Earth Eng. doi:10.1007/s10518-013-9463-2

De Stefano M, Tanganelli M, Viti S (2014) Variability in concrete mechanical properties as a source of in-plan irregularity for existing framed structures. Eng Struct 59:161–172. ISSN: 0141–0296. doi:10.1016/j.engstruct.2013.10.027

EC 8-3 (2005) Design of structures for earthquake resistance, part 3: strengthening and repair of buildings, European standard EN 1998-3. European Committee for Standardization (CEN), Brussels

Fardis MN (2009) Seismic design, assessment and retrofitting of concrete buildings: based on EN-Eurocode8. Springer, Dordrecht

Franchin P, Pinto PE, Rajeev P (2007) Confidence factor? J Earthq Eng 14(7):989–1007

Franchin P, Pinto PE, Rajeev P (2009) Confidence in the confidence factor. In: Proceedings of "Eurocode 8 perspectives from the Italian standpoint workshop, Napoli, pp 25–38

Itaca (2008) Database of the Italian strong motions data. http://itaca.mi.ingv.it

Jalayer F, Iervolino I, Manfredi G (2008) Structural modeling: uncertainties and their influence on seismic assessment of existing RC structures. Structural safety

Magliulo G, Maddaloni G, Cosenza E (2007) Comparison between non linear dynamic analysis performed according to EC8 and elastic and non linear static analyses. Eng Struct 29(07):2893–2900

Magliulo G, Ramasco R, Capozzi V (2012) Seismic performance of R/C frames with overstrength discontinuities in elevation. Bull Earthq Eng 10:679–694

Mander JB, Priestley MJN, Park R (1988) Theoretical stress-strain model for confined concrete. J Struct Div ASCE 114:1804–1826

Marano GC, Quaranta G, Mezzina M (2008) Hybrid technique for partial safety factors calibration. In: Proceedings of Reluis2Rm08 "Valutazione e riduzione della vulnerabilità sismica di edifici esistenti in c.a.", Roma, 29–30 maggio, Cosenza E, Manfredi G, Monti, G (eds). Polimetrica International Scientific Publisher, ISBN 978-88-7699-129-5, pp 109–120

Masi A, Vona, M (2009) Estimation of the in-situ concrete strength: provisions of the European and Italian seismic codes and possible improvements provisions of the European and Italian seismic codes and possible improvements. In: Proceedings of "Eurocode 8 perspectives from the Italian standpoint workshop, Napoli, pp 67–77

Masi A, Vona M, Manfredi V (2008). A parametric study on RC existing buildings to compare different analysis methods considered in the European seismic code (EC8-3). In: Proceedings of 14th WCEE, Beijing, China

Monti G, Alessandri S, Goretti A (2007) Livelli di conoscenza e fattori di confidenza. XII Convegno ANIDIS, L'ingegneria sismica in Italia

Rajeev P, Franchin P, Pinto PE (2010). Review of confidence factor in EC8-Part 3: a European code for seismic assessment of existing buildings. In: International conference on sustainable built environment ICSBE 2010, Kandy, Sri Lanka

Seismosoft (2006) Seismostruct version 5.2.2 – a computer program for static and dynamic nonlinear analysis of framed structures. Available online from URL: www.seismosoft.com

Chapter 18
Evaluating the Efficiency of Recent Nonlinear Static Procedures on the Seismic Assessment of an Asymmetric Plan Building

A. Belejo and R. Bento

Abstract The Seismic Assessment of an asymmetric plan building is performed through an assemblage of recent Nonlinear Static Procedures (NSPs); some of them are extensions of known NSPs. Among these methods are included two multi-mode methods: Modal Pushover Analysis (MPA) and Improved Modal Pushover Analysis (iMPA); the Extended N2 which considers the higher modes effects in both plan and elevation; and the 3D Pushover wherein each step derives from a different known NSP in order to obtain the most reliable results.

The seismic response of an asymmetric plan building is studied considering both components of ground motion acting simultaneously. The seismic assessment of the building is performed in terms of pushover curves, top displacement ratios, lateral displacements profiles, interstorey drifts, normalized top displacements and shear forces. Such seismic quantities are compared with the results obtained by means of Nonlinear Dynamic Analysis (NDA).

18.1 Introduction

Over the years, NSPs have been constantly modified and improved in order to overcome the drawbacks and inaccuracies discerned in the studies previously performed. The capture of the torsional behavior of the buildings, the influence of higher mode effects and the load pattern considered are included among the most common issues faced by the scientific committee in the application of NSPs in buildings.

A. Belejo • R. Bento (✉)
ICIST, DECivil, Instituto Superior Técnico, Universidade de Lisboa,
Avenida Rovisco Pais 1, 1049-001 Lisbon, Portugal
e-mail: rita.bento@tecnico.ulisboa.pt

G.P. Cimellaro et al. (eds.), *Computational Methods, Seismic Protection, Hybrid Testing and Resilience in Earthquake Engineering*, Geotechnical, Geological and Earthquake Engineering 33, DOI 10.1007/978-3-319-06394-2_18,

© Springer International Publishing Switzerland 2015

The original version of MPA was created by Chopra and Goel (2002) and is a complete version of multi-mode pushover analysis. It is a multi-run method, where several pushover curves are obtained from different load patterns proportional to each mode of vibration. The final response is attained combining the results corresponding to each pushover curve using an appropriate combination rule. In 2004 the application was extended to the case of plan asymmetric buildings (Chopra and Goel 2004), and also a modified approach assuming higher modes as elastic was proposed (Chopra et al. 2004). The MPA has been permanently improved and updated until the most recent version which is an adaptation to consider both components of ground motion acting simultaneously in buildings, developed by Reyes and Chopra (2011a, b, 2013).

In 2008 Paraskeva et al. introduced an improved version of the MPA procedure (iMPA) for application in bridges, which was published after (Bento and Pinho 2008; Paraskeva and Kappos 2010). The aim of iMPA was to overcome the weakness of the control node localization and the invariability of the lateral force distribution. In buildings, the node control position is not, in general, an issue; on the other hand the lateral load redistribution considered in iMPA, taking into account the deformed shape of the structure in inelastic regime may be a valid alternative in order to improve results when the structure exhibits inelastic behavior.

Some attempts to consider the redistribution of inertia forces after structure yields were already suggested for a planar frame structure by Jianmeng et al. (2008); and recently this methodology was tested in 3D asymmetric plan buildings (Belejo and Bento 2014).

Also Fajfar and his team, who proposed the original N2 method (Fajfar and Fischinger 1988) which is recommended in Eurocode 8 (CEN 2004), continued to develop their method through extensions applied to the original version in order to consider the torsion effects (Fajfar et al. 2005), the higher mode effects (Kreslin and Fajfar 2011) and most recently, both effects simultaneously (Kreslin and Fajfar 2012). The extensions are based on the assumption that the structure remains in the elastic range when vibrating in higher modes, therefore the seismic demands are obtained combining the results obtained through a simple Pushover analysis with corrective factors obtained through linear response spectrum dynamic analysis.

Recently due the available set of methods, studies were performed in order to understand which the best approach for the seismic assessment of plan asymmetric buildings, considering the most known procedures. Thus, as result of this study, in Bhatt (2012) the 3D Pushover was proposed, also called as Extended Adaptive Capacity Spectrum method in Bhatt and Bento (2014). This method has as sources the known methods: ACSM (Casarotti and Pinho 2007) regarding to the lateral distribution applied and type of pushover curve obtained, CSM proposed in FEMA440 (Freeman 1998; Freeman et al. 1975; ATC 2005) to obtain the damping considered in the reduced spectrum and consequently the peak displacement and Extended N2 in order to capture the torsional behavior.

The objectives of this paper are to evaluate the individual efficiency of the methods mentioned before in their most recent versions when applied to an asymmetric plan structure. The reliance of the results obtained by the NSPs is evaluated through comparison with Nonlinear Dynamic Analyses (NDAs) for two different levels of seismic intensity.

18.2 Nonlinear Static Procedures

18.2.1 Modal Pushover Analysis for Asymmetric-Plan Buildings (MPA)

The MPA considers a conventional force based pushover analysis based on the vibration modes of the structure. In each run, a different load pattern proportional to the considered vibration mode of the structure is applied, and the results computed from each run are combined in order to obtain the final results. The complete methodology as a whole is described step by step in Chopra and Goel (2004).

The load pattern applied in the scope of MPA for asymmetric plan buildings includes two lateral forces and torque at each floor level as explained in Chopra and Goel (2004). However, in order to substitute the torque a different loading can be applied in the building in the nodes with mass assembled, normalizing the modal displacements of each node for both directions to the maximum modal displacement of the structure and multiplying by the respective mass.

Since both components of ground motion are considered acting simultaneously, the process is repeated for both orthogonal directions for all the modes considered. After obtaining the seismic response due both components of ground motion, they are combined by the SRSS multi-component combination rule to determine the seismic response of the structure.

18.2.2 Improved Modal Pushover Analysis (iMPA)

The iMPA procedure is a two-phase method wherein the deformed shape obtained in the first phase of the method, when the structure is responding inelastically to the considered earthquake level, leads to the load pattern, which is applied in the second phase. The steps of the second phase are the same as in MPA for each mode, but considering the new load pattern.

The iMPA, was originally created for bridges by Paraskeva and Kappos (2010), and tested in buildings by Belejo and Bento (2014).

Considering both components of ground motion acting simultaneously in buildings, in the first phase seismic responses are computed for both components of the ground motion separately for each mode, and in the second phase two more analyses are performed per mode, one for each component. Similar as MPA, SRSS combination rule is used in order to obtain the total seismic response of the structure.

In order to estimate member forces, when combining the seismic response of each mode would lead to forces that exceed the capacity of the elements in cases where both ends of an element deform into the inelastic range, by analyzing the plastic hinge rotations. To overcome this disadvantage, the extension proposed by Reyes and Chopra (2011a) to calculate member forces, is applied in both multi-mode procedures.

18.2.3 Extended N2

The extension of N2 method herein applied is the most recent which takes into account the higher mode effects in both plan and elevation. It corresponds to extended versions of the original N2 method, which is described in Eurocode 8, in order to overcome the torsional problem in asymmetric plan structures and simultaneously considering the higher mode effects, which affects high-rise buildings or buildings irregular in height. This version intends to handle both issues by adjusting the pushover results, computed with the original N2 method, by means of correction factors based on linear dynamic response spectrum procedures, as described in Kreslin and Fajfar (2011).

The method is applied separately, and the results obtained for both directions are combined through SRSS combination rule.

18.2.4 3D Pushover

The 3D Pushover method (Bhatt 2012; Bhatt and Bento 2014) was intended to overcome the problems of a simple pushover analysis using known methods in each step of the procedure. The selection of the method was performed in order to apply the best procedure in each step with the purpose of obtain the most reliable results. The most common issues in performing a pushover analysis are the invariability of the lateral load, the damping associated to the seismic action and the torsional behavior capture. In order to overcome all these problems, all studies performed along the time until this proposal were considered, and the approach which leads to better results was chosen for each step, and combining all steps, a new NSP was created. The methods by which 3D Pushover is based, are essentially the ACSM (Casarotti and Pinho 2007), following its guidelines regarding to the lateral load application; the CSM (Freeman 1998; Freeman et al. 1975) following FEMA 440 (ATC 2005) guidelines in considering the damping associated to the seismic action and to obtain the peak displacement and the extended N2 (Fajfar et al. 2005) to capture the torsional behavior of the structure. The procedure as a whole is described in Bhatt (2012).

After a short description of the procedures, the variants of each method are summarized in Table 18.1.

Table 18.1 Summary of studied nonlinear static procedures

	3D Pushover	Extended N2	MPA	iMPA
Pushover analysis type	Adaptive displacement based	Conventional forced-based	Multi-mode conventional force-based	Multi-mode conventional force-based: two phases
Load pattern	Adaptive displacements loading	First mode proportional loading	All significant modes proportional loading	1st Phase: MPA
	Direction of translational motion	Direction of translational motion	Both directions	Proportional to the deformed shape correspondent to the peak deformation obtained in the 1st phase
				Both directions
Capacity curve(s)	One per direction		One per mode in each direction (dominant direction of motion)	1st Phase: MPA
				2nd Phase: one per intensity (in both directions) of seismic ground motion for all modes in each direction
Demand curve	Elastic viscous damping-based reduced spectrum	Inelastic ductility-based reduced spectrum		
Additional corrective factors	Ratio between the normalized roof displacements obtained by an elastic response spectrum analysis and by the pushover analysis $(\Pi_{RSA}/\Pi_{PA}) \geq 1$			None
		Ratio between the normalized interstorey drifts obtained by an elastic response spectrum analysis and by the pushover analysis $(\Pi_{RSA}/\Pi_{PA}) \geq 1$		

18.3 Case Study

In this work, the case study analyzed is a bi-asymmetric plan nine-story steel building (Fig. 18.1), the same analyzed in Belejo and Bento (2014).

All floors present the same height of 3.96 m and all structure shows 9.14 m spans. The identified columns and the girders that connect them are characterized as Moment Resisting Frames (MRF), whereas gravity frames whose only function is to support the gravity loads compose the remaining structure. Member sizes are governed by drift instead of strength requirements and are defined in Reyes (2009). Due to the lack of available models to define the panel zones in the software used in this work, braced frames were introduced in the alignments C1–C8, C3–C9, C9–C12 and C14–C18 in order to obtain the same modal characteristics of the building studied by Reyes (2009). The translational masses considered in 1st–8th floors are 1,212 tones and 1,074 tones in the roof.

18.4 Modeling Issues

The building was modeled in SeismoStruct v6.0 (SeismoSoft 2006), a downloadable fiber element based finite element software. The model was built using space frames assuming the centerline dimensions. All sections were defined with 100 fibers and each fiber was characterized by the material relationship.

Hysteretic damping is implicitly included in the nonlinear fiber model formulation of the inelastic frame elements. In order to take into account the possible non-hysteretic sources of damping, it was modeled by Rayleigh damping with its two constants selected to give 2 % damping ratio at the fundamental period of vibration T1 and a period of 0.2T1, following the work of Reyes (2009). According to Priestley and Grant (2005), the non-hysteretic damping represents the energy dissipation due to phenomena like friction between structural and non-structural

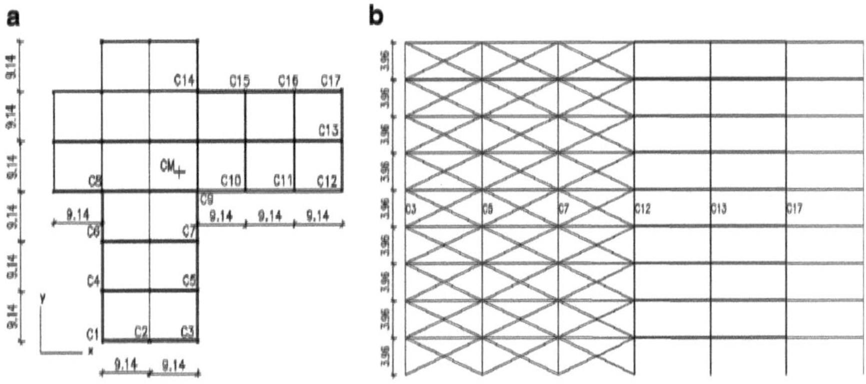

Fig. 18.1 Case study; (**a**) Plan view; (**b**) Lateral view, dimensions in [m] (Belejo and Bento 2014)

members, energy radiation through the foundation, etc., and which is mobilized during the seismic response of the structure. The scientific and engineering community still does not have definitive answers about the type and values of viscous damping used to represent such energy dissipation.

A simplified bilinear stress-strain relationship with 3 % of strain hardening was assumed for steel, based on Byfield et al. (2005) exhibiting an average yield strength around 248 MPa and an ultimate strength of 400 MPa.

Nodal Constraints were modeled with a Penalty Functions option with exponent 10^7 in order to take into account the rigid diaphragm effect. The mass of each floor was applied lumped in the nodes, according to the respective tributary area.

18.5 Seismic Features

Seven ground motion records were randomly selected from the set of records used by Reyes which criteria were defined in Hancock et al. (2006). All records were matched to the seismic hazard spectrum with 2 % probability of exceedance in 50 years.

Table 18.2 shows the Earthquakes and respective station of the records considered.

In the records considered, SeismoMatch v2.0.0 (SeismoSoft 2008) was used to match them to the hazard spectrum for the period range between 0.2T1 and 2T1. SeismoMatch is an application capable of adjusting earthquake accelerograms to match a specific target response spectrum. The method used for spectral matching adjusts the time history in the time domain by adding wavelets to the acceleration time-series as described in Hancock et al. (2006).

The mean spectrum of each component and the overall mean spectrum are shown in Fig. 18.2 as well as all matched spectra and the Seismic Hazard spectrum.

In order to reduce the time of analysis, an interval between the build-up of 5 and 95 % of the total Arias intensity (Bommer and Martínez-Pereira 1999) is considered.

Due to the uncertainty of knowing the position of the buildings relatively to the components of the records, all records were assigned to the building in two different ways: X component of the record according to the X component of the building and Y component of the record assigned to the Y component of the building; and the opposite, i.e. the X component of the record assigned to the Y direction of the

Table 18.2 Ground motion records considered

Earthquake name	Station name
Northridge 1994	Beverly hills
Duzce 1999	Bolu
Kobe 1995	Snishi – Akashi
Superstition Hills 1987	Poe
San Fernando 1971	LA Hollywood
Imperial Valley 1979	Sahop Casa Flores
Hector Mine 1999	Hec

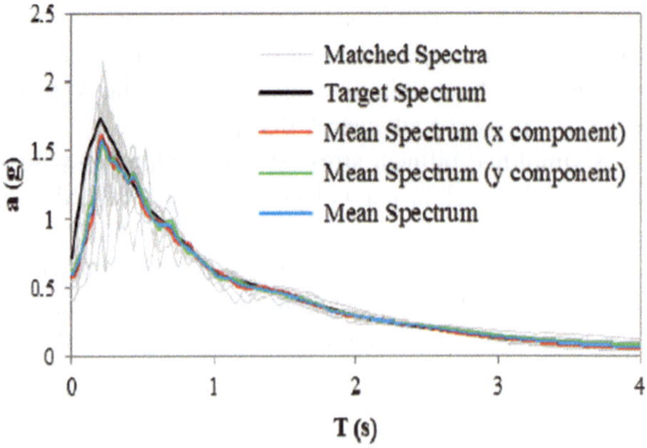

Fig. 18.2 Seismic hazard spectrum and the median response spectra of seven scaled ground motions (X and Y directions and overall)

building and the Y component of the record assigned to the X component of the building. Therefore the final seismic response is determined by the mean of the 14 results obtained. Consequently, for each intensity level, the spectrum used to compute the peak deformation in NSPs, corresponds to the mean spectrum obtained from the 14 records (two components for each ground motion).

18.6 Numerical Results and Discussion

In this section, the seismic response of the building obtained through the NSPs and NDAs, is shown in terms of pushover curves, top displacements ratios, lateral displacement profiles, interstorey drifts, normalized top displacements and Shear Forces for two different levels of seismic intensity, considering both components of ground motion acting simultaneously.

The modal properties of the building are displayed in Table 18.3, which shows the periods and the effective modal mass percentages in both X and Y directions (Ux and Uy) for the two first triplet of modes (6 modes).

The first mode of the building is characterized by torsion motion, the second mode shows translation along both axes, but predominantly in X direction, and the third mode has translational behavior in Y direction coupled with torsion; which means torsional flexibility in both directions. The second triplet of modes (4th to 6th modes) assumes the same order regarding to the nature and directions of motion when comparing with the first triplet.

These two triplets of modes were selected in order to estimate the seismic demands in both directions for the two multi-mode methods. In such procedures, for each mode, only pushover curves in the dominant direction of motion are considered: the pushover curves in the X direction were considered for the 2nd mode and in Y direction for the 1st and 3rd modes for both triplets of modes.

Figures 18.3 and 18.4 display the pushover curves obtained for the MPA (and the 1st phase of iMPA) for each mode considered together with peak displacements obtained for all intensities of ground motion considered, wherein two different

Table 18.3 Periods (in seconds) and effective modal mass percentages of the studied building

Mode	Period (s)	[Ux] (%)	[Uy] (%)
1	1.86	8.0	26.7
2	1.70	59.9	20.5
3	1.57	12.7	33.0
4	0.68	1.1	3.1
5	0.59	7.3	3.5
6	0.55	2.3	4.2

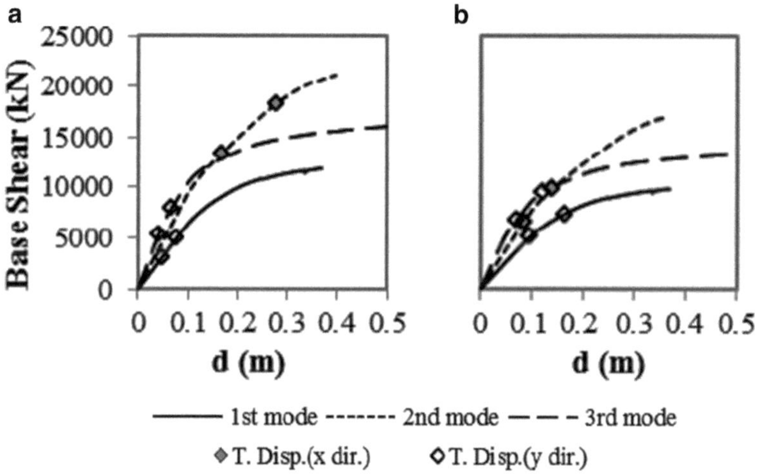

Fig. 18.3 Pushover curves of the 3 modes in MPA procedure: (**a**) 1st triplet of modes, X direction; (**b**) 1st triplet of modes, Y direction

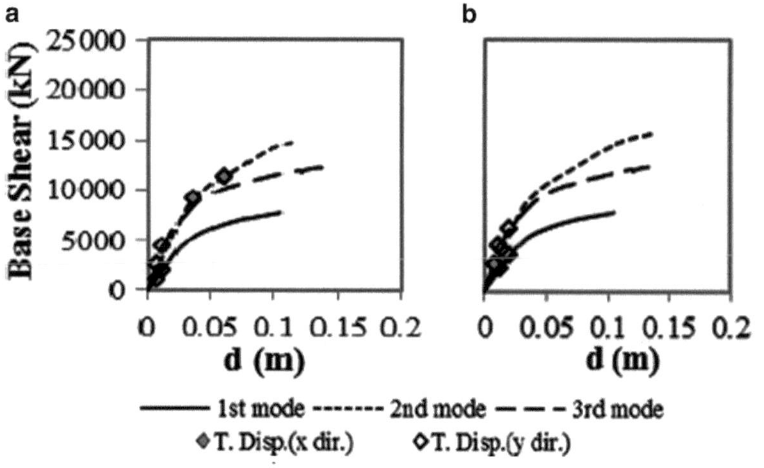

Fig. 18.4 Pushover curves of the 3 modes in MPA procedure: (**a**) 2nd triplet of modes, X direction; (**b**) 2nd triplet of modes, Y direction

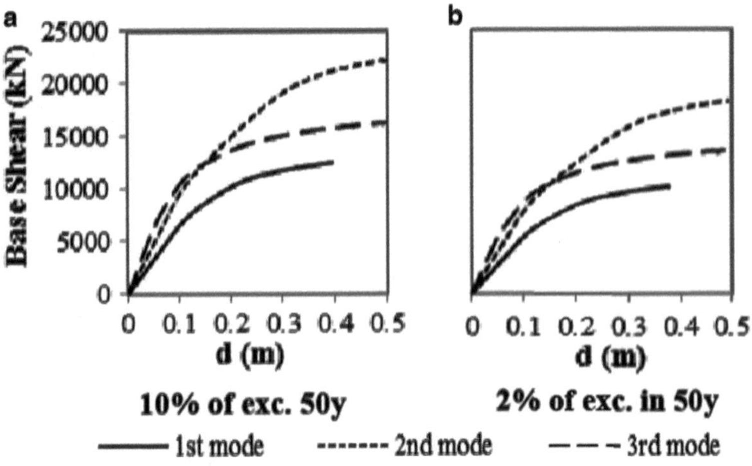

Fig. 18.5 Pushover curves of the 2nd phase of iMPA procedure in X direction

intensities are tested: the first intensity corresponds to a 10 % probability of exceedance in 50 years and the second one to a 2 % probability of exceedance in 50 years.

As mentioned in Sect. 18.2, iMPA is a double-run method, in which the final lateral load pattern for each mode is dependent of a first peak of displacement obtained. This means that for each intensity of seismic action and direction considered, a second pushover curve is achieved, from which seismic response is captured. This phase of the method was not performed for the second triplet of nodes for the reason that the higher mode equivalent SDOF systems do not contribute much to the inelastic response when the structure reaches the peak deformation in the first phase as shown in Fig. 18.3b, and that the errors arising from elastic computation in calculating the response of higher-mode equivalent SDOF systems can be neglected (Gupta and Kunnath 2000). For the first triplet of modes, pushover curves from the second phase of the method are shown in Figs. 18.5 and 18.6.

In extended N2, pushover curves are obtained by applying a lateral load proportional to the 1st mode shape in each direction (Fig. 18.7). Whereas the capacity curves obtained through 3D Pushover are derived from displacement adaptive Pushover analysis in separated directions, which are displayed in Fig. 18.8.

From the curves plotted in Figs. 18.7 and 18.8, as those obtained according to the three first modes in multi-mode methods, one can conclude that the building shows different behavior for both intensities studied: transition between elastic and inelastic behaviors when considering 10 % probability of exceedance in 50 years, and inelastic behavior for 2 % probability of exceedance in 50 years.

Taking into account the Pushover Curves plotted in Figs. 18.3, 18.4 and 18.5, all the seismic demands are obtained.

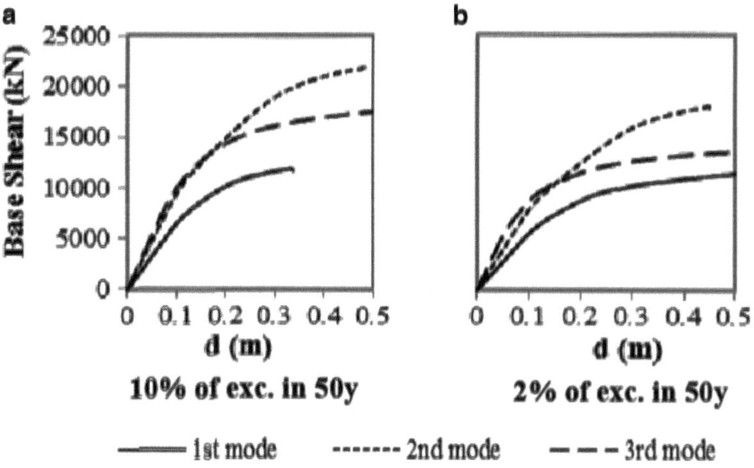

Fig. 18.6 Pushover curves of the 2nd phase of iMPA procedure in Y direction

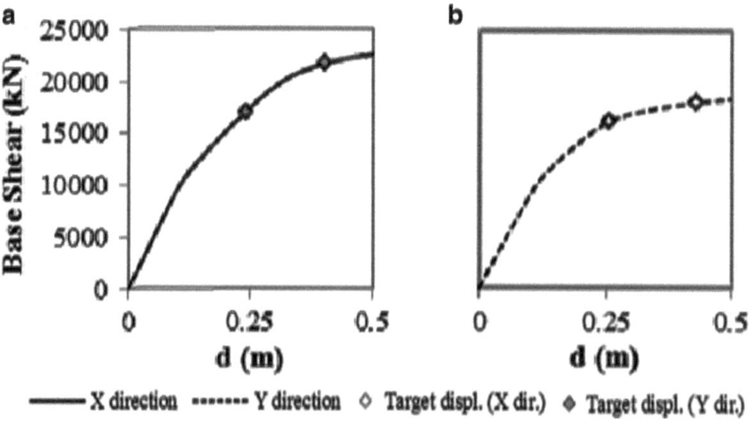

Fig. 18.7 Pushover curves in X and Y direction, respectively, in extended N2

Displacement ratios between the values obtained with the analyzed NSPs and the corresponding mean estimates coming from NDAs are computed (18.1). The NSPs must never lead to underestimated results, therefore these ratios should always be higher than 1.

$$\text{Top Displacement ratio} = \frac{\text{NSP's top displacement}}{\text{NDA mean top displacement}} \tag{18.1}$$

The nonlinear dynamic results obtained are used to compare with NSPs results. Therefore, by this analysis, one would desire such ratios to tend to unity, which means that the NSPs would match to the NDA mean results. These ratios, defined in terms of top displacements in the center of mass, are plotted in Fig. 18.9.

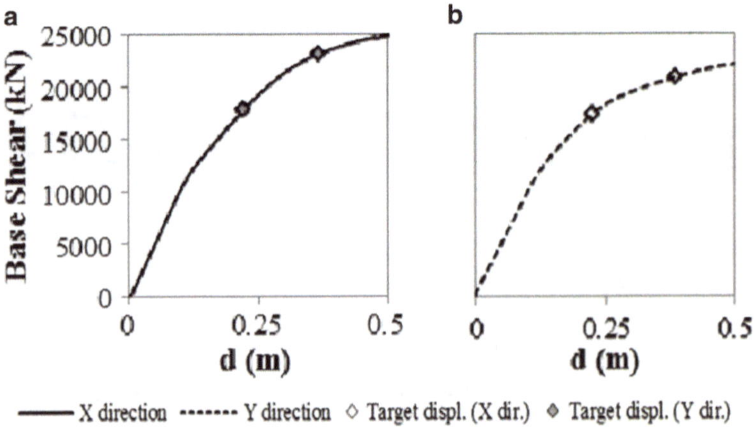

Fig. 18.8 Pushover curves in X and Y direction, respectively, in 3D Pushover

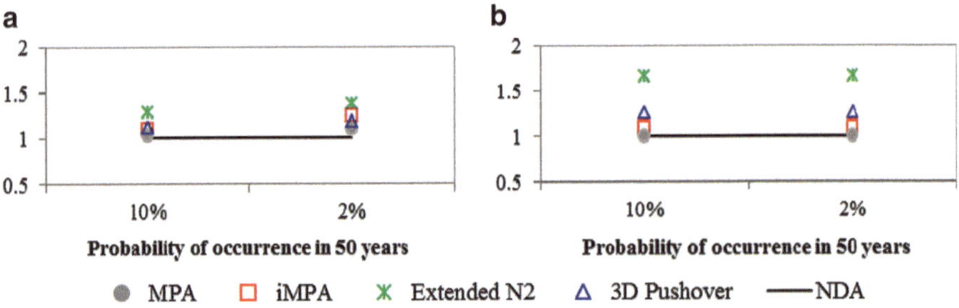

Fig. 18.9 Top displacement ratios in the center of mass: (**a**) X direction; (**b**) Y direction

The Extended N2 is the procedure with highest top displacements values and it is justified by the factor applied to take into account the higher mode effects, which increases considerably the top displacements. On the other hand, the top displacements obtained in the center of mass by the other procedures match with accuracy with the ones obtained through NDAs.

The lateral displacement profiles and interstorey drifts were obtained in center of mass and in edge columns of the building (columns C1 and C17) and are displayed in Figs. 18.10 and 18.11 respectively.

When focusing on the lateral displacement profiles and interstorey drifts obtained, the 3D Pushover and the multi-mode methods generally lead to smaller values when compared to Extended N2, however they generally lead to accurate results. Extended N2 overestimates the results for the two intensities of ground motion in both directions.

In terms of lateral displacements profiles, the multi-mode methods show good accuracy, where Improved MPA is slightly more conservative than MPA, more noticeable in the inelastic range of the structure. Extended N2 overestimates the

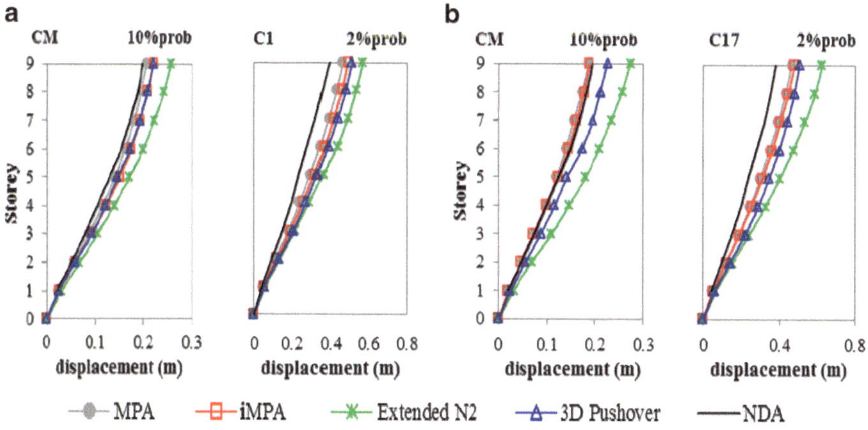

Fig. 18.10 Lateral displacement profiles: (a) X direction; (b) Y direction

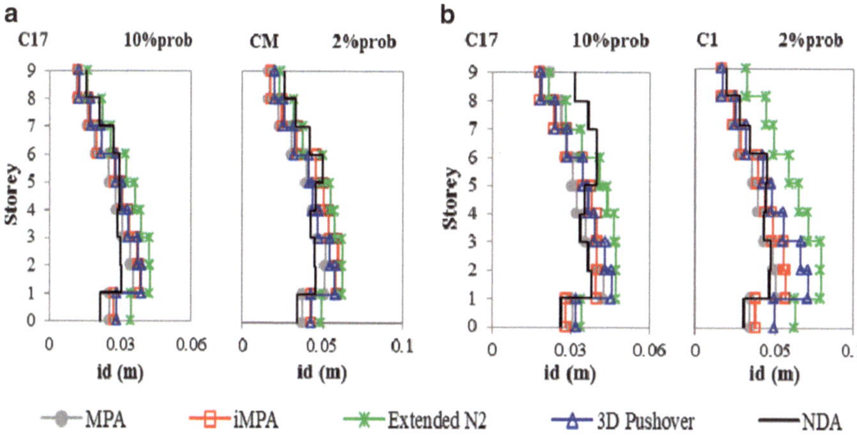

Fig. 18.11 Interstorey drifts: (a) X direction; (b) Y direction

lateral displacements in all columns, for both intensities considered. The lateral displacements obtained with 3D Pushover match perfectly with NDA in X direction, and are shown as a little conservative in Y direction.

Regarding to the Interstorey Drifts obtained, all methods lead to conservative results in terms of maximum values obtained. Generally the results obtained by 3D Pushover and multi-mode procedures are very close among them and achieve a good approximation to NDA results in the most of the stories for all situations. In few cases interstorey drifts in the upper stories are not well captured by these methods, mainly in Y direction. On the other hand, Extended N2 is able to capture the drifts in the upper stories and show conservative results for the other stories in both directions for the intensities studied.

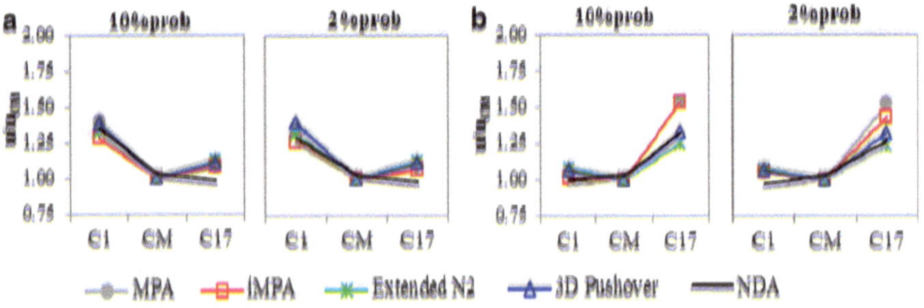

Fig. 18.12 Normalized top displacements: (**a**) X direction; (**b**) Y direction

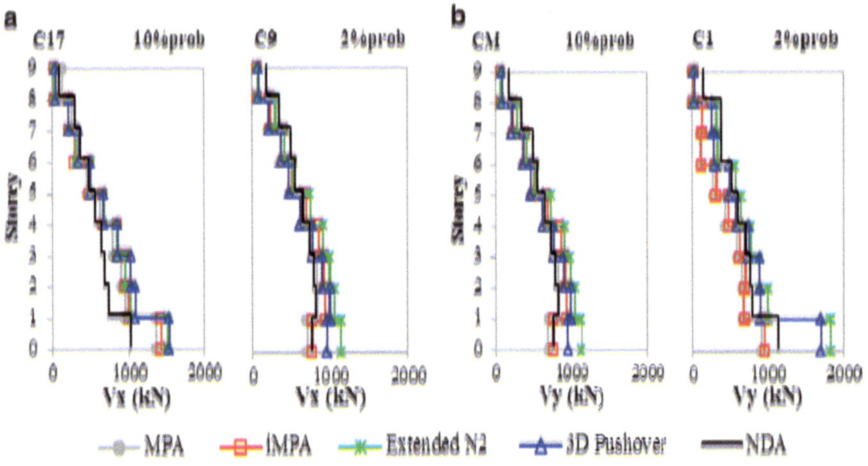

Fig. 18.13 Shear forces: (**a**) X direction; (**b**) Y direction

In order to study the torsional behavior of the building, the trend of normalized top displacements is analyzed and the results are shown in Fig. 18.12. This measurement is obtained by normalizing the edge displacement values with respect to those of the center of mass. The torsional response in NDAs is taken from the stage of the analysis correspondent to the maximum top displacement (in absolute value) in the center of mass.

All studied methods show great accuracy in the capture of the torsional amplification in the edge sides of the building in inelastic regime for both directions. Only the multi-mode procedures show conservative results in the flexible side of the building in Y direction.

Respecting to Shear Forces, the extension to MPA proposed by Reyes and Chopra (2011a) to estimate internal forces in the structure, was herein applied for the multi-mode procedures and when the elements deform into the inelastic range.

In addition to the columns that show displacement results, the column C9, which is close to the center of mass, was added to, in order to obtain a more widespread behavior of the structure in terms of Shear. Hence Shear forces were obtained in both directions of the building and results are shown Fig. 18.13.

All methods provide a good approximation in terms of shear forces for both intensities and in both X and Y directions, as displayed in Fig. 18.13. However 3D Pushover and Extended N2 present slight conservative results in the first stories where the maximum values of Shear Force are achieved.

The shear capacity of the studied columns, calculated by Eurocode 3, is far from being achieved in all columns analyzed.

18.7 Conclusions

In this paper, the nonlinear static procedures MPA, Extended N2, 3D Pushover and an improved version of MPA (iMPA), were applied in order to evaluate their respective individual performance. With this as the main objective, all the aforementioned methods were applied to an asymmetric nine-storey plan building, considering both components of seismic motion acting simultaneously. The results obtained were herein compared with the ones evaluated by means of Nonlinear Dynamic time-history Analyses.

According to all results achieved, one can conclude that all these recent methods or extensions to methods, which had been proposed in the past, lead to very accurate results, as far as this steel building concerned. Since the studied building is torsional flexible in both directions, the capture of its torsional behavior was the most concerning achievement and the results obtained regarding to the capture of the torsional behavior by all methods matched with accuracy the NDA results. Other considerable fact is the height of the building, which some higher modes of vibration are relevant to the seismic performance; however these methods were effective in overcome this issue as well, i.e. taken into account the higher modes of vibration effects.

Notwithstanding the effectiveness of all methods, MPA and iMPA seemed to present the best approach to NDA in terms of lateral displacement profiles and interstorey drifts wherein generally iMPA is slightly more conservative between both methods. The 3D Pushover and mainly Extended N2 generally overestimate these results. On the other hand, in terms of torsional behavior of the building, Extended N2 and 3D Pushover are closer to NDA. Finally, one can say the shear forces values are quite close among all methods and also fit with NDA results.

Having four different approaches which lead to good results, the choice of the method in order to perform an eventual seismic assessment of an asymmetric plan building would be probably sustained by the less time-spending required. In fact, to apply Extended N2 or 3D Pushover, an extra dynamic response spectrum analysis is required and in the case of multi-mode methods, pushover analysis per mode has to be performed, and specifically in iMPA, doubled time-consuming is needed when compared with MPA. Nevertheless, it is important to note that, to apply the 3D Pushover, it is important that the software used is able to perform an adaptive analysis, which is not a common feature of the finite element programs usually used to perform nonlinear static analysis.

Finally it is important to highlight that these methods have already been applied to other asymmetric plan buildings and the same conclusions have been reached about the efficiency of the procedures on the seismic assessment of the buildings.

Acknowledgments The authors would like to acknowledge the financial support of the Portuguese Foundation for Science and Technology through the research project PTDC/ECM/100299/2008. In addition, the authors would like to thank Prof. Juan Carlos Reyes for all the support and material provided whenever it was requested.

References

ATC, Applied Technology Council (2005) Improvement of nonlinear static seismic analysis procedures. FEMA440 report, Redwood City

Belejo A, Bento R (2014) Improved modal pushover analysis in seismic assessment of asymmetric plan buildings under the influence of one and two horizontal components of ground motions. In: Soil dynamics and earthquake engineering (submitted)

Bento R, Pinho R (2008) 3D pushover 2008: nonlinear static methods for design/assessment of 3D structures. IST Press, Lisbon

Bhatt C (2012) Seismic assessment of existing buildings using nonlinear static procedures (NSPs) – a new 3D pushover procedure. PhD dissertation, IST, Lisbon

Bhatt C, Bento R (2014) The extended adaptive capacity spectrum method for the seismic assessment of plan asymmetric buildings. Earthq Spectra 30(2):683–703

Bommer JJ, Martínez-Pereira A (1999) The effective duration of earthquake strong motion. J Earthq Eng 3(2):127–172

Byfield M, Davies J, Danalakshmi M (2005) Calculation of the strain hardening behaviour of steel structures based on mill tests. J Counstractional Steel Res 61:133–150

Casarotti C, Pinho R (2007) An adaptive capacity spectrum method for assessment of bridges subjected to earthquake action. Bull Earthq Eng 5(3):377–390

CEN, Comité Européen de Normalisation (2004) Eurocode 8: design of structures for earthquake resistance. Part 1: general rules, seismic actions and rules for buildings. EN 1998-1:2004, Brussels

Chopra AK, Goel RK (2002) A modal pushover analysis procedure for estimating seismic demands for buildings. Earthq Eng Struct Dyn 31:561–582

Chopra AK, Goel RK (2004) A modal pushover analysis procedure to estimate seismic demands for unsymmetric-plan buildings. Earthq Eng Struct Dyn 33:903–927

Chopra AK, Goel RK, Chintanapakdee C (2004) Evaluation of a modified MPA procedure assuming higher modes as elastic to estimate seismic demands. Earthq Spectra 20(3):757–778

Fajfar P, Fischinger M (1988) N2 – a method for non-linear seismic analysis of regular buildings. In: Proceedings of the 9th World conference in earthquake engineering, Tokyo-Kyoto, Japan, vol 5, pp 111–116

Fajfar P, Marusic D, Perus I (2005) Torsional effects in the pushover-based seismic analysis of buildings. J Earthq Eng 9(6):831–854

Freeman S (1998) Development and use of capacity spectrum method. In: Proceedings of the sixth U.S. national conference on earthquake engineering, Seattle

Freeman S, Nicoletti J, Tyrell JV (1975) Evaluation of existing buildings for seismic risk – a case study of Puget Sound Naval Shipyard, Bremerton, Washington. In: Proceedings of U.S. national conference on earthquake engineering, Ann Arbor, pp 113–122

Gupta B, Kunnath S (2000) Adaptive spectra based pushover procedure for seismic evaluation of structures. Earthq Spectra 16(2):367–391

Hancock J, Watson-Lamprey J, Abrahamson NA, Bommer JJ, Markatis A, McCoy E, Mendis R (2006) An improved method of matching response spectra of recorded earthquake ground motion using wavelets. J Earthq Eng 10(S1):67–89

Jianmeng M, Changhai Z, Lili X (2008) An improved modal pushover analysis procedure for estimating seismic demands of structures. Earthq Eng Eng Vib 7(1):25–31

Kreslin MP, Fajfar P (2011) The extended N2 method taking into account higher mode effects in elevation. Earthq Eng Struct Dyn 40:1571–1589

Kreslin MP, Fajfar P (2012) The extended N2 method considering higher mode effects in both plan and elevation. Bull Earthq Eng 10:695–715

Paraskeva TS, Kappos AJ (2010) Further development of a multimodal pushover analysis procedure for seismic assessment of bridges. Earthq Eng Struct Dyn 39(11):211–222

Priestley MJN, Grant DN (2005) Viscous damping in seismic design and analysis. J Earthq Eng 9(SI1):229–255

Reyes JC (2009) Estimating seismic demands for performance-based engineering buildings. PhD dissertation, Berkeley

Reyes JC, Chopra AK (2011a) Evaluation of three-dimensional modal pushover analysis for asymmetric-plan buildings subjected to two components of ground motion. Earthq Eng Struct Dyn 40:1474–1495

Reyes JC, Chopra AK (2011b) Three-dimensional modal pushover analysis of buildings subjected to two components of ground motion, including its evaluation for tall buildings. Earthq Eng Struct Dyn 40:789–806

Reyes JC, Chopra AK (2013) Three-dimensional modal pushover analysis of unsymmetric-plan buildings subjected to two components of ground motion. In: Lavan O, De Stafano M (eds) Seismic behaviour and design of irregular and complex civil structures. Geotechnical, geological and earthquake engineering, vol 24. Springer, Dordrecht, pp 203–217

SeismoSoft (2006) SeismoStruct – a computer program for static and dynamic nonlinear analysis of framed structures. Available from: www.seismosoft.com

SeismoSoft (2008) SeismoMatch – an application capable of adjusting earthquake accelerograms to match a specific target response. Available online from http://www.seismosoft.com

Index

G.P. Cimellaro et al. (eds.), *Computational Methods, Seismic Protection, Hybrid Testing and Resilience in Earthquake Engineering*, Geotechnical, Geological and Earthquake Engineering 33, DOI 10.1007/978-3-319-06394-2,
© Springer International Publishing Switzerland 2015

MIX
Papier aus verantwortungsvollen Quellen
Paper from responsible sources
FSC® C105338

If you have any concerns about our products,
you can contact us on
ProductSafety@springernature.com

In case Publisher is established outside the EU,
the EU authorized representative is:
Springer Nature Customer Service Center GmbH
Europaplatz 3, 69115 Heidelberg, Germany

Printed by Libri Plureos GmbH
in Hamburg, Germany